Cainozoic geology and landscape of north-east Scotland

This memoir provides a synthesis of the Cainozoic (Palaeogene, Neogene and Quaternary) geology of a large part of the coastal lowlands of north-east Scotland stretching from Elgin to Inverbervie and portrayed on the 'Drift' editions of eleven 1:50 000 geological sheets. The nature, origin and distribution of the superficial deposits are described, together with their use as a resource and as foundation materials. It draws on the results of surveys undertaken over the past 30 years or so, and acts as a guide to the large archive of more detailed information held by the British Geological Survey. It also incorporates and synthesises much information stemming from university research, and is co-authored by leading research workers out with BGS.

The complexity and variety of the Quaternary succession in the district is unrivalled in Scotland particularly for the record it contains of cold stages represented by periglacial deposits and the presence of sediments with a biological or pedological record of interstadial and interglacial climates. The widespread preservation of ancient peneplanation surfaces associated with deeply weathered rock and enigmatic flint and quartzite gravels is unique in north-west Europe. These features, together with a significant offshore stratigraphical record, provide unique information illustrating the effects of Pleistocene glaciation and Cainozoic landscape evolution. The area offers a broad range of future research and teaching opportunities that can draw on the definitive account presented here.

Much of north-east Scotland was peripheral to the main centres of ice accumulation in the west of Scotland during the Pleistocene glaciations resulting in minimal glacial erosion. It repeatedly witnessed the interaction between sluggish, cold-based ice flowing from an inland source and relatively fast, low-gradient ice streams flowing over deformable sediments offshore. A partial record of at least three major glaciations and intervening ice-free periods is preserved. During the last (Main Late Devensian) glaciation, the district was crossed by several distinct ice streams that partially retreated and then re-advanced on more than one occasion. This resulted in the deposition of complex sequences of glacial, glaciofluvial and glaciolacustrine deposits of varied provenance together with of a range of glacitectonic phenomena including numerous glacial rafts derived from the floor of the Moray Firth. Large rafts of Jurassic clay are particularly common. The pattern of deglaciation is also recorded in an extensive network of meltwater channels, and in the raised beach, estuarine and glaciomarine deposits, formed as a result of the considerable glacio-isostatic depression of the ground. Following a period of low sea level in the early Holocene, the Main Postglacial Transgression resulted in marine inundation of the river estuaries and the deposition of estuarine silts, sands and clays, locally over peat.

Cover photograph
Crovie Village (D6116). Sandstones and conglomerates of the Lower Devonian Crovie Group crop out on the foreshore and are down-faulted against Dalradian rocks of the Macduff Slate Formation that form the headland at the far end of the village. The landslip scar beyond the red telephone box exposes fine-grained deposits of the Kirk Burn Silt Formation of the Banffshire Coast Drift Group.

Frontispiece   This winter image of north-east Scotland was acquired by the Landsat Enhanced Thematic Mapper sensor from an altitude of about 700 km, in December 2001. The image is a false colour composite, with bands 7, 4 and 1 displayed in red, green and blue. Bands 7 and 4 record infrared solar radiation reflected by the land surface, and highlight rocks/soil and vegetation, respectively, while band 1 records reflected visible blue light.

Pink areas represent areas of snow and bright greens represent active vegetation. Coniferous forests are displayed as dark greens and moorland appears brown. Bare fields, rock and cultural features are seen as tones of red, while lowland areas appear in blue hues. Water is black due to absorption of solar radiation by the spectral ranges displayed.

The Black Burn and Blackhills glacial drainage channels are conspicuous in the vicinity of Elgin, as are the channels associated with the Tore of Troup, east of Banff. The former inselberg of Mormond Hill dominates the low ground between Fraserburgh and Peterhead and the Insch and Alford basins show up particularly well.

BRITISH GEOLOGICAL SURVEY

J W MERRITT
C A AUTON
E R CONNELL
A M HALL
J D PEACOCK

Cainozoic geology and landscape evolution of north-east Scotland

CONTRIBUTORS

J F Aitken
D F Ball
D Gould
J D Hansom
R Holmes
R M W Musson
M A Paul

Memoir for the drift editions of 1:50 000 geological sheets 66E Banchory, 67 Stonehaven, 76E Inverurie, 77 Aberdeen, 86E Turriff, 87W Ellon, 87E Peterhead, 95 Elgin, 96W Portsoy, 96E Banff and 97 Fraserburgh (Scotland)

Edinburgh: British Geological Survey   2003

iv

ISBN 0 85 272463 2

Bibliographical reference

MERRITT, J W, AUTON, C A, CONNELL, E R, HALL, A M, and PEACOCK, J D. 2003. Cainozoic geology and landscape evolution of north-east Scotland. *Memoir of the British Geological Survey*, Sheets 66E, 67, 76E, 77, 86E, 87W, 87E, 95, 96W, 96E and 97 (Scotland).

Authors

C A Auton, BSc, CGeol
J W Merritt, BSc, CGeol
British Geological Survey, Edinburgh

E R Connell, BA, FGS
Department of Geography and the Environment, University of Aberdeen

A M Hall, BSc, PhD
Fettes College, Edinburgh

J D Peacock, BSc, PhD, FRSE
Honorary Research Associate,
British Geological Survey, Edinburgh

Contributors

D F Ball, BSc, MSc
D Gould, BSc, PhD, CGeol, MIMM
R Holmes, BSc
R M W Musson, BSc, PhD
British Geological Survey, Edinburgh

J F Aitken, BSc, PhD
Badley, Ashton and Associates Ltd, Winceby House, Winceby, Horncastle, LN9 6PB

J D Hansom, BSc, PhD
Department of Geography and Topographic Science, University of Glasgow, Glasgow, G12 8QQ

M A Paul, BSc, PhD
Civil and Off-shore Engineering Department, Herriot-Watt University, Edinburgh, EH14 4AS

CONTENTS

TABLES

PLATES

MAPS

Acknowledgements

This memoir was compiled by J W Merritt from published information (as cited) and contributions of unpublished text and information. Chapters in this memoir have been written by the following authors.

One   Introduction   J W Merritt with contributions from D Gould on solid geology

Two   Applied geology   C A Auton with contributions from D F Ball on hydrogeology, J D Hansom on coastline stability, R M W Musson on seismicity and seismic hazard and M A Paul on ground stability and foundation conditions

Three   Landscape evolution   A M Hall

Four   Palaeogene and Neogene deposits, weathering and soil development   A M Hall with contributions from C A Auton

Five   The Quaternary period   J W Merritt with contributions from R Holmes on the offshore record and E R Connell, A M Hall and J D Peacock on the present model of glaciation

Six   Quaternary deposits   J W Merritt with contributions from the other authors

Seven   Geomorphological features   J W Merritt with contributions from the other authors

Eight   Quaternary lithostratigraphy and lithostratigraphical correlations   J W Merritt with contributions from the other authors

Information sources   C A Auton

Appendix 1   Important localities
Introduction   J W Merritt
1   Teindland, near Elgin   A M Hall
2   Boyne Limestone Quarry   J W Merritt and J D Peacock
3   Castle Hill, Gardenstown   J D Peacock and J W Merritt
4   King Edward   A M Hall and J D Peacock
5   Crossbrae Farm, Turriff   A M Hall
6   Howe of Byth Quarry   A M Hall
7   Kirkhill and Leys quarries   A M Hall and E R Connell
8   Oldmill Quarry   J W Merritt and E R Connell
9   Philorth valley, Fraserburgh   C A Auton
10   Ugie valley   J D Peacock and E R Connell
11   St Fergus (Annachie)   J D Peacock, A M Hall and E R Connell
12   Sandford Bay   E R Connell
13   Windyhills   C A Auton
14   Moss of Cruden   A M Hall
15   Ellon (Bellscamphie)   A M Hall and E R Connell
16   Kippet Hills, Slains   C A Auton and E R Connell
17   Errollston, Cruden Bay   E R Connell and J D Peacock
18   Mill of Dyce   J F Aitken and C A Auton
19   Strabathie   J F Aitken
20   Nether Daugh, Kintore   J F Aitken
21   Rothens, Monymusk   J F Aitken
22   Nigg Bay, Aberdeen   E R Connell
23   Loch of Park   C A Auton
24   Balnakettle   C A Auton
25   Knockhill Wood, Glenbervie   C A Auton
26   Burn of Benholm   C A Auton

Appendix 2   Sand and gravel resources   C A Auton

Appendix 3   Results of shallow geophysical surveys   C A Auton

Maps   J W Merritt and C A Auton

This memoir was edited by A A Jackson, J R Mendum, E A Pickett and D G Woodhall. Figures were produced by BGS Cartography in Murchison House and Keyworth.

Notes

Throughout this publication the word 'district' refers to the area covered by the eleven 1:50 000 geological sheets that are described (Figure 1b).

National Grid references are within the 100 km squares NJ, NK and NO and are given in square brackets. Enquiries concerning geological data for the district should be addressed to the Manager, National Geoscience Information Centre, British Geological Survey, Murchison House, West Mains Road, Edinburgh, EH9 3LA

In this volume and on the CD that accompanies it, colours and colour codes have been described using the Munsell colour scheme, either in the Munsell Soil Color Chart (1992 revised edition) or the Rock Color Chart (1984), distributed by the Geological Society of America.

PREFACE

Pressures on present-day environments are continually increasing. In areas of population and industrial growth such as the coastal zone between Stonehaven and Peterhead, the need for a thorough understanding of the local natural resources, hazards and ground conditions is paramount. A major aspect of this need is a comprehensive knowledge and understanding of the geological, geomorphological and environmental changes that have occurred over the past few million years, and which were responsible for the present distribution of drift deposits and landforms. To this end the British Geological Survey is directing research towards the Quaternary period, particularly concentrating on major centres of population, as well as surveying neighbouring rural and wilderness areas where relatively little geological information is currently available.

This memoir is a synthesis of the Cainozoic (Palaeogene, Neogene and Quaternary) deposits over a wide area of north-east Scotland and represents a departure from the previous survey styles. It focusses on the nature, origin and distribution of the drift deposits, their use as a resource and as foundation materials. As such it represents part of the large geoscience database that BGS holds for the UK that is available to provide solutions to geological problems as well as underpinning the scientific understanding of glacial processes, landforms and deposits. It describes areas that have been resurveyed by BGS over the last few decades in the coastal areas of north-east Scotland and portrayed on the Drift or Solid-and-Drift editions of eleven 1:50 000 geological sheets. Much of the mapping was undertaken as commissioned research specifically for sand and gravel resource appraisal or as part of environmental geology portfolios. Although the drift deposits on some of the sheets have been documented elsewhere in BGS publications, most have not been described systematically.

Several classic concepts of the British Quaternary have been formulated in north-east Scotland over the past 150 years, including the pioneering research work of Thomas Jamieson and James Croll during the 19th century. Croll first suggested that ice ages were caused by changes in the amount of solar irradiance received at the poles as a result of changes in shape, tilt and wobble of the Earth's orbit around the Sun. This memoir summarises and builds on a wealth of research work published in the literature, particularly in publications of the Quaternary Research Association and Scottish Natural Heritage. It includes an analysis of how the landscape of north-east Scotland evolved throughout the Cainozoic era and it summarises our present understanding of the Quaternary events that have affected the district. Several research workers outwith the British Geological Survey have contributed to this memoir, in particular A M Hall and E R Connell, testifying to the co-operation of university academics with BGS's core programme. This collaboration together with the combination of publically funded and commissioned research furthers understanding of British geology and enhances the national database of earth science information available to the public.

David A Falvey, PhD
Director

British Geological Survey
Kingsley Dunham Centre
Keyworth
Nottingham
NG12 5GG

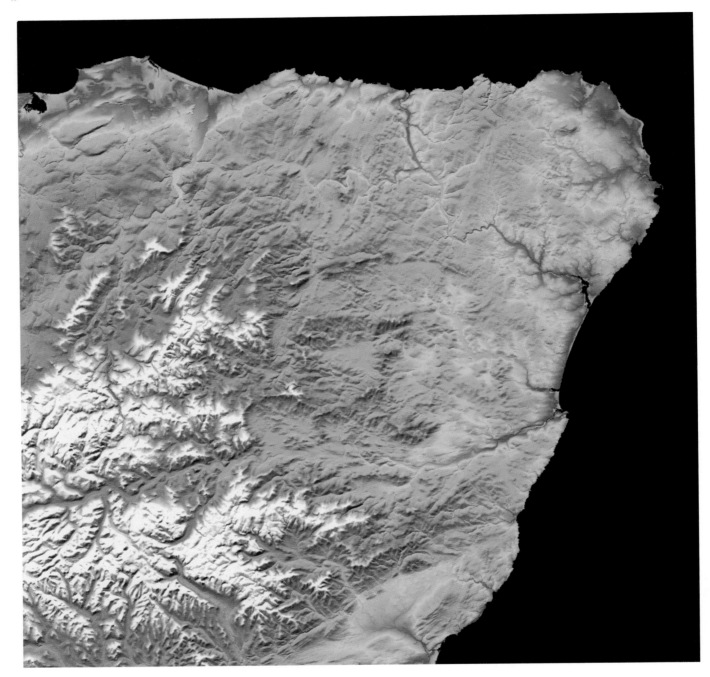

Figure 1a Ground surface created from a digital terrain model made by the Centre for Ecology and Hydrogeology, Wallingford, from Ordnance Survey panorama data.

ONE

Introduction

The region described here (and referred henceforth as 'the district') spans the coastal hinterland and Buchan Plateau of north-east Scotland stretching from Elgin, on the southern coast of the Moray Firth, to Inverbervie, on the North Sea coast (Figure 1). This account summarises the Quaternary geology (Table 1) presented on the Drift (or Solid-and-Drift) editions of the 1:50 000 Series sheets 66 Banchory, 67 Stonehaven, 76E Inverurie, 86E Turriff, 87W Ellon, 87E Peterhead, 96W Portsoy, 96E Banff and 97 Fraserburgh. It also encompasses sheets 77 Aberdeen and 95 Elgin, for which modern memoirs including full descriptive accounts on the Quaternary are available. Details of the coverage and availability of maps, memoirs, sheet explanations and other publications are given in *Information Sources*.

The aim of this memoir is to consider landscape evolution throughout the Cainozoic era (Table 2), as well providing a description of the Quaternary deposits

Figure 1b
Topography of the district and locality map showing the boundaries of the eleven geological sheets. Additional topographical details are given on maps 1 to 11.

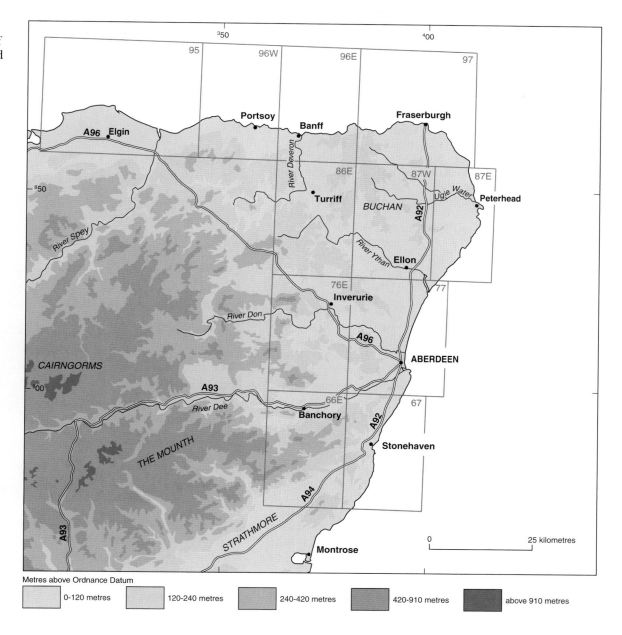

Metres above Ordnance Datum

| 0-120 metres | 120-240 metres | 240-420 metres | 420-910 metres | above 910 metres |

Table 2 Division of the Cainozoic.

Era	Period (System)	Epoch (Series)	Age (of base Ma)
CAINOZOIC	QUATERNARY	Holocene	0.01
		Pleistocene	2.4
		Pliocene	5.3
(Tertiary*) NEOGENE		Miocene	23.8
		Oligocene	33.7
PALAEOGENE		Eocene	54.8
		Paleocene	65.0

* This familiar term is now obsolete, but is used where it would be inappropriate to be more specific

shown on the above-mentioned maps. Many elements of the landscape were created prior to the Quaternary, and furthermore, important deposits of Palaeogene to Neogene age have been shown as drift deposits on some maps. Place names mentioned in the text and a selection of geomorphological features, deposits and other Cainozoic phenomena are shown on Maps 1 to 11 at the back of this publication.

The coastal lowlands of north-east Scotland flank the Grampian Highlands, which rise south-westwards towards the Cairngorm Mountains. Much of the district lies below the 250 m contour and is characterised by a series of ancient plateau surfaces eroded across a wide variety of rock types. The predominant surface is the rolling Buchan Plateau (Figure 1) lying between 60 and 150 m OD (Chapter 3). Only the most resistant quartzitic rocks form distinct hills, such as the conspicuous Mormond Hill (234 m OD), between Fraserburgh and Peterhead, and the more subdued Hill of Dudwick to the north-east of Ellon. The Buchan Plateau is crossed by a broad, gently undulating ridge that trends south-westwards from the coast at Troup Head (Sheet 96E) and which is developed mainly on relatively flat-lying Old Red Sandstone lithologies and Dalradian slates (Figure 2). Other similar trending ridges, some 220 to 310 m high, are formed mainly of steeply dipping Dalradian quartzite, and reach the high, rugged coastline of the Moray Firth between Banff and Portknockie. These ridges merge westwards, towards the River Spey, into a dissected plateau standing between 180 and 265 m OD. It is developed across Old Red Sandstone lithologies, Dalradian flaggy micaceous psammites and semipelites and more massive quartzites.

The Buchan Plateau is overlooked on its south-western margin by Bennachie (528 m OD) (Sheet 76E), a local granite landmark that has become part of the folklore of the region (Plate 1). Farther south, other large hills of granite also dominate the scenery inland from Aberdeen, including the Hill of Fare (471 m OD), Brimmond Hill

(266 m OD) and Kerloch (534 m OD). Many of the rocks of the district have been deeply weathered and these weakened strata have been eroded into wide basins, for example the basic igneous rocks around Maud and Insch (Figure 2). However, some basic rocks are relatively fresh and form isolated hills, such as the broad ridge that extends from Pitgavenny Hill (236 m OD) towards Belhelvie. Another belt of high ground largely underlain by granite forms the eastern continuation of The Mounth. It is bounded to the south by the Highland Boundary Fault and to the north by the Dee valley, which is a major topographic corridor descending from the Gaick Plateau and the Cairngorm Mountains to the west of the district. The extreme south of the district falls within the gently undulating vale of Strathmore, which is underlain mainly by Old Red Sandstone lithologies and associated late Silurian to Devonian volcanic rocks. Although Strathmore is generally low lying, conglomerates near the Highland Boundary Fault form ridges parallel to the fault and volcanic rocks form hills such as Hill of Bruxie (216 m), south-west of Stonehaven.

Plate 1 Bennachie (D4331).View of Bennachie taken from the north-east showing the two tops, Mither Tap (left) and Craigshannoch (right), capped by tors formed of the Bennachie Granite. The lower ground is underlain by sandy till of the East Grampian Drift Group resting on norites of the Insch Basic Intrusions.

The coastal communities of the district owe their existence mainly to the fishing industry, whereas those inland have relied on farming. Both industries have been in decline in recent years. Beef production was once prevalent, but arable farming has become increasingly important following the application of modern methods of land drainage to the notoriously poorly drained and stony soils of the region. Indeed, much of the Buchan Plateau now resembles parts of East Anglia following the widespread removal of boulders and stone walls. The production of aggregates makes an important contribution to the local economy, but the large stone quarrying industries formerly centred on the granites of Aberdeen and Peterhead have gone. Peat is still worked commercially around Strichen, but all of the clay-based brick and tile works in the district have closed. Forestry has become increasingly important, especially in the west of the area, but provides relatively little employment. Tourism is a steady, albeit mainly seasonal industry.

Without doubt, the North Sea oil- and gas-related industries are the mainstays of the present economy of northeast Scotland. However, while they have provided much-needed employment and wealth, they have brought about, or speeded up, irrevocable demographic and other changes across the region. This is especially so in the vicinities of Aberdeen and Peterhead where there has been a significant increase in population during the past 25 years or so. Large numbers of people have moved in from farther south. House prices have risen substantially and the increased demand for housing has placed pressures on local authorities to release land for building, some of which is not entirely suitable for such purposes (e.g. prone to flooding). Wage inflation has occurred causing the declining traditional activities to become uneconomic, especially the more marginal areas of agriculture where many hill farms have been bought to become residences and 'hobby' farms. There has been a dramatic increase in commuting across the entire region placing pressure on the infrastructure and leading to new road building and widespread road improvement projects.

SOLID GEOLOGY

A simplified map of the solid geology of north-east Scotland is shown in Figure 2. The nature of the bedrock that underlies the Pleistocene deposits has exerted some measure of control on their development and form. The hardness of the differing types of bedrock, their weathering profiles, propensity to jointing, and overall response to subglacial deformation are all important factors that have controlled the pattern of glacial erosion and deposition. Resistant rock types such as granite and quartzite generally form positive topographical features that have diverted the ice, while readily releasing large erratic clasts into tills. The gabbros and ultramafic rocks generally form areas of negative relief, but like the granites, they have widely spaced joints and have readily yielded large erratics boulders, many of which were originally 'corestones' within saprolites (Chapter 4). Their distribution can be used in the reconstruction of ice movement directions.

Easily fractured and finely jointed lithologies are also readily incorporated into ice sheets. Graphitic pelite and felsite are two examples of particularly common and widespread clasts, although neither are particularly resistant rock types. In general, the nature of the till in upland areas that formed beneath sluggish, possibly cold-based ice sheets strongly reflects the underlying bedrock.

The bedrock of the district ranges in age from late Precambrian to Early Cretaceous. Deposits of Palaeogene to Neogene age also occur (Chapter 4). An outline of the solid geology is given below, but further details can be found in Stephenson and Gould (1995) and Craig (1991). A wider view of the Precambrian Dalradian succession may be found in Harris et al. (1994, pp.43–50).

Dalradian

A large part of the district is underlain by crystalline metamorphic rocks of the Neoproterozoic Dalradian Supergroup. These metamorphic rocks represent original sedimentary rocks, now much altered by heat and pressure. Sandstones are thus represented by quartzites and quartz-feldspar-granulites (psammites), shales by various types of mica-schist and slate (pelites), and rocks intermediate between these two by quartz-mica-schists (semipelites). Limestones are now crystalline (marble), and originally muddy gritty sandstones have been weakly metamorphosed to become cleaved metagreywacke and schistose grit. The Dalradian sediments were deposited in a large intracontinental basin and later on a continental margin with small offshore oceanic basins.

An almost continuous succession from the upper part of the Grampian Group to a high level in the Southern Highland Group is exposed along the Moray Firth coast between Buckie and Macduff. The lowest part of the succession, assigned to the Grampian Group, consists of well-bedded quartzite and psammite that is mostly thickly bedded (Cullen Quartzite Formation), and was deposited in shallow water. The succeeding Appin Group was also deposited under shallow marine conditions, but the amount and grain size of the clastic material progressively decreased, and increasing quantities of limestone and organic material were deposited. The Lochaber Subgroup is represented by thinly bedded, flaggy psammite and semipelite; it becomes more calcareous upwards, eventually passing into conspicuous tremolite-rich semipelites, calc-silicate rocks and impure limestones (Findlater Flag and Cairnfield Flag formations). The overlying Ballachulish Subgroup consists of semipelite and pelite, commonly graphitic, interbedded with limestone units up to 35 m thick (Mortlach Graphitic Schist and Tarnash Phyllite and Limestone formations). The Blair Atholl Subgroup is represented by the Fordyce Limestone Formation, another interbedded pelite and limestone.

The Islay Subgroup at the base of the succeeding Argyll Group is marked by a discontinuous bed of meta-diamictite, the 'Boulder Bed', which records a glacial event correlated with the widespread Neoproterozoic Varanger glaciations of the North Atlantic region. Associated psammite and quartzite units (Arnbath and Durn Hill formations) are succeeded by mixed semipelitic,

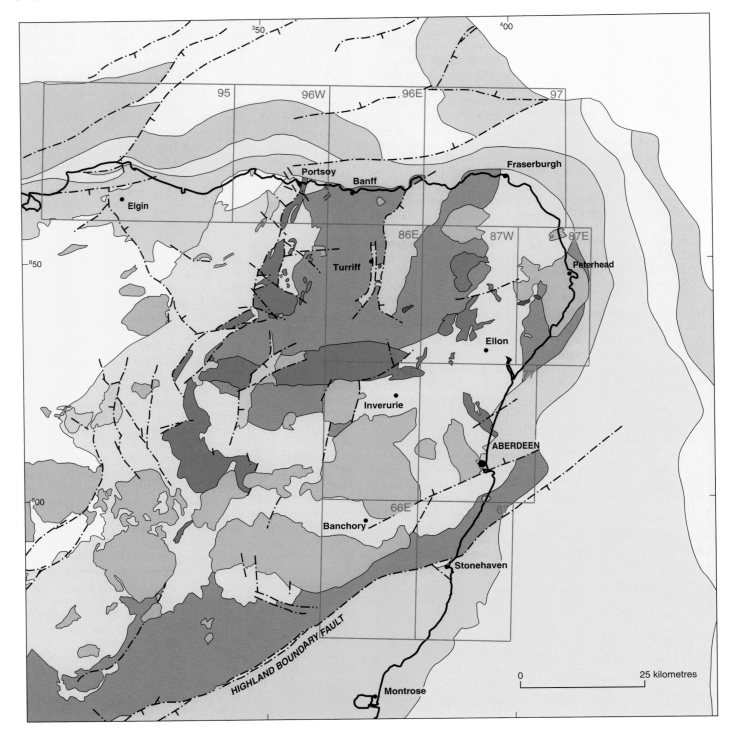

Figure 2   Solid geology of the district.

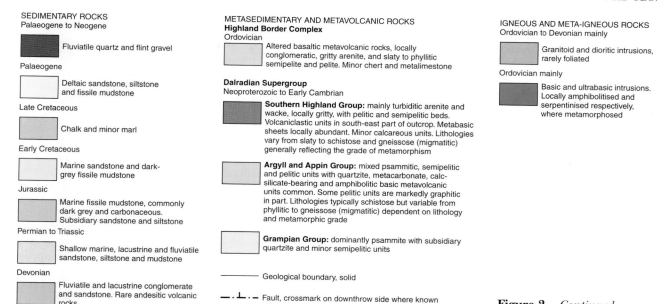

SEDIMENTARY ROCKS
Palaeogene to Neogene

Fluviatile quartz and flint gravel

Palaeogene

Deltaic sandstone, siltstone
and fissile mudstone

Late Cretaceous

Chalk and minor marl

Early Cretaceous

Marine sandstone and dark-
grey fissile mudstone

Jurassic

Marine fissile mudstone, commonly
dark grey and carbonaceous.
Subsidiary sandstone and siltstone

Permian to Triassic

Shallow marine, lacustrine and fluviatile
sandstone, siltstone and mudstone

Devonian

Fluviatile and lacustrine conglomerate
and sandstone. Rare andesitic volcanic
rocks

METASEDIMENTARY AND METAVOLCANIC ROCKS
Highland Border Complex
Ordovician

Altered basaltic metavolcanic rocks, locally
conglomeratic, gritty arenite, and slaty to phyllitic
semipelite and pelite. Minor chert and metalimestone

Dalradian Supergroup
Neoproterozoic to Early Cambrian

Southern Highland Group: mainly turbiditic arenite and
wacke, locally gritty, with pelitic and semipelitic beds.
Volcaniclastic units in south-east part of outcrop. Metabasic
sheets locally abundant. Minor calcareous units. Lithologies
vary from slaty to schistose and gneissose (migmatitic)
generally reflecting the grade of metamorphism

Argyll and Appin Group: mixed psammitic, semipelitic
and pelitic units with quartzite, metacarbonate, calc-
silicate-bearing and amphibolitic basic metavolcanic
units common. Some pelitic units are markedly graphitic
in part. Lithologies typically schistose but variable from
phyllitic to gneissose (migmatitic) dependent on lithology
and metamorphic grade

Grampian Group: dominantly psammite with subsidiary
quartzite and minor semipelitic units

——— Geological boundary, solid

— . ⊥ . — Fault, crossmark on downthrow side where known

IGNEOUS AND META-IGNEOUS ROCKS
Ordovician to Devonian mainly

Granitoid and dioritic intrusions,
rarely foliated

Ordovician mainly

Basic and ultrabasic intrusions.
Locally amphibolitised and
serpentinised respectively,
where metamorphosed

Figure 2 *Continued*

pelitic and calcareous rocks (Easdale Subgroup: Castle Point Pelite and Portsoy Limestone formations), and then by striped, thinly bedded psammites and semipelites (Crinan Subgroup: Cowhythe Psammite Formation). The overlying Boyne Limestone Formation (Tayvallich Subgroup) includes a 200 m-thick unit of 'relatively pure' limestone exposed on the coast section, surrounded by calc-silicates and calcareous semipelites.

The Southern Highland Group is locally unconformable on the Argyll Group west of Whitehills. The basal Whitehills Grit Formation is characterised by calcareous psammites and semipelites, while the overlying Macduff Slate Formation consists of micaceous semipelite and pelite with abundant units of gritty psammite. The group is characterised by turbidites, and was deposited from density currents in deeper water than the rest of the Dalradian Supergroup. The group includes the Macduff Boulder Bed, which preserves a record of distal glaciomarine sedimentation in the form of ice-rafted dropstones and related bedding-deformation structures (Stoker et al., 1999). The Macduff Slate Formation may extend from the Neoproterozoic into the Cambrian and possibly even Ordovician, but the evidence for this is contradictory and inconclusive.

To the east of Macduff and in the more southerly parts of the district, only the Southern Highland Group (Collieston, Glen Effock and Glen Lethnot formations) and the upper part of the Argyll Group (Aberdeen, Ellon and Strichen formations) are represented. The rocks are dominantly psammites and semipelites, with rare calc-silicate beds. Significant areas of limestone outcrop (Deeside Limestone Formation) are confined to the area immediately south-east of Banchory.

Caledonian orogeny and magmatism

The Dalradian rocks were folded and metamorphosed during the Caledonian orogeny, which reached its acme at about 490–470 Ma. Four phases of folding and deformation, D_1 to D_4, are recognised, but are rarely all present in any one place. In the district, the large-scale effects of the earliest phase, which produced the Tay Nappe, are seen in the Collieston area and south of Aberdeen, where the strata are regionally inverted. The abundant upright, open to tight, northerly plunging early folds on the Banffshire coast section may also relate to this early phase. The large-scale structure of the area north of Aberdeen and east of Portsoy is dominated by the later 'open', upright north-north-east-trending Turriff Syncline and the parallel Buchan Anticline to the east. They are D_3 or possibly D_4 in age. South of Aberdeen, gently dipping, inverted strata of the Southern Highland Group are tilted to the vertical along the Highland Border Downbend. This is a late regional east-north-east-trending monoform; its axial trace runs a few kilometres north of the Highland Boundary Fault. Major zones of ductile shearing occur both west of Portsoy (probably a continuation of the Keith Shear Zone), and in and just east of Portsoy (the Portsoy Shear Zone). They are also found between Inzie Head and Ellon, and along the northern and southern boundaries of the Insch basic intrusion. The Keith Shear Zone is associated with early granite intrusions and may itself reflect an early, Proterozoic structure. Other shear zones are closely associated with the emplacement of the basic and ultrabasic intrusions and are probably of early Ordovician age, although they may also reflect older lineaments in basement gneisses at depth.

To the east of Portsoy and to the north of Aberdeen, the regional metamorphism was of a relatively low-pressure type (Buchan metamorphism), and the characteristic minerals formed in pelites and semipelites were andalusite and cordierite. To the west of the Portsoy Shear Zone and south of Aberdeen, regional metamorphism took place at higher pressures, and the characteristic minerals to form in pelitic rocks were garnet, staurolite and kyanite.

Large volumes of basic magma were intruded into the Neoproterozoic metasedimentary rocks at about 470 Ma. It crystallised as several large layered bodies of basic and ultrabasic rock. The Insch, Haddo House, Arnage, Maud, Belhelvie and Portsoy masses lie within the district, and the nearby Huntly and Knock masses have also provided erratics to the till of the area. The commonest lithologies are serpentinite (derived by hydration of ultrabasic rocks), olivine gabbro, olivine norite, and gabbros and norites (both cumulate and granular types). Troctolite, monzonite and syenite are more restricted in outcrop. Contact metamorphism associated with the basic intrusions produced hornfelses rich in sillimanite and cordierite, and also produced characteristic 'contaminated' and 'hybrid' rocks such as cordierite-bearing norite. Locally these metamorphic effects appear to have reinforced the regional pattern giving rise to areas of semipelitic and psammitic gneisses.

A phase of late Neoproterozoic (about 600 Ma) acid plutonic intrusion produced the Windyhills and Boggierow granitic intrusions, which now lie along the Keith Shear Zone, and are deformed by it. Acid magmas were also intruded about 470 Ma and at 425–395 Ma. The earlier intrusions, confusingly termed the 'Older Granites', are grey muscovite-biotite granites (S-type) and are typically foliated. This group includes the Aberchirder, Strichen, Forest of Deer, Kemnay, and Aberdeen granites. The later intrusions show calc-alkali differentiation trends (B-type) and are more voluminous. They comprise an earlier group of quartz-diorites, tonalites and granodiorites, typically grey in colour, which include the Torphins, Crathes, Balblair and Clinterty intrusions, and a slightly later group of biotite-granites, generally pink and in some cases megacrystic, which show a highly evolved geochemistry. These include Peterhead, Bennachie, Hill of Fare and Mount Battock. The distant Cairngorm, Ballater and Glen Gairn plutons also belong to this suite.

A suite of minor intrusions is associated with the 420–390 Ma plutons. They are widely distributed, but their outcrops are too small to be shown on smaller scale maps. The principal lithologies are felsite, microgranite, quartz-feldspar porphyry, microdiorite, and the lamprophyres, spessartite and vogesite. The more acid lithologies can form hard and resistant blocks that have yielded erratics and, in places, form a conspicuous though minor component of sand and gravel deposits.

Ordovician, Devonian and Carboniferous

A very narrow, discontinuous strip of early Ordovician rocks, the Highland Border Complex, crops out just north-west of the Highland Boundary Fault. They include metabasalts, slaty pelites, gritty psammites and rare impure limestones.

Devonian rocks formed in arid intermontane basins following the Caledonian orogeny. They are found as outliers in the East Grampian area to the north of the Highland Boundary Fault, around Elgin, in the Turriff basin and in several smaller outliers including one found recently on the Moss of Cruden, near Peterhead (Appendix 1).

The Lower Devonian Crovie Group crops out mainly along the eastern side of the Turriff Basin. It consists of sandstones and conglomerates deposited on river floodplains and alluvial fans. The conglomerates underlying parts of Aberdeen are also assigned to the Lower Devonian. The unconformably overlying Middle Devonian rocks belong to the Inverness Sandstone Group. In the Turriff Basin, they form the Gardenstown Conglomerate Formation. This consists of a sequence of breccias and conglomerates, but also includes the Findon Fish Bed. Inland, the conglomerates are commonly deeply weathered. To the southeast of Elgin, and in outliers as far east as Deskford, the Spey Conglomerate Formation at the base of the Inverness Sandstone Group comprises a basal breccia overlain by a conglomerate with rounded cobbles mainly of quartzite set in a loose sandy matrix. It is overlain by red-brown and purple-brown flaggy sandstones of the Fochabers Sandstone Formation, which also contains beds of red- and purple-brown and green-grey mudstone. The unconformably overlying Upper Devonian Nairn Sandstone Formation, a gritty to pebbly arkose with scattered clasts of micaceous mudstone, crops out in the Alves–Elgin area.

To the south of the Highland Boundary Fault, between Edzell and Stonehaven, there is a thick sequence of late Silurian to early Devonian mainly fluviatile rocks that was deposited adjacent to the rising Highland block. The basal Stonehaven Group consists largely of sandstones with some siltstone and a thin andesitic lava flow. The Dunnottar–Crawton and Arbuthnott–Garvock groups here consist largely of conglomerate, derived partly from older metamorphic and igneous rocks and partly from contemporaneous lavas. Flows of andesite and mugearite are conspicuous in the coastal section. The uppermost Strathmore Group consists of sandstones and siltstones, some notably red, interspersed with fans of conglomerate, where large Devonian river systems drained the embryonic Highlands.

A suite of east–west-trending quartz-dolerite dykes of latest Carboniferous age occur throughout the district, though they are most abundant in the southern part. The dykes are typically wider than an earlier felsite to lamprophyre suite of Siluro-Devonian age, and large blocks are commonly found as erratics close to these dykes.

Mesozoic

Mesozoic sedimentary rocks are widely preserved offshore and reflect the stability of the land masses in the East Grampians and adjacent sea areas since Permian times. Permian and Triassic rocks are preserved only in the extreme north-west of the district, in a strip from Burghead to Lossiemouth. The Upper Permian to Lower Triassic Cutties Hillock and Hopeman sandstones comprise aeolian sandstones with large-scale dune-bedding and sparse fluviatile deposits. The overlying fluviatile sandstones and conglomerates of the Burghead Beds are succeeded by the Upper Triassic Lossiemouth Sandstone, which is also of aeolian origin. This grades upwards into the Stotfield Cherty Rock, a calcrete-dominated sandstone with chert concretions. Jurassic rocks have been proved at depth near Lossiemouth. They consist of basal calcareous mudstones (or marls) overlain by a rhythmic sequence of sandstones, siltstones and mudstones, and succeeded by coarse-grained, kaolinitic sandstones. A hitherto unknown outlier of Lower

Cretaceous glauconitic sandstone has been found recently on the Moss of Cruden (Hall and Jarvis, 1994; Appendix 1). No Upper Cretaceous rocks are known in north-east Scotland, but evidence of a Late Cretaceous transgression is given by the presence of nodular flints at the base of the Buchan Ridge Gravel Member (Chapter 4), which contains abundant flint pebbles yielding Cretaceous fossils.

A thick and more complete succession of Mesozoic rocks occupies the Moray Firth basin. Triassic sandstones rest unconformably on sandstones and conglomerates of the Old Red Sandstone and are overlain by sandstones, black shaly mudstones and siltstones of Early Jurassic to Early Cretaceous age. The Lower Cretaceous rocks also include glauconitic and calcareous sandstones. Upper Cretaceous chalk and calcareous mudstones crop out on the sea bed at the eastern end of the Moray Firth and parallel to the North Sea coast some 15 km offshore from Peterhead.

SUMMARY OF CAINOZOIC GEOLOGICAL HISTORY

Palaeogene and Neogene (formerly the Tertiary)

The geological record of this period of time on the mainland is largely one of landscape evolution by subaerial erosion and the development of deep weathering profiles. The Paleocene was marked by considerable volcanic activity along the western seaboard of Scotland. Uplift of as much as 1.5 km was associated with this event, causing an eastward tilting of peneplains formed across north-east Scotland during the Late Cretaceous and earlier. Cretaceous sediments in the developing Inner Moray Firth Basin and on the Buchan Plateau were exposed to erosion and many of the major valleys of the region were initiated as river systems (Chapter 3). They became established on the tilted peneplains and drained towards the deepening Moray Firth and North Sea basins, where sedimentation was rapid. For example the chalk is overlain by a thick succession of Paleocene sandstone, mudstone and siltstone with beds of lignite, which crop out in the eastern North Sea basin (Gatliff et al., 1994). The kaolinitic flint and quartzite-rich deposits of the Buchan Gravels Formation preserved on isolated hilltops in central Buchan were laid down in the catchments of such rivers, their present position resulting from subsequent topographical inversion (Chapter 3). A warm, humid climate prevailed until the late Miocene, during which time the effects of chemical weathering penetrated deeply below the land surface forming 'clayey gruss' saprolites (Chapter 4). Mechanical weathering became increasingly intense during the Pliocene and early Pleistocene as the climate deteriorated and granular disintegration occurred, particularly of coarse-grained igneous rocks, to form 'gruss' saprolites. Subsequent glacial erosion has failed to remove these ancient regoliths over large areas of the district, and they locally extend to depths of 20 m or more.

In this volume 'Tertiary' is used as an informal term in contexts where it would be inappropriate to be more specific.

Quaternary (Pleistocene and Holocene)

The history of the Pleistocene period is described fully in Chapter 5, where both the onshore and offshore evidence is examined. The Scottish mainland has experienced numerous glacial episodes during the past two million years or so, but only the last glaciation, the Main Late Devensian, has left an appreciable sedimentary record (Table 1). Despite over 150 years of research, the extent of this ice sheet in north-east Scotland remains controversial (Figure 3). Sparse pre-Devensian deposits representing stages back, possibly, to the Anglian occur locally, and many geomorphological features, such as some former glacial meltwater channels, have clearly evolved during several glacial episodes. The Quaternary succession is more complete offshore, where it thickens towards the axis of the North Sea basin.

The Quaternary deposits and features of the district occur in four distinct geomorphological domains and can be related to five distinct ice masses that co-existed during the Late Devensian and earlier glaciations (Figure 4). The eastern Grampian Highlands were mainly affected by relatively sluggish ice that flowed outwards from centres of accumulation in the Cairngorms and Gaick. It deposited mainly thin, sandy diamictons in this domain, followed during deglaciation by meltwater deposits that are mainly confined to major valleys such as the Dee and Spey.

A northern domain was affected by a powerful ice stream that flowed out of the Moray Firth and impinged on the northern coast of the district. It typically laid down dark bluish grey diamictons containing abundant erratics and rafts of Mesozoic rocks and cold-water Quaternary sediments dredged up from the bed of the firth. Ice from the Spey merged with the Moray Firth ice stream to the west of Portsoy. An eastern domain was influenced by ice that flowed north-eastwards from Strathmore and was deflected back onshore to the north of Aberdeen where it left a distinctive suite of reddish brown sediments. These contain clasts of Permo–Triassic sandstone and dolomite and possible Mesozoic calcareous siltstone from the adjacent North Sea basin, in addition to material from the Old Red Sandstone and Devonian volcanic rocks of Strathmore. The deflection was probably caused by Scandinavian ice offshore. Red till is widespread in Strathmore itself, but is relatively thin, demonstrating that this third domain was mainly one of glacial erosion.

The fourth distinctive geomorphological domain is that of the Buchan Plateau. This area was once thought by some to be 'moraineless', but glacial deposits do occur, although they are thin and patchy (Figure 4). Features of glacial erosion are very poorly developed, but they do indicate that the final ice movement was from the west. This observation is supported by the presence of tills with erratics derived from the west, and by the pattern and distribution of glacial drainage channels across the area that indicate westwards retreat of the ice. However, an earlier, south-eastward movement of ice deposited dark bluish grey tills which occur sporadically at depth. The widespread presence of deeply weathered rock in central

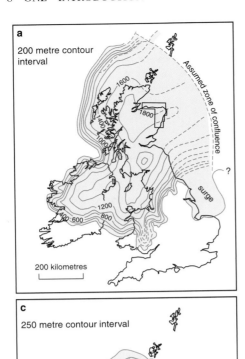

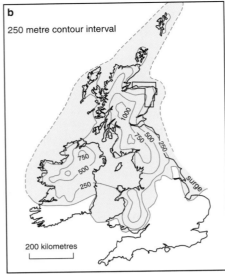

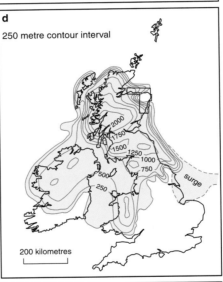

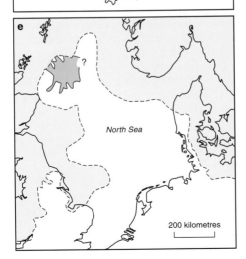

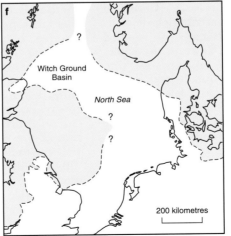

Figure 3 Models of the British Main Late Devensian ice sheet at its maximum extent.

a   Boulton et al. (1977)
b   Boulton et al. (1985)
c   Boulton et al.(1991)
d   Lambeck (1991)
e   Jansen (1976)
f   Ehlers and Wingfield (1991)

Figure 4
Generalised
flow-lines of ice
during the
Main Late
Devensian
glaciation and
profile map of
the five groups
of glacigenic
deposits
mapped in
north-east
Scotland.

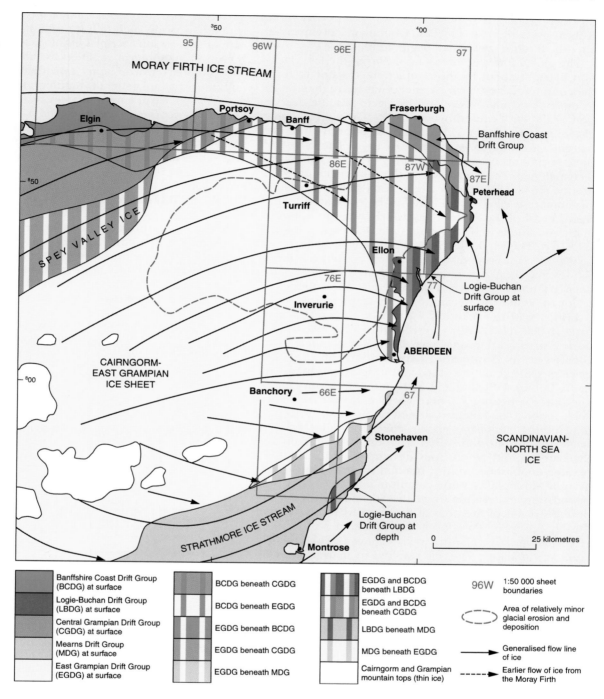

Buchan coupled with the survival of the Buchan Gravels indicates that minimal glacial erosion has occurred there throughout the Quaternary period, probably as a consequence of the ice being sluggish and cold-based in this peripheral part of the mainland. Periglacial deposits and structures such as ice-wedge polygons are relatively common across the Buchan Plateau leading to the belief, now considered unlikely, that it was unglaciated at least during the whole of the Late Devensian (Chapter 5).

Coastal glacigenic deposits commonly overlie tills derived from inland suggesting that the East Grampians ice had retreated from the coast prior to inundation by the Moray Firth and Strathmore ice streams (Figure 4). This view is supported by the presence, in the lower valleys of the Deveron, Ugie, Ythan and Don, of laminated glaciolacustrine deposits that locally interdigitate with the coastal tills, but which generally rest on tills derived from inland. The lakes were dammed by the coastal ice streams and, for some time, meltwaters ponded in the catchment of the Deveron may have flowed into that of the Ythan via a prominant glacial spillway between Turriff and Fyvie. In Strathmore, red deposits are widespread in the lower reaches of the Grampian valleys, but here there is local evidence that the East Grampian ice expanded after the Strathmore ice had begun to retreat.

The global lowering of sea level during the last glaciation brought about by the abstraction of water to form the continental and local ice sheets was more than offset in most of mainland Scotland by glacio-isostatic depression of the land caused by ice loading. As a result some glaciomarine sediments, laid down at the retreating ice margins, are now raised above sea level. The oldest of these sediments occur in the vicinity of the St Fergus gas terminal, near Peterhead, where they have been dated to around 15 000 BP (radiocarbon years before present). Glaciomarine clays near Elgin can be compared to the 'Errol Beds' of the Firth of Tay, which were laid down in arctic conditions while the main Late Devensian ice sheet retreated. The whole district was probably deglaciated by 13 000 BP. However, relative sea level continued to fall for several thousand years, leading to the incision of deep channels that are now buried at the mouths of major rivers like the Dee and Ythan (Chapter 7).

Evidence of the relatively warm Windermere Interstadial (13 000–11 000 BP) is restricted to a few sites where buried organic deposits have been found, such as at Glenbervie, in Strathmore (Appendix 1 Site 25). No glaciers existed in the district during the subsequent Loch Lomond Stadial (11 000–10 000 BP), but many periglacial phenomena were formed during this very cold period. The climate warmed abruptly at the beginning of the Holocene epoch, the present interglacial stage. Intense erosion would have occurred initially on the bare ground before soils developed and vegetation became established. At first, braided rivers flowing across gravelly floodplains were common, but they later stabilised into the mainly single-thread and locally meandering streams of the present day. Relative sea level rose during the Main Postglacial Transgression of the mid-Holocene, peaking at a few metres above OD about 6000 years ago (Chapter 7). This resulted in marine inundation of the lower reaches of river valleys and deposition of estuarine silt, sand and clay, commonly on terrestrial peats and fluvial sediment laid down earlier in the Holocene. There is evidence preserved near Fraserburgh (Appendix 1 Site 9) and in the lower Ythan valley that at least one huge tidal wave (tsunami) hit the coast about 7000 years ago. It was generated by a massive submarine slide on the Norwegian continental margin.

CLASSIFICATION OF DEPOSITS

The Drift editions of most BGS 1:50 000 scale maps covering north-east Scotland show deposits that have been classified using a *morpho-lithogenetic* scheme and identified by standard colours and symbols (Figure 5), for example 'Glaciofluvial ice-contact deposits' in pink or crimson and 'Peat' in brown. This method of classification has proved to be a practical means of mapping deposits cropping out at the surface and it is particularly appropriate for air photo interpretation. The Quaternary deposits classified in this way are described systematically in Chapter 6.

The symbols have been embellished on more recently published maps, such that lithostratigraphical map codes are added as superscripts, lithological codes as prefixes, chronostratigraphical qualifiers as subscripts and inferred depositional environments as suffixes (Figure 5). More subcategories are generally found on the detailed 1:10 000 or 1:10 560 'clean copies' that are available for large parts of the district (see *Information sources*). The symbol scheme has been modified over the years with the result that there are variations in presentation between sheets, although the differences are largely semantic (Table 3).

LITHOSTRATIGRAPHY

The morpho-lithogenetic scheme does have some failings. For example, it does not easily accommodate complicated sequences of deposits or bodies of sediment that contain a mix of lithologies. In order to overcome these difficulties, the more recently published maps depict some deposits that have been classified *lithostratigraphically* in accordance with internationally agreed codes (Hedberg, 1976). The Neogene deposits have also been classified in this way. It is beyond the scope of this

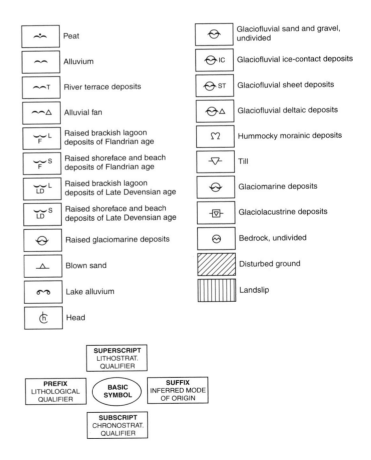

Figure 5 Key to the most common morpho-lithogenetic symbols used on recently published BGS drift maps of the district.

publication to inaugurate a formal lithostratigraphical framework that systematically encompasses all the Quaternary deposits occurring across the district, but a start has been made with the glacigenic deposits (Chapter 8). For example, deposits laid down in association with ice that flowed out of the Moray Firth are placed within the *Banffshire Coast Drift Group* (formerly the 'blue-grey series'), whereas deposits laid down by more local ice from the Grampian mountains are allotted to the *East Grampian Drift Group* (formerly the 'inland series'). The former 'red series' is divided so that deposits laid down by ice flowing through Strathmore (*Mearns Drift Group*) are distinguished from those deposited by ice moving onshore from the North Sea (*Logie-Buchan Drift Group*). Deposits laid down by ice entering the district from the Monadhliath mountains and the Spey valley are placed in the *Central Grampian Drift Group*. Subdivisions at formation, member and bed level are described in Chapter 8 and the type localities of most units are described in Appendix 1. The original names of units are retained whenever possible.

A new mapping-related lithostratigraphical framework for the Quaternary of Britain has been proposed recently by McMillan and Hamblin (2000). For example, fluvial, lacustrine, estuarine, coastal and aeolian deposits are defined geographically within a series of Catchment Groups defined by major river drainage systems. It is proposed to change the ranking and naming of the glacigenic groups described here slightly. The Central Grampian Drift Group may be subdivided in order to

Table 3   Synonyms used on BGS drift maps of the district.

Current use	Obsolete term
Alluvium	River alluvium Floodplain alluvium Alluvium of the first terrace
River terrace deposits	Alluvium of the 2nd, 3rd … terrace
Alluvial fan deposits	Alluvial cone
Till	Boulder clay
Glaciofluvial sheet deposits	Fluvioglacial sand and gravel Fluvioglacial terrace deposits Glacial meltwater deposits … terraced
Glaciofluvial ice-contact deposits	Glacial sand and gravel Glacial meltwater deposits … moundy
Glaciolacustrine deposits	Brick clay Loam
Hummocky glacial deposits	Morainic drift

identify the sandy tills of the Inverness area, which extend towards Elgin.

CHRONOSTRATIGRAPHY

Chronostratigraphy is the definition of internationally agreed boundaries to units of strata (*systems, series and stages*) that correspond to intervals of geological time (*periods, epochs and ages*). The Cainozoic era is divided here into three periods, the Palaeogene, Neogene and Quaternary (the first two formerly known as the 'Tertiary'; Table 2). The Quaternary period embraces two epochs, the *Pleistocene* ('Ice Age') and the *Holocene* (the last 10 ka), but the latter is really the last of a series of relatively short-lived, warm *interglacial* stages separating longer *glacial* stages. The Quaternary has been divided traditionally into climato-stratigraphical stages because climatic change has had a dominant influence on sedimentation. The British chronostratigraphy embraces an alternating sequence of glacial and interglacial stages that have been defined principally in East Anglia (Figure 6; Table 1), but problems have arisen because the geological record there is incomplete. The more comprehensive north-west European scheme has been adopted offshore. The placing of the base of the Quaternary is discussed in Chapter 5. In northern Britain, glacial, periglacial (tundra) and boreal (like central Scandinavia) environments existed during the glacial stages, and both boreal and temperate environments occurred at times during the interglacial stages.

The Pleistocene *Series* is divided formally into Lower, Middle and Upper as shown on Figure 33. The Quaternary *Period* is divided here into Early, Middle and Late as shown on Table 1. Although capital letters are used here for these divisions of the Quaternary, the precise definition and formal status of many of them are not agreed internationally.

The terrestrial record of the Quaternary is fragmentary because later glaciations have removed most of the evidence of earlier events. In contrast, cores taken from deep ocean floors provide a more continuous record. A complete chronostratigraphy has been based on the analysis of calcareous microfossils preserved in these sediments. The changing microfaunal assemblages preserve a record of fluctuating oceanic water temperature (Ruddiman and Raymo, 1988) and the relative proportions of the two common isotopes of oxygen contained in the tests provide a proxy record of global ice volume and sea level (Imbrie et al., 1984). During glacial periods, water is lost from the oceans to form ice sheets. The oceans consequently become relatively enriched in water containing the heavy isotope of oxygen (O^{18}). The oscillating O^{18} content of ocean water (as determined from the carbonate tests of organisms that lived in it) can thus be used as an index of ice sheet growth and decay. The *oxygen isotope stages* thus defined now provide a universal means of dividing the Quaternary (Emiliani, 1954; Shackleton and Opdyke, 1973). The even numbered stages generally refer to cold periods and the odd numbered stages to the warm part of each global glacial–interglacial

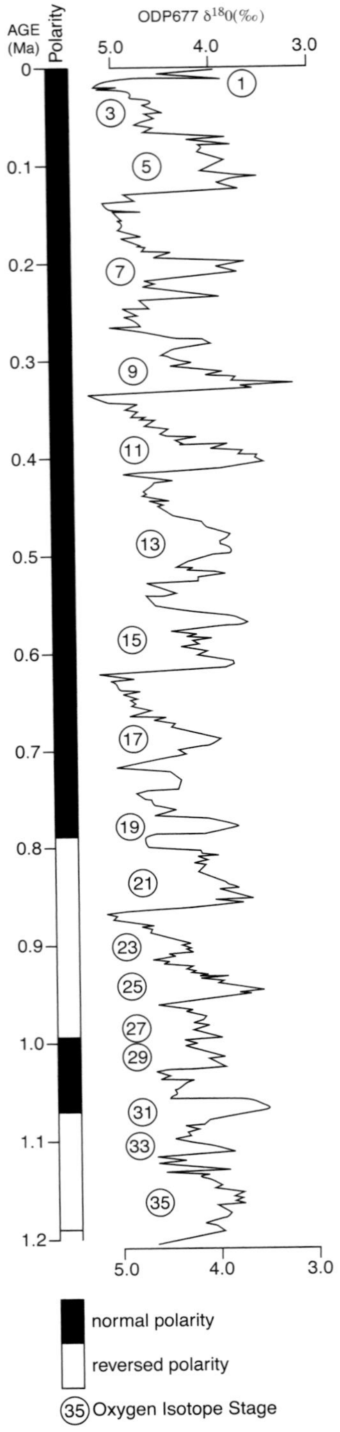

	Age (ka BP)	Oxygen Isotope Stage	Chronostratigraphy: Britain			Chronostratigraphy: NW Europe	
Late Quaternary	10	1	Holocene			Holocene	
Late Quaternary	11	1	Devensian	Late Devensian	Late-glacial * : Loch Lomond Stadial	Weichselian	Younger Dryas Stadial
Late Quaternary	12, 13	2	Devensian	Late Devensian	Late-glacial * : Late-glacial (Windermere) Interstadial	Weichselian	Allerød Interstadial; Early Dryas Stadial; Bølling Interstadial
Late Quaternary	24	3	Devensian	Late Devensian	Dimlington Stadial	Weichselian	Pleniglacial
Late Quaternary	26, 30	3	Devensian	Middle Devensian	'Sourlie Interstadial' (Upton Warren)	Weichselian	
Late Quaternary	40, 50	3	Devensian	Middle Devensian		Weichselian	Denekamp Interstadial; Hengelo Interstadial; Glinde Interstadial; Oerel Interstadial
Late Quaternary	59	4	Devensian	Early Devensian		Weichselian	
Late Quaternary	71	5a	Devensian	Early Devensian	'Brimpton Interstadial'	Weichselian	Odderade Interstadial
Late Quaternary	85	5b	Devensian	Early Devensian		Weichselian	Rederstall Stadial
Late Quaternary	92	5c	Devensian	Early Devensian	'Chelford Interstadial'	Weichselian	Brørup Interstadial
Late Quaternary	105	5d	Devensian	Early Devensian		Weichselian	Herning Stadial
Late Quaternary	116	5e	Ipswichian			Eemian	
Middle Quaternary	128	6	'Wolstonian'			Saale (Warthe)	
Middle Quaternary	186	7	'Ilfordian' (Strensham)				
Middle Quaternary	245	8				Saale (Drenthe)	
Middle Quaternary	303	9	'Hoxnian' (Hoxne)				
Middle Quaternary	339	10					
Middle Quaternary	362	11	'Hoxnian' (Swanscombe)			Holsteinian	
Middle Quaternary	423	12	Anglian			Elsterian	
Middle Quaternary	480	13-19	Cromerian (in part)			Cromerian Complex	
Early Quaternary	780–2400	20+				Bavelian	

* Note: The 'Late-glacial' spans the time between the retreat of the Main Late Devensian ice sheet in an area and the beginning of the Holocene (Gray and Lowe, 1977).

Figure 6 British and north-west European chronostratigraphy (after Gordon, 1997; Bowen et al., 1999) and a representative oxygen isotope and geomagnetic polarity record (ODP 677) (after Shackleton et al., 1990; Bowen et al., 1999). See also Figure 33.

cycle. An exception is Stage 5, which is broken down into 5a–5e where only 5e is generally accepted to represent a full interglacial (Figure 7). The oxygen isotope stages (OIS) are used in this publication in association with the 'long' British stratigraphy onshore and the north-west European scheme offshore (Figure 6), although many correlations with oxygen isotope stages remain uncertain and are the subject of active research.

GEOCHRONOMETRY

Radiocarbon dating

Radiocarbon dating is the principal method for determining the age of organic materials from the present to about 60 000 years ago. The method takes advantage of the natural occurrence of a radioactive isotope of carbon (^{14}C), which is produced in the upper atmosphere by the interaction of cosmic ray neutrons with nitrogen-14. The carbon-14 is taken up by plants during photosynthesis and then passes up the food chain to other organisms. Once an organism dies the ^{14}C in its structure is gradually lost by radioactive disintegration back to ^{14}N. Carbon-14 decays by beta particle emission with a half life of 5730 years. Conventional radiocarbon dating involves the measurement of beta particle transformation and is primarily limited by the amount of raw carbon that is available for counting. The Accelerator Mass Spectrometer (AMS) technique involves counting the carbon atoms rather than the beta particles emitted during decay. AMS radiocarbon dating is generally superior to conventional methods and, importantly, only very small samples (from 2 mg to 5 mg) are required for dating.

Dates quoted in this memoir in the style 12.5 ka* BP are conventional (uncalibrated) radiocarbon years before present (taken as 1950). Dates on shell material should

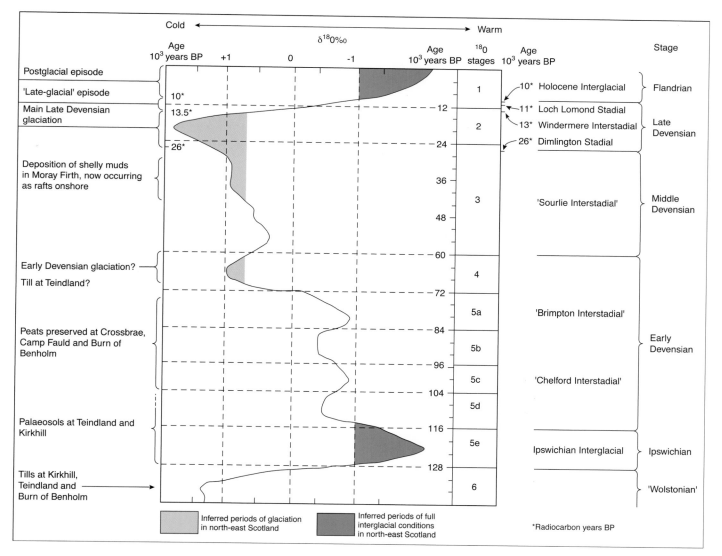

Figure 7 The 'SPECMAP' oxygen isotope curve for the last glacial–interglacial cycle (after Imbrie et al., 1984) with climato-stratigraphical stages and selected events in north-east Scotland.

have been *adjusted* to take into account the 'marine reservoir effect' for British waters, which involves subtracting 405 ± 40 years from the conventional radiocarbon age. It should be noted that calendar (sidereal) ages based on historical records, annual layering in ice cores, tree rings, varve counting etc are somewhat older than conventional radiocarbon ages. **If ages quoted in this publication have been *calibrated* to take this disparity into account, the Calib 3.0 radiocarbon calibration program (Stuiver and Reimer, 1993) has been used, and they are quoted as 'cal yr BP'.**

Amino-acid dating

Amino-acid dating has been used widely in north-east Scotland to date deposits older than the last glaciation. The technique mainly involves the analysis of proteins locked-up in the shells of certain molluscs and tests of foraminiferids (Sykes, 1991). Several time-dependent chemical reactions occur upon death that provide a means for relative dating. The most useful reaction is racemisation, where L-isomers of individual amino acids transform to the D-configuration. North-east Scotland has been important in the development of British *aminostratigraphy*, where amino-acid ratios have been used to rank fossils and their associated sediments according to relative age (for example Figure 43).

Luminescence dating

Luminescence dating is the collective term that covers a variety of dating methods, specifically thermoluminescence (TL) and optically stimulated luminescence (OSL). These methods are based on the principle that naturally occurring minerals like quartz and feldspar can act as dose meters, recording the amount of nuclear radiation that they have been exposed to (Miller, 1990). The total radioactive dose can be calculated by making a series of luminescence measurements, either where the sample is heated (TL) or exposed to a monochromatic or narrow band source of light (OSL). The methods are particularly applicable for dating loess and wind-blown sand, but less reliable for glaciofluvial and glaciolacustrine deposits. Several dates have been obtained from the sequences at Kirkhill Leys, Teindland and Howe of Byth (Appendix 1).

*   ka   thousand years

TWO

Applied geology

The Quaternary sediments of north-east Scotland play a major role in the economic development of the district. Although these deposits are generally thin, they cover some 90 per cent of the land surface onshore and most of the sea bed in the surrounding offshore area. Quaternary strata form the foundations for most urban, and industrial buildings, as well as road and rail links. They are also the source of most of the aggregate used in their construction. Much of the district's drinking water has passed through these sediments and its most fertile soils are developed upon them. Many nature reserves and Sites of Special Scientific Interest (SSSIs), especially those concerned with wetland and dune habitats, result from accumulations of peat and blown sand during post-glacial times. Domestic and industrial waste is disposed by landfill into former workings in Quaternary sands and gravels or tipped on top of undisturbed permeable Quaternary sediments. Knowledge of the extent and nature of these deposits and other types of made ground, and the engineering properties of in situ Quaternary strata and deeply weathered bedrock, is crucial in determining ground conditions for future urban and industrial developments and protecting groundwater resources.

BULK MINERAL RESOURCES

Construction and industrial minerals form an important resource in north-east Scotland. Sand and gravel deposits are by far the most important and reserves with planning permission accounted for 25 per cent of the Scottish total in 1993 (BGS, 1998). The value of naturally occurring aggregate extracted from north-east Scotland was approximately £6 million in 1997. Of this, 431 000 tonnes were used for fill, 350 000 tonnes for concreting sand, 290 000 tonnes as coarse aggregate for concrete and 109 000 tonnes for building sand. In 1997, twenty-two major workings (Figure 8) were notified to BGS as being operational within the district (Cameron et al., 1998). Brick clay and peat also constitute significant bulk resources in north-east Scotland, although there is little current commercial peat extraction and no large-scale brick clay working from these deposits.

A brief general account of the most extensive spreads of sand and gravel in north-east Scotland is given below, but, because of their economic importance, abundance and variability, these resources are more fully described in Appendix 2.

Sand and gravel

Sands and gravels of several types occur in north-east Scotland (Chapter 6). Each type reflects its mode of deposition and the rocks from which it was derived. The extensive glaciofluvial deposits that flank the lower reaches of most of the major river valleys, and extend onto the adjacent inter-fluves, are the principal sources of naturally occurring coarse-grained bulk aggregates. Most were laid down at the close of the Main Late Devensian glaciation (Chapter 5). Less extensive spreads of sand and gravel accumulated around the coast as raised beach and marine deposits (Chapter 7), when sea level was higher than at present (during the Late-glacial and mid-Holocene). Sands and gravels that were deposited by postglacial streams and rivers also commonly underlie floodplains and river terraces, while extensive dunes of blown sand fringe the coast, particularly in the eastern part of the district.

The most important sand and gravel resources occur as moundy and flat topped glaciofluvial deposits. These flank the lower reaches of the valleys of the rivers Don and Dee, and extend onto the adjacent interfluves around the city of Aberdeen (the largest market for aggregate in northern Scotland). However, many of the most attractive deposits around Aberdeen have been extensively worked, and some of the resources close to the city have been sterilised by urban expansion. More extensive spreads of good quality sand and gravel, suitable for most end-uses, occur in the coastal lowlands between Forres and Elgin. They form coarsening-upward sequences that were laid down as fans at the mouths of drainage channels or in temporary ice-dammed lakes, kettled terraced spreads that were laid down on bodies of stagnant ice during deglaciation and gravelly Late-glacial raised beaches. Thick deposits of gravel and sand are locally present beneath river flood-plains, but none is exploited at present, as the bulk of the resource lies beneath the water table.

Thick, easily worked resources of sand and gravel, lying above the water table, form flights of terraces flanking the North Ugie Water, South Ugie Water, and the rivers Lossie, Spey, and Ythan. The Ythan terrace gravels contain a high proportion of boulders, but have been extensively worked on both sides of the river downstream of Methlick. Gravels were also won from several pits sited on the Ugie terraces. Moundy glaciofluvial ice-contact deposits that crop out inland of Fraserbrugh, between Aberdeen and Peterhead, and between Stonehaven and Auchenblae are also major resources. However, these accumulations are commonly discontinuous and individual deposits may vary considerably in thickness and quality over relatively short distances. Typically, they contain waste partings of silt and clay and may be concealed beneath a thin overburden of till.

Most deposits, particularly those in the upland areas, contain high proportions of clasts derived from crystalline resistant rock types (Appendix 2), and after on-site screening, washing, crushing and grading, they produce high-quality aggregates that suit most end-uses (Merritt et

Figure 8
Major sand and
gravel workings
in north-east
Scotland
(1999).

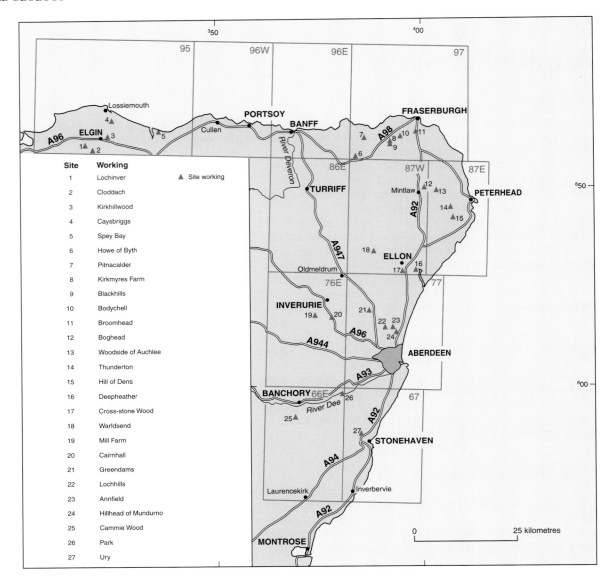

Site	Working
1	Lochinver
2	Cloddach
3	Kirkhillwood
4	Caysbriggs
5	Spey Bay
6	Howe of Byth
7	Pitnacalder
8	Kirkmyres Farm
9	Blackhills
10	Bodychell
11	Broomhead
12	Boghead
13	Woodside of Auchlee
14	Thunderton
15	Hill of Dens
16	Deepheather
17	Cross-stone Wood
18	Warldsend
19	Mill Farm
20	Cairnhall
21	Greendams
22	Lochhills
23	Annfield
24	Hillhead of Mundurno
25	Cammie Wood
26	Park
27	Ury

▲ Site working

al., 1988). However, the spreads of sand and gravel overlying Devonian bedrock, such as those in the south-eastern part of the district, tend to yield weaker aggregates, as they locally contain significant amounts of friable sandstone, mudstone and porphyritic lavas. Much of the sand and gravel in north-east Scotland is used in the production of ready-mix concrete and concrete products, as a source of mortaring and plastering sand, and as fill in civil engineering works (Plate 2). Some sand is used for coated aggregate in road building, but little gravel is used for road surfacing, as crushed-rock aggregate provides clasts with better resistance to abrasion and polishing. Poorer quality sand and gravel is used extensively in the construction of unmetalled roads and tracks.

Potentially workable deposits of coarse-grained aggregate also occur within the Buchan Gravels Formation, an unlithified gravelly deposit of Palaeogene to Neogene age, which crops out between Peterhead and the upper reaches of the Deveron and Ythan valleys (Chapter 4). In

some areas, sands, and less commonly gravels, derived from decomposed igneous, metamorphic and sedimentary bedrock are also worked as fill for untarred roads.

Brick clay

Clays suitable for brick and tile making are widely distributed within the Quaternary deposits of north-east Scotland, and have been worked at almost 20 sites during the last two centuries. Most workings were small-scale and served local markets; no commercial brick or tile making currently takes place. The widely scattered, variable quality and remote nature of many deposits, the small size of reserves at most sites and the relatively high costs of firing superficial clays, all contributed to the decline of the industry.

The only comprehensive evaluation of clays in northeast Scotland suitable for brick, tile and pipe-making was undertaken by Eyles et al. (1946). This work, which forms

Plate 2 Mill of Dyce sand and gravel pit in glaciofluvial deposits on the southern side of the valley of the River Don north-west of Aberdeen (D4795A).

the basis of the following account, provided geological descriptions of all of the worked and potentially workable clay deposits that had been identified in the district prior to the end of the Second World War. It also reported the results of physical and chemical tests on deposits at a selection of clay pits, which were primarily undertaken to establish the suitability of each clay deposit for brick making. The results showed the material to be of variable, but generally rather poor quality, but suitable for the production of drainage pipes, tiles and common bricks.

The data provided by Eyles et al. (1946), together with that gleaned during more recent geological surveys have been used to compile a map of the main sources of brick clay in the district (Figure 9). A total of 29 brick clay sites are recognised, of which 21 have been worked commercially. Eyles et al. (1946) established that common bricks had been made from a wide variety of clayey Quaternary sediments, including, 'boulder clay' (till), 'fluvioglacial clay' (clayey glaciolacustrine deposits), clayey raised marine deposits and alluvium. Workings for brick clay were also recorded in large glacially transported rafts ('erratics') of Jurassic mudstone. Data on the type of clay deposit and its maximum thickness at each site are summarised in Table 4. This shows that clays of marine, glaciomarine and glaciolacustrine origin (Chapter 6) have been the most widely worked for brick making, possibly because they are the least heterogeneous, widespread clayey deposits in the district.

Silty clays of marine origin (Spynie Clay Formation) were formerly worked for brick and tile making in two pits close to the town of Elgin (Figure 9; Map 1). The clay deposits occur close to sea level, beneath shelly sands and peat near Loch Spynie. Considerable resources are thought to be present north of Milltown Airfield [NJ 266 658] and beneath Lossiemouth Airfield [NJ 210 695] (Peacock et al., 1968).

Dark blue-grey clay with wisps of sand was worked for pipes, bricks and tiles in a clay pit at Tochieneal [NJ 521 652] (Map 2), south of Cullen, though test results indicate that mixing sand with the clay would improve its suitability for brick making. Previous accounts are undecided as to whether the worked deposits are part of a large erratic of Jurassic mudstone or a Quaternary marine deposit. It is probable that they are rafts within the Whitehills Glacigenic Formation (Chapter 8). Similar rafted material was also worked in pits at Blackpots [NJ 659 657] near Whitehills (Map 3), and at Plaidy [NJ 730 550], north of Turriff (Map 5).

Greenish grey laminated clays and silts, containing dropstones and marine shells of Late Devensian age, have been worked for brick making in a pit near Annachie [NK 105 529], north of Peterhead (Appendix 1   St Fergus). Much of the resource is concealed beneath alluvium, but 16 m of laminated silt and clay were recorded in a nearby BGS borehole (NK15SW1).

Most recent brick production has been concentrated on Quaternary clay resources between Peterhead and Aberdeen, which are reviewed by Ridgeway (1982). Until the middle part of the 1980s, reddish brown laminated silty clays and associated clayey tills were worked for brick making at the Cruden Bay Brick and Tile Works at Errollston [NK 088 368], south of Peterhead (Appendix 1 Errollston; Map 7). Laminated silts, sands and clays, interbedded with silty clayey diamicton, were recorded to a depth of 6.2 m, in a section at the working pit examined during 1974 by Peacock (1984). A comparable section was recorded in 1944, by Eyles et al. (1946), although the workable deposits were reported to extend

Figure 9
Principal former brick-clay localities in north-east Scotland.

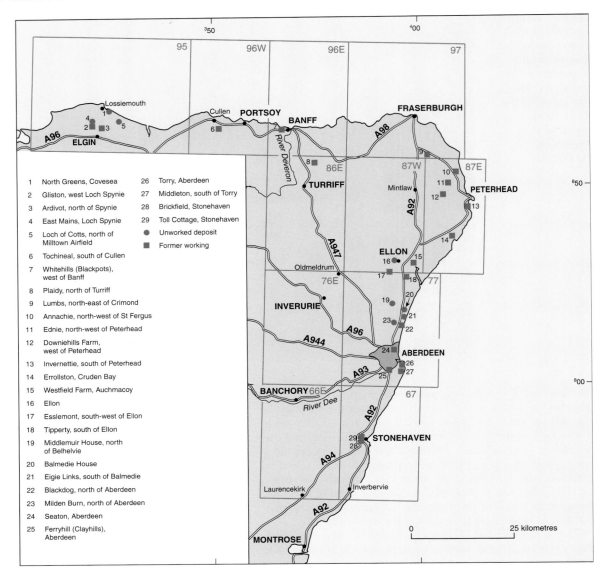

1   North Greens, Covesea
2   Gliston, west Loch Spynie
3   Ardivot, north of Spynie
4   East Mains, Loch Spynie
5   Loch of Cotts, north of Milltown Airfield
6   Tochineal, south of Cullen
7   Whitehills (Blackpots), west of Banff
8   Plaidy, north of Turriff
9   Lumbs, north-east of Crimond
10 Annachie, north-west of St Fergus
11 Ednie, north-west of Peterhead
12 Downiehills Farm, west of Peterhead
13 Invernettie, south of Peterhead
14 Errollston, Cruden Bay
15 Westfield Farm, Auchmacoy
16 Ellon
17 Esslemont, south-west of Ellon
18 Tipperty, south of Ellon
19 Middlemuir House, north of Belhelvie
20 Balmedie House
21 Eigie Links, south of Balmedie
22 Blackdog, north of Aberdeen
23 Milden Burn, north of Aberdeen
24 Seaton, Aberdeen
25 Ferryhill (Clayhills), Aberdeen

26 Torry, Aberdeen
27 Middleton, south of Torry
28 Brickfield, Stonehaven
29 Toll Cottage, Stonehaven
●   Unworked deposit
■   Former working

to more than 24 m depth in places. The test results indicate that the Errolston deposit is suitable for the manufacture of facing bricks as well as common bricks.

Reddish brown clay and silt was also worked by the Cruden Bay Brick and Tile Company in a pit in the valley of the Tarty Burn at Tipperty [NJ 971 268] (Map 9) until the middle part of the 1980s. Eyles et al. (1946) record 5.3 m of laminated clay resting on sand and gravel in the working face exposed in 1944. The form and extent of the deposit, which was investigated by mapping and trial pitting during the 1970s, is illustrated in Munro (1986, fig. 37). At the time, 2 m of unstratified 'red' clay passing downward into 2 m of laminated silt and clay was recorded from the working pit. The laminated unit was seen to overlie sand and gravel resting on red-brown till. In 1944, the clay was only worked for the manufacture of agricultural drain tiles, but facing bricks were produced during the 1980s.

In and around the city of Aberdeen several clay pits were worked in the past; for example Seaton, Blackdog, Ferryhill, and Torry (Map 9). The sites of all but the first mentioned

have long been abandoned and the exact extent of each working is unclear. About 1 m of crudely laminated 'bluish-yellow' clay was exposed in the pit at Seaton Brick Works in 1944, but 4.9 m of similar material, beneath 0.9 m of gravel was recorded from the site by Jamieson (1858). Bricks were produced at Seaton during the 19th century. By 1944, however, the deposit was worked solely for the manufacture of earthenware pottery and horticultural ware, such as flower pots. The worked deposits are assigned to the Tullos Clay Member of the Logie-Buchan Drift Group.

Peat

Widespread commercial exploitation of peat in north-east Scotland has been limited by transportation costs and environmental concerns. Production for horticultural use and as fuel has been concentrated around New Pitsligo and Strichen in Buchan, but resources there are almost exhausted. Other patchy deposits of peat are widespread, but there are few major resources. Most are restricted to

Table 4   Brick clay deposits in north-east Scotland.

Locality	Sheet	Site	Thickness* m	Postulated origin	Published analytical results	Formerly worked	References
1	95(w)	North Greens, Covesea	6	Raised glaciomarine deposit of Spynie Clay Formation	yes	no	Eyles et al., 1946
2	95(w)	Gilston, W of Loch Spynie	13	Raised glaciomarine deposit of Spynie Clay Formation	no	yes	Eyles et al., 1946; Peacock et al., 1968
3	95(w)	Ardivot, N of Loch Spynie	15	Raised glaciomarine deposit of Spynie Clay Formation	no	yes	Peacock et al., 1968
4	95(w)	East Mains, Loch Spynie	20	Raised glaciomarine deposit of Spynie Clay Formation	no	no	Peacock et al., 1968
5	95(e)	Loch of Cotts, N of Milltown Airfield	~	Raised glaciomarine deposit of Spynie Clay Formation	no	no	Peacock et al., 1968
6	96 W	Tochieneal, S of Cullen	4	Erratic raft of Jurassic mudstone and marine deposit	yes	yes	Read, 1923; Eyles et al., 1946
7	96 E	Whitehills (Blackpots), W. of Banff	6	Erratic raft of Jurassic mudstone and marine deposit	no	yes	Read, 1923; Eyles et al., 1946
8	86 E	Plaidy, N. of Turriff	8	Erratic raft of Jurassic mudstone	no	yes	Eyles et al., 1946
9	97(e)	Lumbs, NE of Crimond	~	Till of Banffshire Coast Drift Group	no	yes	Eyles et al., 1946; Peacock, 1983
10	87 E	Annachie, NE of St Fergus	16	St Fergus Silts of Banffshire Coast Drift Group	no	yes	Eyles et al., 1946; Peacock, 1983
11	87 E	Ednie, NW of Peterhead	10	Till and laminated deposits of Banffshire Coast and Logie-Buchan drift groups	no	yes	Eyles et al., 1946; Peacock, 1983
12	87 E	Downiehills Farm, W of Peterhead	16	Till of Logie-Buchan Drift Group	no	yes	Eyles et al., 1946; Peacock, 1983
13	87 E	Invernettie, S of Peterhead	14	Till and glaciolacustrine/marine deposits of Logie-Buchan Drift Group	no	yes	Jamieson, 1858
14	87 E	Errolston, Cruden Bay	24	Till and glaciolacustrine deposits of Logie-Buchan Drift Group	yes	yes	Eyles et al., 1946; Peacock, 1983
15	87 W	Westfield Farm, Auchmacoy	3	Alluvium and (?) glaciolacustrine deposits of Logie-Buchan Drift Group	no	yes	Eyles et al., 1946
16	87 W	Ellon	10	(?) Glaciolacustrine deposits Logie-Buchan Drift Group	no	no	Eyles et al., 1946
17	87 W	Esslemont, SW of Ellon	2	Till and glaciolacustrine deposits of Logie-Buchan Drift Group	no	yes	Eyles et al., 1946
18	77	Tipperty, N of Aberdeen	5	Glaciolacustrine or glaciomarine deposits of Logie-Buchan Drift Group	yes	yes	Eyles et al., 1946; Munro, 1986
19	77	Middlemuir House, N of Belhelvie	2	Glaciolacustrine deposits of East Grampian Drift Group	no	no	Eyles et al., 1946; Smith, 1983
20	77	Balmedie House	2	Glaciolacustrine deposits of Logie-Buchan Drift Group	no	no	Eyles et al., 1946; Munro, 1986
21	77	Eigie Links, S of Balmedie	5	(?) Glaciolacustrine deposits of Logie-Buchan Drift Group	no	no	Munro, 1986; Smith, 1983
22	77	Blackdog, N of Aberdeen	3	Glaciomarine or glaciolacustrine deposits of Logie-Buchan Drift Group	no	yes	Eyles et al., 1946; Munro, 1986
23	77	Milden Burn	5	(?) Glaciolacustrine deposits of Logie-Buchan Drift Group	no	no	Eyles et al., 1946
24	77	Seaton, S of Bridge of Don	5	(?) Glaciolacustrine deposits of East Grampian Drift Group	no	yes	Eyles et al., 1946; Munro, 1986
25	77	Ferryhill (Clayhills), Aberdeen	20	Deposits of the Tullos Clay Member of the Logie-Buchan Drift Group†	no	yes	Eyles et al., 1946; Munro, 1986
26	77	Torry, Aberdeen	7	Deposits of the Tullos Clay Member of the Logie-Buchan Drift Group†	no	yes	Eyles et al., 1946; Munro, 1986
27	77	Middleton, south of Torry	6	Deposits of the Tullos Clay Member of the Logie-Buchan Drift Group†	no	no	Simpson, 1948
28	67	Brickfield, Stonehaven	2	Glaciolacustrine Ury Silts Formation of Mearns Drift Group	no	yes	Eyles et al., 1946; Carroll, 1995a
29	67	Toll Cottage, Stonehaven	~	Glaciolacustrine Ury Silts Formation of Mearns Drift Group	no	yes	~

* Maximum recorded thickness.

† Red-brown clays associated with brown and grey clays.

Table 5 Potential peat resources between Aberdeen and Elgin (based on Fraser, 1948).

Sheet	Name	Type of peat	Condition[†]	Altitude (m)	Area (ha)	Maximum thickness (m)
95	Broken Moan	Basin	33% cut over	275	10	~
96W	Aultmorehill Moss*	Hill	Unexploited	235	80	2.4
96W	Old Fir Hill–Hill of Moss Clashmadin	Hill & Basin	25% cut over	235	60	4.3
96W	Sheil Muir Moss	Basin	Mostly cut over	200	60	~
96W	Hill of Ord	Basin	Mostly cut over	160	8	~
96E	Moss of Clochforbie– Moss of Byth	Basin	All cut over, partly reclaimed	150	~	~
96E	Moss of Fishrie*	Hill	25% removed	190	280	2.4
97	Moss of Bracklemore*	Basin	50% disturbed by cutting	130	150	6.1
97	Windyheads Hill	Basin	All cut over	170	~	~
97	Moss of Blackrigg	Basin	75% cut over	90	30	4.9
97	Mosses of Skelmanae and Auchmacleddie	Basin	All cut over	90	100	4.3
97	Montsolie	Basin	All cut over	100	~	~
97	Mosses of Middlemuir*	Basin	33% cut over	100	400	4.3‡
86E	Cuminestown	Hill and Basin	75% cut over	150	80	2.1
86E	Moss of Swanford*	Basin	Mostly cut over	90	25	2.1
86E	Mosses NE of Fyvie	Basin	All cut over	110	120	1.2
86E	Mosses at Leet, Cairns and Heatherybanks	Basin	All cut over	110	40	2.4
86E	Wartle Moss	Basin	Mostly cut over	120	100	~
86E	Moss of Redhill	Basin	Mostly cut over	150	25	~
87W	Mosses at Cowbog	Hill & Basin	75% cut over	135	120	6.1
87W	Mosses at Nittanshead	Hill & Basin	Mostly cut over	175	180	3.7
87W	Craigculter Moss	Basin	25% cut over	120	40	3.7
87W	Mosses N of New Leeds	Basin	Mostly cut over	75	60	4.3
87W	Mosses of Auchleuchries and Muirtack	Basin	All cut over	75	60	~
87W	Moss of Dudwick	Hill	All disturbed by cutting	105	65	~
87W	Mosses of Elrick and Annochie	Basin	All cut over	110	~	~
87E	Mosses of Crimond and Logie	Basin	50% cut over	60	160	4.3
87E	St Fergus Moss	Basin	50% disturbed by cutting	45	220	6.1‡
87E	Rora Moss	Hill	50% cut over	55	400	2.7
87E	Mosses of the Cruden Hills	Hill & Basin	20% cut over	115	600	2.4‡
87E	Moss of Kinmundy	Basin	Mostly cut over	75	55	4.3
87E	Moss of Lochlundie	Basin	70% disturbed by cutting	75	240	6.1
76E	Skene Moss	Basin	75% cut over	100	30	3.5‡
76E	Mosses of Air and Lochside	Basin	75% disturbed by cutting	100	60	4.3‡
76E	Springhill Moss	Basin	50% cut over	75	25	3.7
76E	Luechar Moss	Basin	90% cut over	75	35	~
76E	Red Moss at Candyglirach	Basin	66% cut over	75	240	3.0‡
76E	Mosses of Blacknose and Lochmuir*	Basin	All disturbed by cutting	70	260	1.9
76E	Hill of Fare	Hill & Basin	Wide area removed	250	10	4.6
77	Burreldale Moss	Basin	75% cut over	150	120	4.3‡
77	Moss of Logierieve	Basin	All cut over	60	30	~
77	Moss of Pettymuck	Basin	All cut over	60	12	~
77	North Moss of Ardo	Basin	Mostly cut over	75	15	~
77	Mosses of Ardo, Wardhillock and Harestone	Basin	Extensively cut over	70	20	4.6
77	Red Moss between Belhelvie and Dyce	Basin	60% cut over	90	200	3.7‡
77	Grandholm Moss	Basin	All cut over	70	20	~

*   Deposit extends onto adjoining sheet.
†   Refers to the condition of the peat evaluated at the time of Fraser's study; many of the deposits have been worked extensively on a piecemeal basis since 1948. Cut over — indicates that the top layer of peat was partially or completely extracted; deeper layers may remain intact.
‡   Cross-section of moss and representative peat profile in Glentworth and Muir (1963).

relatively inaccessible spreads of hill peat, many of which are covered by forestry (Chapter 6). Areas of basin peat occur on lower ground, but the bulk of these have been worked to the water table, or are preserved as nature reserves and Sites of Special Scientific Interest (SSSIs). The use of peat as fuel by farmers and crofters has not been important in the recent past, but persists on a small scale. In western parts of the district it has a specialised application in whisky distilling.

The first systematic assessment of the peat deposits in Scotland began during the Second World War (Fraser, 1943), with the aim of evaluating their potential uses as fuel and in agriculture. The results from the first of a planned series of surveys, dealt with the deposits of Aberdeenshire, Banffshire and Morayshire (Fraser, 1948) and are summarised in Table 5; subsequent systematic surveys were abandoned in the immediate postwar period. Assessments recommenced in 1949, but concentrated only on major deposits with a view to their potential use as fuel for peat-fired power stations (Department of Agriculture and Fisheries for Scotland, 1962). These surveys evaluated resources in south-west Scotland, the Western Highlands and Islands, Central Scotland, and Caithness, Shetland and Orkney (Department of Agriculture and Fisheries for Scotland, 1964, 1965a, b, 1968). None of the deposits in north-east Scotland was included in these surveys, but the overall potential of Scottish peat for electricity generation was reviewed by Dryburgh (1978).

Fraser (1948) identified 46 areas of peat in north-east Scotland that were considered potentially workable (Table 5), but none of the deposits on 1:50 000 sheets 66E and 67 was investigated. In general terms, the hill peats are extensive, but thin, with much of the resource forming hags above the water table. The basin peats are thicker, but less extensive and predominantly water-saturated; many are relict and now eroding.

By the mid 1940s, most of the 4845 hectares (ha) of peat identified by Fraser, between Aberdeen and Elgin had been worked in a piecemeal fashion for fuel. In the 1960s, eight of the major peat mosses between Aberdeen and Fraserburgh (Table 5) were examined in detail by the Soil Survey (Glentworth and Muir, 1963). In general, the extent of the peat and amount of cutting that had occurred at each site, recorded in the later survey, were comparable to the figures given in Fraser (1943). The detailed later survey showed that maximum thickness of peat at each site was generally greater than that recorded by Fraser. The average thickness of peat that remained in cut areas, ranged between 25 per cent of maximum thickness recorded at Moss of Air, to 60 per cent of the maximum at St Fergus Moss. Recent geological mapping indicates that piecemeal working has continued, particularly up to the late 1970s, and many of formerly identified resources are now too thin and waterlogged to be attractive for commercial extraction.

Commercial extraction of peat around New Pitsligo and Strichen reached its peak in the early 1990s. Seven to eight thousand tonnes per year is currently extracted from Lambhill Moss, by the Northern Peat and Moss Company, although remaining reserves here are small. Some 15 to 20 per cent of production is sold throughout Scotland for domestic fuel; the remainder is exported, via Fraserburgh, to southern Sweden, for use as fuel in a district heating plant. Commercial extraction of peat from St Fergus Moss commenced during 1998, with 7–8000 tonnes being produced in the first year of working; most of the material is also exported to Sweden for use as fuel, but some peat for horticultural use is also produced.

More than 3000 ha of peat has been mapped on Sheet 66E Banchory, most of which occurs as extensive hilltop spreads on the watershed between the catchments of the River Dee and Water of Feugh and the rivers and burns that drain south-eastwards into Strathmore. The hill peat is generally less than 2.5 m thick and much of it rests either on thin till or directly on the dominantly granite bedrock. The most notable spreads occur on the uplands, north and south of Glen Dye (1500 ha), between Little Kerloch [NO 674 874] and Leachie Hill [NO 739 853] (900 ha) and around Little Sheil Hill [NO 796 917] (350 ha). The hill peat has been worked on a piecemeal basis in several places, but much of the resource is currently sterilised by extensive forestry plantations. Minor deposits of thin, waterlogged basin peat are present around Loch of Park, on the northern edge of Sheet 66E.

About 400 ha of basin peat occurs in the northern part of Sheet 67 Stonehaven within rock basins and in the floors of glacial drainage channels. Significant deposits (160 ha) occur in basins, linked by drainage channels, between Broomhill [NO 844 901] and Forester's Croft [NO 873 883], and at Red Moss (120 ha), near Netherley [NO 855 934]. Mapping indicates that, apart from the deposit at Red Moss, the basin peat is generally thinner (1–2 m) than that commonly found north of Aberdeen, and it has not been worked as extensively.

HYDROGEOLOGY

Across the whole of north-east Scotland groundwater occurs within a wide range of superficial deposits as well as in bedrock (Robins, 1990). Storage capacity and permeability are the main controlling factors determining both the volume of groundwater stored within the aquifers and the rate of flow.

Alluvial deposits

The deposits with the largest storage and highest permeability are the coarse alluvial gravels within the main river basins (Figure 10). The rivers Spey and Dee contain significant volumes of alluvial gravels below the water table; the Spey deposits are of particularly high permeability. Other rivers, notably the Deveron around Banff, and the Lossie, near Elgin, also have well developed alluvial gravel deposits and there are many other minor watercourses containing smaller amounts of water-saturated granular alluvium.

River gravels beneath valley floors typically contain large amounts of groundwater relative to the volume of the deposits. The gravels have greater lateral continuity than the more heterogeneous deposits, such as glaciofluvial sand and gravel, resulting in more predictable flow routes for groundwater in the alluvial sequences. They are also generally characterised by high rates of through flow.

Figure 10
Shallow
groundwater
occurrence.

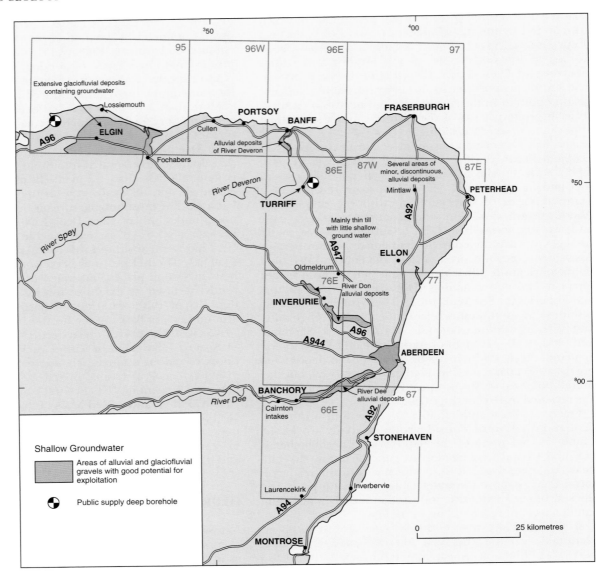

The relatively homogeneous nature of alluvial deposits and the presence of shallow water tables under the river floodplains means that there is considerable potential for exploitation of this type resource for supply. The potential for recharge to alluvial aquifers is also high, because they are located in valleys and, as surface run-off moves towards the main watercourses, a significant proportion of it infiltrates into the ground. Alluvial deposits act as an important storage medium for groundwater prior to its discharge (sometimes after several years storage) as base flow to rivers.

Glaciofluvial deposits

Glaciofluvial sand and gravel deposits are much more widespread than alluvial sediments, particularly along the Moray coast, but they commonly have a lower overall permeability. The heterogeneous nature of many glaciofluvial deposits also results in complex groundwater flow paths, for example via gravel-filled channels, many of which are discontinuous.

Glaciofluvial sheet deposits commonly form terraces flanking the alluvium of the major river systems. They form higher ground on the valley sides and mostly lie above the water table. Glaciofluvial ice-contact sand and gravel forms extensive moundy deposits and much of this material also occurs above the local water table. The sandy nature of the soils developed on these terraced and moundy deposits means that recharge by rainfall to the water table is high where they occur, and both play an important role in the water cycle by providing base flow to rivers and wetland areas.

Blown sand

Blown sand deposits are very well sorted and have high intergranular storage capacity, but are generally of very limited thickness and lateral extent. Where underlain by peaty or clayey material above sea level, the basal parts of blown sand deposits are commonly saturated with groundwater and relatively high values for hydraulic conductivity are present. However, the deposits are restricted to coastal areas and the total volume of groundwater stored within them is generally quite small in comparison with that stored in other, less permeable material. The rate of recharge to these deposits is very high, with no surface runoff present, except where peaty interbeds form perched water tables.

Till

Till deposits are widespread across the district, most of them being relatively sandy. Owing to its compaction, higher clay content and heterogeneous nature, till contains less groundwater than the more porous sandy and gravelly deposits and, overall, its hydraulic conductivity is significantly lower. It is important, however, on a regional scale, for determining the movement of shallow groundwater. For example, many springs and seepage lines are present within this type of deposit, where sandier horizons and gravelly interbeds crop out. These can help sustain relatively high rates of groundwater flow. Springs and seepage lines are also commonly developed at the base of till units, particularly where they overlie less permeable fresh bedrock. Recharge to till deposits can be high where sandy horizons are present and more than 200 mm of rainfall per year can infiltrate to the water table.

Bedrock

Fresh pre-Palaeozoic igneous and metamorphic bedrock generally has a low permeability, but where shallow zones have been subjected to weathering, fractures may be enlarged allowing groundwater to move relatively freely. Flow and storage are dependent entirely on the presence of secondary voids such as joints and fractures within fault zones. This results in highly complex, unpredictable flowpaths. The Devonian and Permian sandstone aquifers near the Moray coast contain greater volumes of groundwater than the older rocks owing to their greater porosity and intergranular permeability. In spite of this, a large proportion of the total groundwater flow uses secondary voids to move through the aquifer.

Groundwater quality

Shallow groundwater quality in the district is variable, although many shallow aquifers contain potable groundwater. Chemical quality depends to a large extent on the nature of the surface deposits. For example, groundwater in the Spey gravels is normally of good quality, but where overlain by waterlogged layers of peat or silt, significant local concentrations of iron and manganese may occur. This is due to the groundwater having a relatively low content of dissolved oxygen and higher than normal acidity, causing the dissolution of these elements. Elsewhere, the effects of intensive agricultural practices have influenced the quality of the groundwater, with the presence of locally high concentrations of nitrate derived from fertilisers. This is particularly noticeable around the Ythan estuary. Groundwater is particularly vulnerable to pollution where the water table is less than 5 m below the ground surface, and where sand or gravel deposits are present below the soil layer. In general terms, groundwater is of very good quality in bedrock aquifers.

Groundwater exploitation

Groundwater from springs as well as shallow and deep sources is abstracted for many hundreds of private domestic supplies in north-east Scotland. In addition, public supply sources include two deep boreholes and many shallow bores into alluvial aquifers. The industrial usage of groundwater has, to date, been limited to maltings, distilleries and agriculture-related activities, and the absence of high-yielding aquifers in urban areas such as Aberdeen has hindered development. However, several water boreholes were sunk into Devonian sandstones in the Elgin area, between 1990 and 1998. This relatively unexploited source appears to have the potential for further development.

Historically, groundwater has provided a safe, reliable water supply to many communities and individual dwellings across the area. The absence of public supply networks across much of the more rural parts of the district has necessitated the exploitation of hundreds of springs in many different settings. Where natural occurrences of groundwater were absent, shallow wells were constructed, commonly to depths of 4 to 10 m. Most farms had private wells, many of which exploited gravelly interbeds in till or the more permeable zone at rockhead. Where high-yielding springs were present, communities would organise a limited water distribution network. The larger towns, such as Peterhead and Aberdeen, relied on groups of springs to supply large numbers of people. The Peterhead spring supply was still in use until after the Second World War and was based around a group of seven springs on the Hill of Longhaven, 5 km south-west of the town.

Beneath Aberdeen, superficial aquifers with more than 30 m of saturated granular material were exploited for factory supplies in the 19th century. The variable nature of the Quaternary deposits and indifferent water quality restricted usage, especially adjacent to the tidal sections of the rivers, where the shallow groundwater is brackish. The main supply for Aberdeen presently comes from intakes on the River Dee, including one at Cairnton [NO 665 965].

The nature of groundwater exploitation began to change towards the end of the 18th century when deep boreholes were first drilled into bedrock. Depths ranged from 10 m to greater than 100 m. However, the absence of widespread, highly permeable groundwater resources in the bedrock of the district has meant that relatively few deep boreholes have been drilled across north-east Scotland. Examples of public supply boreholes (Figure 10) include those at Turriff, where a Middle Devonian conglomerate aquifer is exploited, and at Burghead, where the local supply is sourced from Triassic, Permian and Devonian sandstones, but almost all the supply comes from the Permian strata. Deep boreholes in the Moray area produce low to moderate supplies for distilleries and maltings.

Owing to their favourable hydraulic characteristics, shallow superficial aquifers remain the most heavily exploited type of aquifer for public supply. The River Spey groundwater scheme, located immediately to the south of Fochabers, is the latest scheme to abstract groundwater from gravel deposits and was commissioned in 1996. The main well field lies just outside the district, but is a good example of large-scale groundwater development. A total of 40 boreholes, some up to 20 m deep, have been sunk within the river gravels on the eastern side of the Spey. Each is capable of an abstraction rate of 10 litres per second. Recharge to the aquifer is by leakage from the Spey into the gravel beds between the boreholes and the river, by lateral groundwater movement from the valley sides and by direct infiltration of rainfall on the floodplain. Another example of alluvial aquifer exploitation occurs in the valley of the River Deveron, south of Banff. This is a much smaller abstraction scheme than on the Spey at Fochabers. It uses a horizontal infiltration design to collect water from the gravel, rather than a vertical borehole.

The exploitation of natural springs for public supply is now rare within north-east Scotland, with most supplies having a surface origin. A total of 20 springs are maintained by the North of Scotland Water Authority including one in Moray, eleven in Gordon and eight in Banff and Buchan. Owing to continual improvements in supplies, many of these springs may be decommissioned, either to comply with European requirements on water quality, or because of the replacement of several previously separate supplies by a single regional network.

PLANNING CONSIDERATIONS AND CONSERVATION ISSUES

One of the most fascinating attributes of the landscape of north-east Scotland is the degree to which the topography reflects the interplay between rock type, structure and geomorphological processes, the latter largely driven by climate changes throughout the Cainozoic. As a consequence, the district contains many sites where geological sequences are preserved; these contain evidence critical to understanding the landscape evolution not only of north-east Scotland, but also northern Britain and north-west Europe during the Cainozoic. This latest part of Earth history is particularly important, as it provides the environmental setting within which the whole of human cultural and economic development has taken place. It is only during the last thirty years, however, that the importance of the Cainozoic geological record of the frequency and rate of naturally occurring environmental changes has been generally recognised. It provides base-line data against which anthropogenic degradation of the environment, such as temperature changes caused by increased CO_2 emissions, elevated sea levels, or vegetation changes due to farming intensification, can be gauged.

Decisions as to which Cainozoic sites should be conserved are based upon guidelines that try to encapsulate the range of scientific interest at each site. Some are designated as SSSIs on geological and geomorphological grounds. Short descriptions of several of these SSSIs are given in Appendix 1. Most of the key Quaternary sites were included in the Geological Conservation Review (GCR), initiated by the Nature Conservancy Council in 1977 and are fully described in Gordon and Sutherland (1993).

Several important Quaternary sequences are also preserved within SSSIs, National Nature Reserves (NNRs), Royal Society for the Protection of Birds (RSPB) reserves and local nature reserves, which were originally designated mainly on the basis of their botanical and wildlife importance. An example is the NNR covering the Sands of Forvie, north of Newburgh; this covers an extensive area of active and stabilised dunes of blown sand up to 60 m in height (Ritchie, 1992). The reserve was first established in 1959. The area was designated a SSSI in 1971, for its coastal plant communities, colonies of sea birds (it is home to Britain's largest population of Eider duck), as well as its coastal geomorphology. Holocene sediments on the banks of the Ythan estuary, within the SSSI, include a thin bed of grey sand, deposited by tsunami waves from the second Storegga slide. It occurs beneath blown sand, peat and clay, in boreholes at Waterside [NK 007 267] on the eastern side of the estuary (Smith, 1984; Smith, et al., 1983, 1999). This 'tsunami deposit', which was generated by a massive submarine landslide on the Norwegian continental margin, is analogous to that recorded from the cores taken in the Philorth valley near Fraserburgh (Appendix 1), about 40 km farther north.

Important Quaternary biogenic sequences are sometimes recorded from waterlogged peat mosses and lake basins, which are primarily conserved for their modern flora and fauna. For example, the Loch of Park, northeast of Banchory, is a local nature reserve that was established to protect its population of breeding birds, its reed beds and Alder woodland. Pollen records and radiocarbon ages, determined from Holocene and Late-glacial peats and organic muds recovered in cores taken from the lake basin, have been critical in establishing climatic and vegetational changes in north-east Scotland since deglaciation (Vasari, 1977). The Loch of Park succession is described in more detail in Appendix 1.

Knowledge of the nature and distribution of Quaternary sediments are fundamental to decisions made regarding the economic and environmental well being of the district. Planning in the hinterland of Aberdeen, for example, not only requires forecasts of demand for housing, infrastructure and industrial construction, but also needs to ensure that adequate supplies of aggregates are available and that new building developments are not sited on the most attractive remaining aggregate resources. It was partly to this end that the detailed assessments of sand and gravel resources, described in Appendix 2, were undertaken between 1979 and 1990.

Development planning also benefits from knowledge of ground stability and foundation conditions and seismic hazard, which are discussed below, and groundwater vulnerability, described in the preceding section on hydrogeology. These and other factors, such as coastline evolution were initially evaluated for the hinterlands of Aberdeen and Peterhead, in two projects funded by the Scottish Office. The results of these two studies are presented in Smith (1983) and Peacock (1983).

GROUND STABILITY AND FOUNDATION CONDITIONS

Engineering properties of glacigenic sediments

It is well known that glacial deposits often present problems to the ground engineer owing to their inherent variability and the complex sequences in which they lie. Observations on modern glaciers have enabled the development of a process-based model (Boulton and Paul, 1976) that has been found applicable to several areas of lowland Britain (Paul and Little, 1991). This model distinguishes erosion and transport processes, which are responsible for the grading and plasticity of the sediments, from depositional and postdepositional processes, which are responsible for the packing and strength of the materials. At the larger scale, the model identifies a number of recurrent *landform-sediment associations*, which describe the geomorphology and facies architecture of the glacigenic sequences as a whole.

Based on the analysis of commercial site investigation data held at the British Geological Survey, it is possible to classify foundation materials in the study area into three broad assemblages.

i   Relatively thin (usually less than 5 m) glacigenic sequences, which lie on bedrock and may be weathered or glacially tectonised. They usually extend to surface, and comprise coarse glacial tills, sands and gravels. These sequences are normally found in upland areas and may be locally extensive.

ii Thick (5–10 m or more), generally complex glacigenic sequences of interdigitated tills, glaciofluvial sands, and glaciolacustrine/glaciomarine silts and clays, typically found as local valley or low ground embayment infills.

iii Thick (over 10 m), postglacial sequences of clays, silts and sands, together with interbedded peats, which overlie glacigenic sequences, found locally in bedrock depressions.

Assemblage i is widespread across the area and appears to be present in the outcrop of each of the drift groups shown in Figure 4 and described in Chapter 8. It seems common within the outcrops of the Central Grampian, East Grampian and Banffshire Coast drift groups, where it has been encountered in excavations for several road improvement schemes. Assemblage ii, by contrast, is more restricted in extent, being confined to the coastal lowlands and the lower parts of the adjacent valleys. To the west and north of Aberdeen excavations in the Don and Ythan valleys have shown local development of this assemblage involving sediments of the Logie-Buchan Drift Group. Similarly, excavations in the Feugh valley to the south-west of Aberdeen have revealed analogous sequences in sediments of the East Grampian Drift Group. Assemblage iii has been reported only rarely, mainly from coastal settings. In the Peterhead area, excavations to 20 m depth have revealed peats, sands and gravels resting on glacial deposits of both the Banffshire Coast and Logie-Buchan drift groups. Around St Fergus, similar sequences have been reported extending to 15 m depth.

Most of the geotechnical properties described below deal with glacigenic successions, dominated by glacial diamictons (tills) and sampled from shallow boreholes and trial pits. A more detailed study of engineering properties of glacial sediments from sites throughout north-east Scotland (Ramsay, 1999) shows, however, that at any particular location, site investigations are required to adequately characterise the geotechnical properties of specific Quaternary sequences. Geotechnical information is not presented for the Central Highland Drift Group owing to the sparse, and unrepresentative coverage of site investigation data within its area of outcrop in the district.

PARTICLE SIZE   Ternary composition plots (Figure 11) show that the glacial and glaciofluvial deposits within all of the drift groups are broadly sandy gravels, with an admixture of fine (silt and clay) particles. They are thus classified as clast-dominant to well-graded diamictons in the scheme of McGown and Derbyshire (1977). In detail, the sediments of the East Grampian Drift Group have a relatively constant proportion of fines (15–20%), although the proportions of sand and gravel can vary greatly. The sediments of the Mearns Drift Group appear indistinguishable from those of the East Grampian Drift Group on granulometric criteria alone. By contrast, sediments of the Logie-Buchan and Banffshire Coastal drift groups differ granulometrically both from those of the other two and from one another. Those of the Logie-Buchan Drift Group contain up to 40 per cent fines and around 30 per cent each of sand and gravel, whereas those of the Banffshire Coast Drift Group, although sandy, generally contain less gravel (less than 40 per cent) and have a fines content in the range 20 to 40 per cent.

LIQUID AND PLASTIC LIMITS   Glacial tills show a relationship between their liquid limit (LL) and plasticity index that arises from their characteristic particle-size distributions, which are themselves, in part, the result of crushing processes and subsequent modification (Dreimanis and Vagners, 1971; McGown, 1971), and also reflect the provenance of the deposits. This relationship is expressed on the plasticity chart as a straight line (the T-line of Boulton and Paul, 1976). Plasticity charts for glacial sediments from four of the north-east Scotland drift groups considered here (Figure 12) illustrate that they conform to this linear relationship. Comparison of these charts shows a number of differences between the drift groups, which reflect the differences in their grain-size and source material. Sediments of the East Grampian Drift Group (Figure 12a) are of relatively low plasticity (LL 20–30%) and values are scattered around the T-line, probably owing to their low fines content, and the possible inclusion of some weathered bedrock in the 'till' category. The field of values for sediments sampled from the Mearns Drift Group (Figure 12b) overlaps that for samples from the East Grampian Drift Group, which is expected from the apparently similar granulometry and clay-size mineralogy of both sets of samples.

The sediments of the Banffshire Coast Drift Group (Figure 12b) also plot along the T-line. Those from the Lhanbryde area are of generally low plasticity (LL 25–35%), which reflects their sandy character and incorporation of material from the local sandy bedrock. In fact, the tills at Lhanbryde are probably representative of the Central Grampian Drift Group. Those from St Fergus have a generally greater range of liquid limits (25–50%), possibly reflecting their derivation from offshore clayey strata, and also as a result, in some instances, of a concentration of fines by resedimentation processes. In a similar manner, sediments of the Logie-Buchan Drift Group (Figure 12a) also have generally high liquid limits (25–50%), presumably as the result of their large fines content. The probable inclusion of a number of glaciomarine and glaciolacustrine silts and clays, clayey deformation tills and resedimented deposits ('flow tills') in this group, in addition to lodgement tills, may be responsible for the spread of points below the T-line.

The mineralogy and grading of the sediments exert the fundamental control on their plasticity. Figure 12 indicates that the sediments of the Logie-Buchan Drift Group are the most plastic of the four groups examined and those of the East Grampian Drift Group the least. Hall and Jarvis (1995) have reported the clay mineralogy of glacial sediments in the Ellon area, and have shown that the Hatton Till Formation of the Logie-Buchan Drift Group is dominated by illites and mixed illite smectites. By contrast, the Pitlurg Till Formation of the Banffshire Coast Drift Group, near its southern limit, contains a mixture of illites, illite-smectite and kaolinite. Although their composition has not been reported in the literature, the East Grampian and Mearns drift groups, which contain largely locally derived lodgement and deformation tills, may be expected to be dominated by kaolinite and quartz flour. This is supported by a study of the clay mineralogy of soils developed on the Logie-Buchan and Mearns drift groups (Glentworth et al., 1964), which showed that the latter contain higher proportions of kaolinitic clays and vermiculite and lower proportions of illite. If all of these results, taken together, are broadly representative of the groups as a whole, then at a given clay content, sediments of the Logie-Buchan Drift Group should behave more plastically than those of the Banffshire Coast Drift Group. The sediments of the East Grampian and Mearns drift groups will be less plastic still. The results shown in Figure 12 confirm these comparisons.

WATER CONTENT AND LIQUIDITY INDEX The in situ water content of a fine-grained sediment is determined by its stress history (the sequence of load changes in terms of effective stress) and is thus indicative of its geological origin and subsequent postdepositional history. In general, subglacial sediments have experienced larger loading events than those from ice-marginal settings and so, for a given composition, their water contents are normally lower. This is not to imply that the stress history of a glacial sediment is a simple function of the superincumbent weight of ice: it is now recognised (e.g. Boulton and Dobbie, 1993) that pore-water pressure plays a complex role in the control of subglacial stress and that for ice-marginal sediments, episodes of drying or freezing

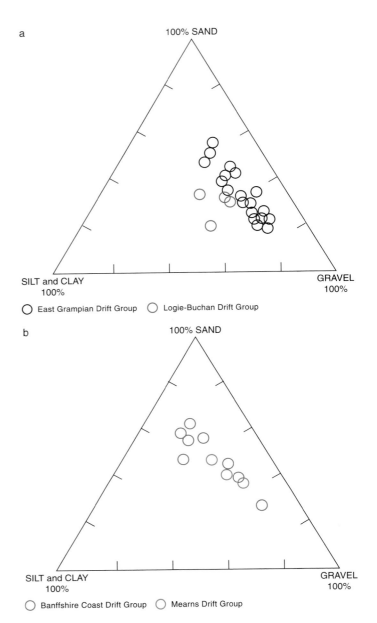

Figure 11 Grain size plots for four of the drift groups in north-east Scotland.

are often the most significant events in their stress histories (Boulton and Paul, 1976).

For a given stress history, the water content of a sediment is determined by the quantity and type of its clay content, which is reflected by its liquid and plastic limits. They are generally compared in terms of their *liquidity index*, a normalised value of water content that is relative to their liquid (LI=1) and plastic (LI=0) limits. A review of glacigenic deposits (Paul and Little, 1991) has shown that, in many areas of Britain, subglacial lodgement tills usually have liquidity indices in the range − 0.1 to − 0.5, whereas resedimented ice-marginal diamictons normally have higher liquidity indices in the range 0.5 to 0. Thus, sediments of ice-marginal origin appear to be typically wetter and softer than their sub-glacial counterparts, a not unexpected situation, in view of the differing stress histories of these classes of sediment.

Comparison of liquidity indices across the four drift groups for which adequate data exits (Figure 13), shows significant differences. The sediments of the East Grampian and Banffshire Coast drift groups have relatively low liquidity indices (mostly in the range 0 to -1.0), which is consistent with a subglacial origin, whereas those of the Logie-Buchan Drift Group are higher (+ 0.5 to − 0.3), which suggests a mixed suite of both subglacial and ice-marginal deposits. The results from the Mearns Drift Group are more difficult to interpret, but may also suggest a bimodal distribution with both subglacial and ice-marginal components.

UNDRAINED SHEAR STRENGTH The undrained shear strength of a remoulded clay sediment is determined by its liquidity index according to a logarithmic relationship (Skempton and Northey, 1952). At the plastic limit (LI=0) remoulded clays have an undrained strength around 170 kPa; at LI=0.5 their strength is around 7 kPa and at LI= − 0.3 it is around 400 kPa, although the latter is difficult to determine accurately and may be reduced by fissuring. Thus naturally remoulded sediments, such as glacial tills, may, in the case of subglacial tills, be expected to have undrained strengths around 200 to 400 kPa and, in the case of ice-marginal flow tills, to have undrained strengths around perhaps 50 to 100 kPa. Data from two drift groups (Figure 14) supports these suggestions. Sediments of the Banffshire Coast Drift Group, of presumed subglacial origin, possess strengths up to 300 kPa with a modal value of 100 to 150 kPa. Those from the Logie-Buchan Drift Group, formed in a more ice-marginal setting, have strengths in the general range 50 to 150 kPa with a modal value 50 to 75 kPa. In both cases there is an expected relationship between liquidity index and undrained strength, although the absolute values of strength are less than those expected, possibly as a result of fissuring or other difficulties in testing.

RELATIONSHIP TO DEPOSITIONAL AND POSTDEPOSITIONAL HISTORY As discussed in Chapter 8, the glacigenic deposits of north-east Scotland are divisible into five drift groups on the basis of lithology and provenance (Figure 4). They were deposited by three major ice streams, involving both subglacial and ice-marginal processes. The engineering properties of the sediments within each group generally reflect their differences in composition, their mode of deposition, and the degree and type of postdepositional modification (weathering, cryoturbation, solifluction etc) that has affected each unit.

Although the data is somewhat limited and scattered geographically, a general model emerges when the area is considered as a whole. The ice streams deposit subglacial till at their base, with sedimentation of flow till and melt-out till (including associated glacio-aqueous sediments) at their margins, especially where ice locally abutted against reverse slopes. This leads to the observed association of low liquidity index and high undrained shear strength in assemblages from coastal regions, where basal tills dominate, and to the association of higher liquidity index and lower shear

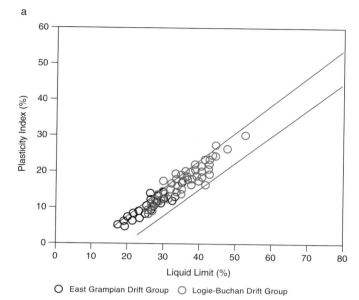

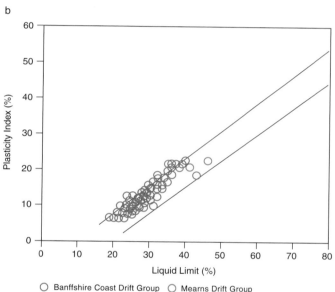

Figure 12 Plots of Plasticity Index versus Liquid Limit for four drift groups in north-east Scotland.

strengths in marginal valleys plugged by mixed sedimentary infills.

The diamictons of the East Grampian Drift Group are mostly subglacial in origin and, being derived from weathered crystalline or arenaceous bedrock, have a low liquidity index and low plasticity. The liquidity index data suggest that subglacial (deforming bed) tills also dominate the Banffshire Coast Drift Group. The Logie-Buchan Drift Group contains both subglacial and ice-marginal elements, as implied by the liquidity index and shear strength data. It suggests that ice that moved inland from the North Sea deposited basal tills along the coast, but there was supraglacial (flow till) sedimentation into the valleys at its western margin. At the few sites examined here, the Mearns Group appears to contain diamictons deposited in both ice marginal and subglacial settings.

Deeply weathered bedrock, periglacial and postglacial deposits

The widespread development in north-east Scotland of deep weathering profiles on bedrock has important implications for ground stability conditions in the district. The nature, distribution and age of these weathering covers are discussed fully in Chapter 4.

The postdepositional remobilisation of glacial diamictons has been widespread within the district, but it can only be recognised at a few sites, notably where the remobilised material rests on Windermere Interstadial and Loch Lomond Stadial organic sediments. These sites demonstrate the former instability of low-angle till slopes, particularly during the Loch Lomond Stadial and their identification is important because such sites are liable to be rendered unstable by engineering works. The nature and distribution of these landslipped and soliflucted deposits, and postglacial materials such as peat, blown sand and alluvium are discussed more fully in Chapter 6.

Made, landscaped and worked ground, and landfill

Only the largest, well-defined areas of made and worked ground are shown on the published 1:50 000 scale geological maps of the district. Many of the smaller areas are only shown on 1:10 000 and 1:25 000 scale geological maps dealing with the Drift deposits. The coverage of these larger scale maps is shown in Figure 51. Most of the examples of made, landscaped and worked ground, quoted below, come from recently surveyed sheets (77, 76E, 67 and 67) in the southern part of the district. Knowledge of the extent of these deposits and workings in the remainder of the district is, at best, patchy and requires revision. However, many of the observations on types of workings and man-made deposits appear typical of similar excavations and areas of made ground in the less well known northern parts of the district.

Made ground is ubiquitous within some parts of the larger built up areas, but many of the deposits are less than 1 m in thickness and their precise nature and distribution is difficult to determine by traditional geological mapping alone. The composition, thickness and extent of individual deposits can only be determined in areas where adequate site investigation records exist. For example, site investigations indicate that thick man-made deposits overlie undisturbed bedrock and Quaternary strata around the harbours of the larger coastal settlements, with jetties and breakwaters commonly constructed on reclaimed ground. Several of the larger road and railway embankments shown on Sheet 66E Banchory and Sheet 77 Stonehaven are underlain by made ground up to 15 m thick.

Embankments are generally composed of naturally occurring Quaternary sediments (soils, tills, sands and gravels etc) which have been excavated and mixed together; embankment drains are infilled with coarse crushed rock aggregate. Made ground in urban areas

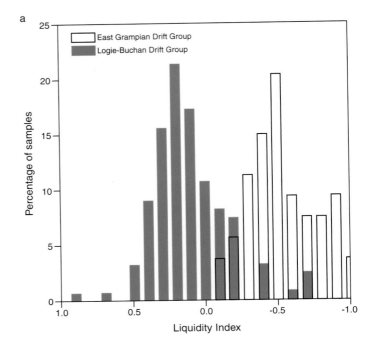

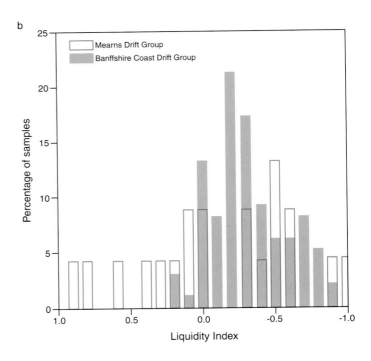

Figure 13 Histograms of Liquid Index of four drift groups in north-east Scotland.

(particularly areas of housing and industrial regeneration) is commonly composed of mixtures of artificial inert material (builders rubble, ash and slag) and disturbed Quaternary sediments. In major civil engineering projects made ground is commonly composed of graded hard rock aggregate from local bedrock quarries, which has been faced and capped by reinforced concrete.

Spoil heaps associated with bedrock quarrying and sand and gravel workings are widespread in rural areas. Piles of granite blocks up to 20 m in height have accumulated on the valley sides adjacent to disused granite quarries on Sheet 76E Inverurie around Tillyfourie [NJ 645 124] and Raimoir [NJ 702 003], and the working quarry at Corrennie [NJ 641 119]; many of the blocks are several metres in diameter. The quarries themselves are many tens of metres deep. Similar spoil tips have also developed around the quarry in the basic/ultrabasic igneous intrusion at Belhelvie [NJ 944 181] on Sheet 77 Aberdeen. Spoil associated with sand and gravel working is generally finer grained and commonly comprises mixtures of overburden (generally topsoil and till) and waste partings (tills, silts and clays) together with boulders. Much of this material is stored within worked-out areas of active pits or has been used to backfill worked-out and abandoned sites.

The most widespread sand and gravel workings in the district are found on Sheet 77 Aberdeen between Corby Loch and the coast north of Bridge of Don (Appendix 2). Some pits were excavated to depths of more than 20 m. Most are now disused and many have been backfilled with inert domestic and industrial waste; some less inert domestic refuse, agricultural waste and chemicals used in paper making have also been placed in landfills. Most of these landfill sites are in highly permeable strata and only the more recent are lined. The waste is commonly covered by less permeable spoil from the former workings which is often capped by replaced topsoil. Partial infilling of abandoned areas within active pits is commonplace, with final reinstatement occurring when extraction has ceased. Several large landfill sites have been landscaped and returned to agriculture. Disused sand and gravel workings have been landscaped and incorporated into golf courses, turned into nature reserves or built over. For example, industrial units have been erected in open disused pits at Blackdog [NO 957 143] and Upper Tarbothill [NO 951 134].

Red clayey diamicton (Hatton Till Formation) has been dug at a site near Teuchan [NK 0839 3896] since 2000 for burying domestic waste at the new Stonyhill landfill site, south-west of Peterhead.

Many abandoned bedrock quarries have been used as landfill sites. For example, a quarry in felsitic microgranite at Kirkhill [NK 011 528] has been infilled with domestic waste from nearby Peterhead. Similar tipping of domestic waste has taken place in Burnside Quarry [NJ 775 126], a disused quarry in the Crathes Granodiorite, south-east of Kintore. Tipping of inert domestic and building waste on top of lowland peat mosses has been widespread also, particularly north of Aberdeen. Many tips are now disused and have been graded and landscaped, but several remain active, for example the large tip [NJ 916 457] near Maud. Numerous smaller enclosed depressions in bedrock and till have been completely infilled by farmers, with rocks cleared from nearby fields, subsoil and farm waste. The fill is covered by top soil and the ground returned to agriculture. Around Aberdeen, larger scale excavations and hollows in bedrock and till have been infilled, for example around Westhill [NJ 830 870], Loirston Loch [NJ 940 010] and Portlethen [NO 920 970], where extensive building developments have taken place on the reclaimed ground.

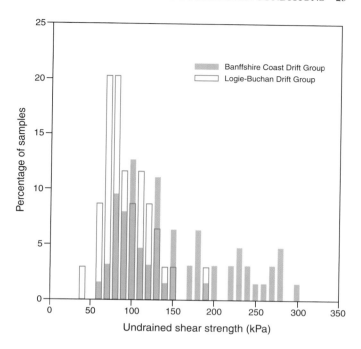

Figure 14 Histograms of Undrained Shear Strength for the Banffshire Coast and Logie-Buchan drift groups.

Many of the recent large-scale industrial developments on the outskirts of Aberdeen, at Bridge of Don, Dyce and Nigg, for example, have involved widespread modification of the ground surface, with millions of tons of loose material being removed or remodelled using mechanical excavators. The extent of these spreads of landscaped ground is currently being mapped as part of ongoing continuous revision of Sheet 77. Smaller areas of landscaped ground associated with industrial and commercial developments around Portlethen and housing developments in Stonehaven are shown on Sheet 67. A housing development on the southern bank of the River Dee, 1 km upstream of the Bridge of Dee [NO 697 952], and the Invercannie Water Treatment Works, on the northern bank, 2 km farther upstream are both sited on landscaped ground associated with former sand and gravel workings.

Landslips

Landslips are widely developed in coastal areas where steep rock cliffs are being eroded by the sea. Slips are widespread along the rocky shoreline of the Moray Firth, between Buckie and New Aberdour and along the North Sea coast south of Aberdeen, where resistant Dalradian metamorphic rocks and Devonian sandstones and conglomerates reach the sea. Small slips are widely developed in the Kirkburn Silt Formation, especially where the deposits are exposed on the sides of deeply incised valleys that reach the coast around Troup Head [NJ 826 673], immediately east of Crovie [NJ 809 658] and in the vicinity of Macduff (Plate 3; Map 3). Slips are also developed in reddish brown clayey diamictons of the Mill of Forest Till Formation and clays of the Ury Silts Formation at the top of steep cliffs between Aberdeen and Stonehaven.

Evidence of the mobilisation and probable solifluction of glacial diamictons is widespread in the lowlands of north-east Scotland. Few of the landslips are extensive enough to be shown on the published 1:50 000 geological maps, but some are shown on 1:10 000 and 1:25 000 scale geological maps of the district. Many of these small slips are wholly developed within Quaternary sediments. Dating of organic material incorporated in one example, at Knockhill Wood near Glenbervie, suggests that the earliest movement occurred either during the Loch Lomond Stadial (about 10–11 ka ^{14}C years BP) or at the beginning of the Holocene (Appendix 1 Knockhill Wood).

Another good example of an old landslip was observed in 1995 during the realignment of the A941 road near Rothes [NJ 277 498]. The excavations revealed lenses of organic sediment containing wood fragments within a unit of interbedded grey clayey silt, pale greyish brown sandy silt and sand. The unit rested on gravel and was overlain by 2 m of reddish brown, pebbly sandy diamicton. A radiocarbon age of about 11 110 BP (Table 8) from a fragment of willow demonstrated that disturbance observed towards the top of the organic sequence postdated retreat of ice from the area. Furthermore, the presence of sand blocks within the silt unit indicated that the disturbance and mass movement took place across at least partly frozen sediment. This, together with the degree of disturbance, suggests that the slip was a detachment slide of an 'active layer' that had formed under permafrost conditions during the Loch Lomond Stadial (compare with Ballantyne and Harris, 1994).

A number of other sites have been reported where solifluction deposits rest on organic sediments dated to the Windermere Interstadial and Loch Lomond Stadial. Examples include Garral Hill, near Keith (Godwin and Willis, 1959), Woodhead, Fyvie (Connell and Hall, 1987), Moss-side Farm, Tarves (Clapperton and Sugden, 1977) and sites near New Byth (information from G Whittington, St Andrews University, 2002). These sites demonstrate the instability of low-angle till slopes during the Loch Lomond Stadial and the widespread occurrence of foot-slope accumulations of periglacial diamicton. The frequency of former active layer detachment slides in the region is unclear, but they are liable to be rendered unstable by engineering works.

The largest landslip affecting solid rocks in the district is developed in steeply dipping, interstratified semipelites and psammites of the Southern Highland Group on the western side of Gamrie Bay (Map 3); the slip is almost 800 m long and up to 300 m wide. Another major slip in Dalradian gritty psammites and semipelites occurs in the 150 m-high cliff [NJ 798 649] immediately south-east of Morehead. The largest area of landslip in the southern part of the district occurs on the southern flank of Strathfinella Hill (Sheet 66E), north of Westmoston [NO 683 761]. Five individual slips have been recognised (Carroll, 1995a), all of which are developed in Lower Devonian Strathfinella

Plate 3 Back scar of a landslip in the Kirk Burn Silt Formation near Macduff (P104100).

Conglomerate. The slipped material, which can be up to 70 m thick in places, extends for a distance of some 300 m along the hillside. Each slip has a prominent cuspate back scarp and reorientation of bedding within the slipped masses suggests that the slips are principally rotational failures.

COASTLINE STABILITY

The present and future stability of any coastline depends on its geological context, supply of sediment, energy environment (wave, tide and wind) and the altitude at which this energy is delivered (sea level), together with any secondary effects resulting from human interference.

Much of the coastal edge of north-east Scotland is composed of hard rock cliffs that are highly stable in the short term at least. On the east coast, cliffs of resistant Old Red Sandstone sedimentary rocks and Dalradian metamorphic rocks extend from Inverbervie to Aberdeen, beyond which they give way to a low, sandy coastline backed by dunes between Aberdeen Bay and the Sands of Forvie. To the north of Cruden Bay the cliffs are composed of Peterhead Granite. The north coast is characterised by extensive cliffs formed of resistant Dalradian rocks, broken by a lower, sandy and rocky coastline at the Loch of Strathbeg and Rattray Head and by gravel beaches in Spey Bay, extending from Buckie to Lossiemouth. Permian sandstone cliffs between Lossiemouth and Burghead give way within Burghead Bay to gravel and sand beaches and spits that extend to Nairn and beyond. Elsewhere, on both coasts, beach development is restricted to small and stable cliff-foot bays. The cliffed sections of the coast are stable and hard, restricting the movement of the cliff-foot beaches that are consequently also stable. It is in the softer and low-lying sections of the coast that instability is an issue for planners and engineers.

Although tidal currents are relatively weak in the area, those along the eastern coast move southward and serve to reduce the northward wave-induced drift of sediment. However, a low net drift of sediment to the north still occurs (Hydraulics Research, 1997). Wave energies are highest close to Cairnbulg Point [NK 035 661] (Map 4) and generally decline to the west and south (Natural Environment Research Council, 1998). As a result, Cairnbulg Point represents a sediment drift divide (Hydraulics Research, 1997), although sediment exchanges are restricted on the indented and cliffed sections of both coasts. On the north coast, in Spey and Burghead Bays to the west of Buckie, there is considerable wave-induced westerly longshore movement of sediment. Consequently, these stretches of coastline are affected by erosion in the east (up-drift) sections and by accretion in the west (down-drift). Elsewhere, the overall pattern of coastal sediment movement is locally affected where rivers carry high sediment loads into the sea. This occurs at and beyond the mouths of the Don, Ythan and Ugie; it is particularly, significant where the Spey enters the sea.

The history of sea level change is also important to coastal stability (Chapter 7, Figure 48). The coast of the district has been subject to fluctuating sea level following deglaciation some 15–13 000 years ago, but about 7000 years ago a sea level rise culminated at about 8 m above OD, which allowed large quantities of glacigenic sediment to be moved onshore. This input of sediment is reflected in the impressive suites of raised beaches that are found in virtually all of the low lying coastal areas, but especially in the hinterlands of Spey Bay, Burghead Bay and backing the present-day beaches adjacent to Rattray Head and Forvie. The continuous isostatic rebound of the crust (Figure 49) to its present altitude led to the elevation of these beaches, but also to the reduction of the offshore supply of sediment that once fed them. In such conditions, beaches adjust to reduced sediment supply by internal reorganisation of sediments and by enhanced up-drift erosion to fuel down-drift accretion, a process that is evident at present in the eastern parts of Spey and Burghead Bays and in the southern part of Aberdeen Bay.

Present sea level change in the district is a combination of global rises in sea level and local crustal rise following deglaciation. Present crustal rise in north-east Scotland is about 0.5 mm per year (Shennan, 1989), although this increases to 1 mm per year around Nairn, while global sea level is rising at about 1–2 mm per year (Gornitz, 1995). Thus a small relative sea level rise appears to be under way in the east (Smith et al., 1999; see Appendix 1, Site 9 Philorth Valley) while the west is in approximate balance. However, as global sea level is modelled to be subject to an accelerated rise of about 30 cm over the next 50 years (Hill et al., 1999), this state of approximate balance may be set to change. By 2050, the area may experience a relative sea level rise of 25–29 cm (Hill et al., 1999).

Placing sea-level changes into the context of present and future coastal stability, the low lying sand and gravel coasts are undergoing localised erosion at present. Where longshore drift is significant, this is manifest by a tendency for erosion at one end of the beach to fuel deposition at the other. In the west, in embayments such as at Spey Bay and Burghead Bay, this erosion is associated with reduced levels of sediment supply and is possibly exacerbated by a slowly rising sea level. Where longshore drift is insignificant, such as at the Rattray Head beaches and at Balmedie, the shoreline is either stable or undergoing limited movement landwards by means of frontal erosion. In some places, close to towns and villages, the human response to this situation has been to defend the eroding sections of shoreline. If sea walls are used, such as at the south end of Aberdeen Bay, they commonly have the effect of further reducing the supply of sediment to the beaches and accelerating erosional responses elsewhere. As a result, the short and medium term future of such areas is either enhanced instability or the imposition of an artificial, constructed stability (in the form of a sea wall) which, in the long term, may be unsustainable. Where beach recharge mechanisms are used, such as at the mouth of the Spey, the short and medium term future is for the maintenance of ongoing natural coastal processes under an assured sediment supply regime, and, is consequently more sustainable in the long term (Gemmell et al., 1996).

By the year 2050 there may also be problems of coastal flooding of low-lying land as a result of the projected 25–29 cm sea level rise. Areas most at risk from future flooding are low-lying ground especially that in the east of the district and ground adjacent to river exits, whose flood waters may exacerbate any marine effect. Flooding of the lower parts of Aberdeen, together with the lower parts of coastal towns such as Banff could be experienced (Barne et al., 1996). On the other hand, a rising sea level will enhance the tendency for estuaries and inlets to infill from landward sediment sources and places such as the exits of the rivers Lossie, Deveron, Ugie, Ythan, Don and Dee may experience not only rapid infilling, but also a higher flood frequency.

SEISMICITY AND SEISMIC HAZARD

The area covered by this memoir is remarkable in that it is more or less completely free from earthquakes. This is equally true of both the modern period (i.e. from 1970 onwards) for which there are good instrumental records available of small events, and for the historical period for which earthquakes are known principally from written descriptions. The published UK earthquake catalogue (Musson, 1994), which concentrates on events larger than magnitude 3 ML (Richter Local Magnitude), contains no earthquakes at all for this area, onshore or offshore. Because of the cultural and historical importance of Aberdeen, the availability of historical records is relatively good and it is quite clear that the absence of reported earthquakes represents a real absence of seismicity and not just a data gap. One can estimate, from a knowledge of the historical sources, that there certainly have been no significant earthquakes along the east coast of this area since at least 1750, and along the north coast since about 1800.

Examination of the BGS instrumental database (Walker, 1998) reveals a very few small events, imperceptible except to instrumental recording. These include an event of magnitude 1.4 ML about 10 km east of Elgin on 11 April 1985, and one of magnitude 2.0 ML offshore, about 23 km north of Rosehearty, on 24 June 1980. This is earthquake activity at a trivially low rate, even by the intraplate standards of the UK.

The reason why the east coast of Scotland should be so much less seismic than the west coast remains obscure. It has been noted that the locus of Scottish seismicity correlates rather closely with the distribution of maximum ice load during the last glacial advance (e.g. Musson, 1996), but this is at best only a partial explanation.

There are records of earthquakes having been felt in this area, going as far back as the turn of the 16th century. However, these have all been the distant effects of large earthquakes occurring elsewhere, principally in the Inverness area, around Comrie, Perthshire, and in the North Sea (especially the Viking Graben area). The Comrie earthquake of 8 November 1608 was felt strongly in Aberdeen, and caused some alarm. The local clergy attributed the earthquake to God's wrath at salmon fishing being conducted on the River Dee on the Sabbath. The Inverness earthquakes of 13 August 1816 and 18 September 1901 were felt throughout the whole of the area in question, at intensities between 4 and 5 EMS (European Macroseismic Scale).

Particular mention should be made of the Viking Graben earthquake of 24 January 1927 (Musson et al., 1986). This earthquake had a magnitude of 5.7 ML and was felt over a wide area of eastern Scotland and western Norway; in the UK the effects were strongest in the Buchan area (intensity 5 EMS). Very slight damage was caused at Peterhead. One ceiling in the southern part of the town was damaged and a concrete wall at nearby Keith Inch was said to have been cracked by the shock. These two single instances seem to be the only cases of earthquake damage ever recorded in the whole area.

LAND USE

Soils

More than 40 different soil types, grouped into 25 soil associations, have been recognised across the district (Walker et al., 1982). Soils from three areas are described in detail in Soil Survey One-inch sheet memoirs: Sheet 95 Elgin (Grant, 1960); Sheets 86 and 96 Banff, Huntly and Turriff (Glentworth, 1954) and Sheets 77, 76 and 87/97 Aberdeen, Inverurie and Fraserburgh (Glentworth and Muir, 1963).

Stony, sandy loams are developed on the reddish brown tills and sandstone bedrock of the Moray Firth coastal lowland and on free-draining till slopes flanking the major river valleys. Gravelly loams have formed on many of the glaciofluvial and hummocky glacial deposits. Sandy, gravelly and clayey loams are widely developed in Buchan, where the proportion of gravel, sand and clay in the soil often reflects the nature of the deeply weathered underlying parent material. Thin, coarse, sandy loamy soil, in many places containing large boulders, overlies many tills developed on granite bedrock or outcrops of decomposed granite. The deeply weathered basic igneous rocks around Insch, together with the glacial deposits derived from them, produce highly fertile sandy or sandy clayey loams, while fertile brown forest soils have formed on the reddish brown tills and Old Red Sandstone bedrock of Strathmore. Immature silty and sandy soils, commonly mixed with shingle, are developed on raised beach deposits and spreads of blown sand. Wet mineral soils have developed on alluvial deposits around estuaries, and are exposed in upland areas where peat has been removed.

Agriculture and forestry

The 1:250 000 scale land capability for agriculture map for eastern Scotland (Soil Survey of Scotland, 1983), shows that much of the land in the district is capable of producing good yields of a narrow range of crops, principally cereals and grass. This is true of the Buchan plateau, the major river valleys, the southern coast of the Moray Firth and Strathmore. In these lowland areas, which represent a major part of the main arable belt of Scotland, arable farming is concentrated on the ground generally lying below 100 m OD. A combination of better climate, high agricultural productivity and good natural harbours has led to the concentration of primary population centres along the coastal margins of the district.

North-east Scotland is justly famous for quality of its beef cattle, many of which are raised on improved pastures on the interfluves between the major rivers. However, each area has its own characteristics, reflecting its differing topography, elevation, climate, bedrock and Quaternary geology. Cattle rearing has been dominant on improved pastures in the foothills of the Grampian Mountains, but sheep farming, shooting and forestry are competing forms of land use on most of the ground above 200 m OD. Apart from the patches of very fertile soil developed on the glacial and fluvial deposits between Insch and Inverurie, there appears to be little regional correlation between agricultural land capability and the underlying geology on the lower lying ground. This may reflect the lack of erosive power of the ice that traversed the lowlands during successive glaciations. This is particularly evident in the north-eastern part of the district, where gentle slopes and thick weathering profiles are preserved on igneous and metamorphic rocks, as well as in the Quaternary deposits derived from them. Variations in land capability on this low, undulating ground are more likely to reflect the degree of soil improvement (drainage and fertiliser application) that has taken place rather than the nature of the underlying strata. Many of the most productive patches of agricultural land between Insch and Inverurie occur on glaciolacustrine deposits. These produce relatively stone-free silty and sandy soils, which, given adequate drainage, are capable of producing a wide range of crops. Another notable area of very fertile soil occurs on reddish brown clayey till, overlying Lower Devonian mudstones of the Cromlix Formation in Strathmore. Horticulture as well as arable farming has developed on these clayey till soils and also on adjacent sandy alluvial soils in the catchment of the Luther Water.

The link between underlying geology, elevation, aspect and climate and land use capability is more evident in the upland areas. It is well seen on the high ground between the valley of the River Dee and Strathmore, where relatively thin, acidic soils and hill peat are developed on till, here containing a predominance of granitic clasts, or directly on decomposed Mount Battock Granite. Much of this land is only suitable for use as rough grazing, though some of the lower hillsides are capable of supporting improved grazing. In practice, much of the high ground is devoted to shooting and summer grazing of sheep, or is covered by extensive conifer plantations. A similar pattern of land use is present on the acid soils developed on the Hill of Fare and Bennachie granites.

SHALLOW GEOPHYSICS AND REMOTE SENSING

Ground geophysical surveys have been undertaken to investigate the three dimensional form of Cainozoic successions in several small areas in the district. Each investigation was designed to answer particular questions concerning the nature of the concealed sequence, in ground that had been investigated previously by drilling and trial pitting, as well as by detailed geological mapping.

Three techniques were used:

i   Conductivity surveying, using a non-contacting (EM31) terrain conductivity meter, was employed on an experimental basis to map the near surface (less than 5 m depth) extent of workable deposits of sand and gravel in 2 km² of ground in the Houff of Ury area, north-west of Stonehaven.

ii   Resistivity soundings, using Offset Wenner and Schlumberger arrays, were taken to investigate the nature and thickness of Quaternary sequences encountered in sand and gravel assessments in the Inverurie–Stonehaven (Auton et al., 1988) and Strachan–Auchenblae–Catterline areas (Auton et al., 1990).

iii   Ground probing radar (GPR) traverses, using a Pulse Echo IV radar system, were undertaken to investigate the thickness and lateral extent of Quaternary deposits and the form of rockhead in the Houff of Ury conductivity survey area (Greenwood and Raines 1994; Greenwood et al., 1995). Traverses were also made to elucidate the sedimentary architecture of the Palaeogene to Neogene Buchan Gravels Formation at the Den of Boddam, Moss of Cruden and Windy Hills sites. At Moss of Cruden, resistivity soundings were made, in conjunction with the GPR survey, to augment earlier resistivity measurements made in the area. The GPR traverses, which produce cross-sections analogous to shallow seismic profiles, used radar frequencies between 25 and 100 m Hz.

Detailed results of the conductivity survey at the Houff of Ury and the resistivity soundings in the Inverurie–Stonehaven, and the Strachan–Auchenblae–Catterline areas are discussed in Appendix 3. The results of the GPR traverses are discussed in Appendix 3. They are also incorporated in the site descriptions of the Moss of Cruden and Windy Hills in Appendix 1.

Satellite imagery

The gross geomorphology of north-east Scotland can best be illustrated using satellite imagery (*Frontispiece*). The image data, which were acquired from about 700 km above the surface of the Earth by Landsat, clearly differentiate the upland areas of the Grampian Highlands and Cairngorm Mountains from the lowlands of Strathmore, Buchan and the Moray Firth coast. The major elements of the postglacial drainage pattern are also clearly visible. Winter Landsat imagery (Landsat Thematic Mapper Band 5, Scene 204–020, acquired in October 1985), characterised by a low sun-angle that highlights subtle geomorphological features, was interpreted to aid the Quaternary mapping of Sheet 87W. Unfortunately, comprehensive winter coverage was not available for the district until recently. Consequently, detailed interpretation of aerial photographs, rather than satellite images, has played the major role in mapping out Quaternary landforms in the district, on both a local and a regional scale.

THREE

Landscape evolution

The lowlands of north-east Scotland are an excellent area for the study of long-term landscape evolution. The erosional impact of the Quaternary glaciations has been modest and a range of erosion surfaces of Palaeogene to Neogene age (Table 2) are preserved (Figure 15). The development of these Tertiary landforms has been studied using evidence from deep weathering profiles and preglacial deposits. On land, evidence from rocks that postdate the Caledonian orogeny is limited. The geological record is confined to remnants of formerly extensive Devonian intramontane basins, Permian dykes and very small outliers of Cretaceous rocks. Much greater thicknesses of Mesozoic and Cainozoic rocks are preserved within the adjacent basins of the Moray Firth and central North Sea. These provide important evidence of environments that prevailed on the adjacent land area.

Knowledge of ancient landscapes in north-east Scotland is important for several reasons. Firstly, the region forms the onshore extension of a structural 'high' that divides the Moray Firth and central North Sea basins offshore. Buchan and its contiguous lowlands to the south and west have periodically provided sediments to both basins and, in turn, have periodically been covered with sedimentary rocks laid down beyond the basin margins. While research relating to hydrocarbon exploration has provided a large amount of new data on the tectonic development of the offshore basins, relatively little is known about events on the adjacent structural high. Secondly, the preservation of preglacial landforms is matched only in a few other formerly glaciated areas and so the region provides a rare opportunity to investigate Palaeogene and Neogene landscapes within the limits of the Quaternary ice sheets. Thirdly, the extensive presence of deeply weathered rock is not a feature normally associated with glaciated landscapes. It is more akin to crystalline terrains in parts of Europe beyond the Quaternary ice sheet limits. Deeply weathered rock is a source of aggregate and provides the parent material for Quaternary deposits and soils. Assessment of its characteristics and distribution is very important in site investigations for major engineering projects. Finally, the limited degree of glacial erosion places the lowlands of north-east Scotland within a select group of similar regions around the North Atlantic, including parts of southern Sweden (Lidmar-Bergstrom, 1982) and islands in the Canadian Arctic (England, 1986), in which it is possible to investigate the processes that have operated under dominantly cold-based ice sheets.

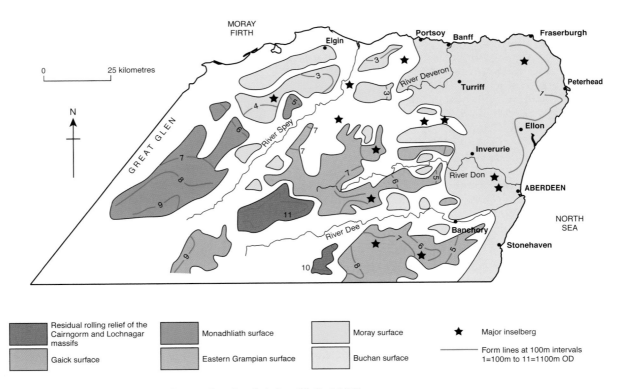

Figure 15   Erosion surfaces in north-east Scotland (after Hall, 1986).

The evolution of the landscape of the lowlands of north-east Scotland is described below. The gross features of the relief, the main hill masses, valleys and plains, first began to emerge in the Palaeozoic and may have taken on much of their present distribution and general form by the end of the Mesozoic. The landscape continued to develop in response to uplift and differential chemical weathering under humid climates throughout the Palaeogene and Neogene, producing a wide variety of preglacial landforms. These landforms and landscapes were then variably modified by the successive ice sheets of the Quaternary period. Descriptions of the Palaeogene and Neogene deposits and related soils is contained in the next chapter.

PALAEOZOIC AND MESOZOIC

Inherited Devonian surfaces

The oldest landscapes preserved in north-east Scotland are of Devonian age. During this period the region was desert and lay in the Southern Hemisphere. It formed the southern boundary of the Orcadian basin at the foothills of a major Grampian mountain chain (Andrews et al., 1990). These mountains were created earlier in the Caledonian orogeny and had already suffered prodigious erosion before Devonian times. At around 470 Ma, numerous large 'Younger Basic' igneous masses were emplaced at crustal depths of around 15 km (Droop and Charnley, 1985; Chapter 1). Between 425 and 395 Ma, final phases of postorogenic magmatism culminated in the emplacement of the 'Newer' Granite plutons at depths of around 5 km (Gould, 1997). Many of these intrusions were already unroofed by Devonian times, as shown by the presence of sandstones and conglomerates of this age resting on exposed basement. Devonian sedimentary rocks overlie the Aberdeen granite, the Belhelvie basic intrusion (Munro, 1986), the Peterhead granite (Hall and Jarvis, 1994) and the Insch–Boganclogh and Morvern–Cabrach basic igneous intrusions (Gould, 1997).

The case of the Bennachie granite is of special interest. The granite was unroofed by about 385 Ma in the Devonian as it apparently contributed pebbles to the Turriff Old Red Sandstone basin (Mackie, 1923). Abundant aplite sheets and the occurrence of intrusive breccias along the lines of the north–south-trending faults show that rocks now exposed represent the uppermost part of the pluton (Figure 16). Furthermore, the pluton extends east and west of the exposed area on Bennachie under thin cover (McGregor and Wilson, 1967). Its present elevation probably reflects Tertiary uplift. An analogous situation occurs at Moss of Cruden. Here a small patch of arkosic Devonian sandstone (Appendix 1 Site 14) rests on Peterhead Granite (Hall and Jarvis, 1994). Overlying the same intrusion nearby is an outlier of Lower Cretaceous Greensand and deposits of the Buchan Gravels Formation (Chapter 4). This association demonstrates that erosion of the Peterhead Granite since the Devonian has been minimal, like Bennachie.

The former extent of Devonian cover rocks may be reconstructed from the present distribution and dip of these rocks and from the composition of the sandstones and conglomerates. The huge volume of clastic sediments that were deposited in the Lower Old Red Sandstone basins of Strathmore indicate that the adjacent eastern Grampian mountains were of considerable relief at this time. North of Aberdeen, the remaining Lower Old Red Sandstone strata were originally laid down on the floor of former intramontane basins. Middle Old Red Sandstone sedimentary rocks occur in the Turriff basin, where they lie with marked unconformity on older sandstones. The Upper Old Red Sandstone is found in coastal areas west of the Spey, but south of the Grampians considerable thicknesses of Upper Old Red Sandstone rest on folded and deeply denuded Lower Old Red Sandstone sequences (Mykura, 1983).

The Lower and Middle Old Red Sandstone form a continuous belt just offshore (Andrews et al., 1990) and occur as widely scattered outliers on the Dalradian basement (Figure 2). Apart from the main outliers of Turriff, Rhynie, Cabrach and Tomintoul, there are several

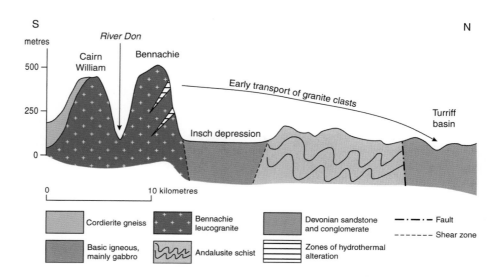

Figure 16   Devonian unroofing and post-Devonian erosion of the Bennachie granite mass (after Hall, 1987).

small outcrops, such as those in upper Strathisla (Hinxman and Wilson, 1902). Other small, concealed outliers may exist. However, an earlier report of an outcrop of Devonian rock in the Feugh basin (Bremner, 1942) appears to have been mistaken for pale orange, sparsely feldsparphyric aplitic microgranite that crops out in the vicinity. Furthermore, the patch of Devonian conglomerate reported from Cruden Bay (Wilson, 1886) appears to be a cemented glacigenic deposit belonging to the Logie-Buchan Drift Group, and another supposed outcrop to the north of Glenlivet seems to be glaciofluvial sand and gravel.

While there is little doubt that the Middle Old Red Sandstone once covered areas well beyond its present outcrops (Sissons, 1967), it is unlikely that cover was continuous away from the current coastal belt, as some have claimed (Bremner, 1942). In the Turriff outlier, dips increase from west to east, conforming to the half-graben basin shape in section. The Turriff, Rhynie and Cabrach basins are bounded to the west by north-east-trending faults. To the east, the Lower to Middle Old Red Sandstone appears to rest with simple unconformity on basement. The basal conglomerates of these and other outliers are dominated by clasts from the local basement (Wilson and Hinxman, 1890; Hinxman and Wilson, 1902; Read, 1923; Peacock et al., 1968; Munro, 1986). These basement rocks were not buried by later deposition as shown by the continued presence of basement-derived clasts in conglomerate bands higher in the successions of the deeper basins (Read, 1923; Peacock et al., 1968). South of Elgin, conglomerates and breccias were deposited against steep hillsides marking the edge of a rugged terrain developed in underlying Moine and Dalradian metamorphic rocks.

The original thickness of Devonian rocks in the various basins is difficult to estimate. Geophysical surveys indicate that a stratigraphical thickness of around 1 km of Devonian sedimentary rocks are preserved in the Turriff basin (Ashcroft and Wilson, 1976). The high porosity and limited development of quartz overgrowths in the Smallburn Sandstone on the Moss of Cruden suggest that postdepositional burial was limited to 2 to 3 km, although part of this burial may have taken place in the Mesozoic.

In general, sedimentation occurred in fault-bounded 'half-grabens' and other topographic basins with erosion of debris from intervening ridges and hills. Sediment transport was generally via a series of north-east- and north-trending valleys towards Lake Orcadie. The strata of the Deskford (Read, 1923) and Strathisla outliers (Geikie, 1878) partly infill two of these ancient valleys. Coarse sandstones and conglomerates indicate high-energy fluvial transport. Rounded and reddened quartzite clasts are abundant and they form a conspicuous component of Quaternary deposits adjacent to the Devonian outliers.

The preservation of sub-Devonian landscapes is likely to be confined to the margins of the Devonian outliers and to the ancient basins and valleys in which these rocks lie (Hall, 1991). Even where Devonian rocks no longer remain, the present subdued topography of the coastal lowlands undoubtedly results largely from post-Devonian planation.

Marine inundation during the Mesozoic

Reconstructions of post-Devonian palaeogeography generally show the lowlands of north-east Scotland as being above sea level for most of the time (Ziegler, 1981). Erosion of this structurally positive area has resulted in the arcuate outcrop of late Palaeozoic sedimentary strata around Buchan (Figure 2). Yet it is clear from the provenance of the Mesozoic rocks lying just offshore, in the inner Moray Firth and west central North Sea, that the lowlands were also inundated periodically during Mesozoic times.

In the Inner Moray Firth Basin, sequences of Jurassic rocks are up to 3.6 km thick, but thin rapidly across faults and basin highs (Andrews et al., 1990). Remnants of Lower Jurassic cover rocks are found in fault basins around Lossiemouth. Marginal marine sands were deposited across much of the Northern Highlands at this time (Hallam and Sellwood, 1976). The significant overstep of the contemporaneous margins of the Inner Moray Firth Basin in the Late Jurassic indicates that adjacent land areas had been reduced to low relief.

Tectonic activity was renewed in the Moray Firth Basin at the Jurassic/Cretaceous boundary (Anderton et al., 1979; Figure 17). Lower Cretaceous strata were once much more extensive both north and south of the basin, but have been removed by Tertiary erosion (Andrews et al., 1990). A small, concealed outlier of late Hauterivian–early Barremian glauconitic sandstone at Moreseat (Hall and Jarvis, 1994) is the only known remnant of this cover (Appendix 1 Moss of Cruden). Greensand clasts are known from the Neogene Buchan Gravel Formation at Windy Hills (Flett and Read, 1921) and Moss of Cruden (Kesel and Gemmell, 1981), which are described in the next chapter. Large blocks of Lower Cretaceous sandstone of late Hauterivian–early Barremian and Aptian age occur also as erratics in glaciofluvial outwash around Cardno, near Fraserburgh (Cumming and Bate, 1933; Map 4). These marine sands of marginal facies must originally have covered most, if not all of Buchan, from the Hauterivian onwards.

The basement rocks in the lowlands of north-east Scotland may not have emerged extensively prior to the Late Cretaceous transgression. That a cover of chalk once existed across Buchan is indicated by the large volume of flint contained within the Buchan Gravel Formation (Chapter 4). Furthermore, small remanié lags of nodular flint occur at the base of the Buchan Gravels at Skelmuir Hill (Bridgland, et al., 1997, 2000) and Moss of Cruden (Hall, 1993). These rest on kaolinised crystalline rocks and place the sub-Cenomanian surface close to current summit levels in the area. As the chalk accumulated at water depths of 100 to 600 m (Hancock, 1975) the former chalk sea must have extended across all low ground in north-east Scotland. The lack of terrigenous debris in the chalk offshore (Andrews et al., 1990) also implies that any areas of high ground were remote from the contemporaneous shoreline.

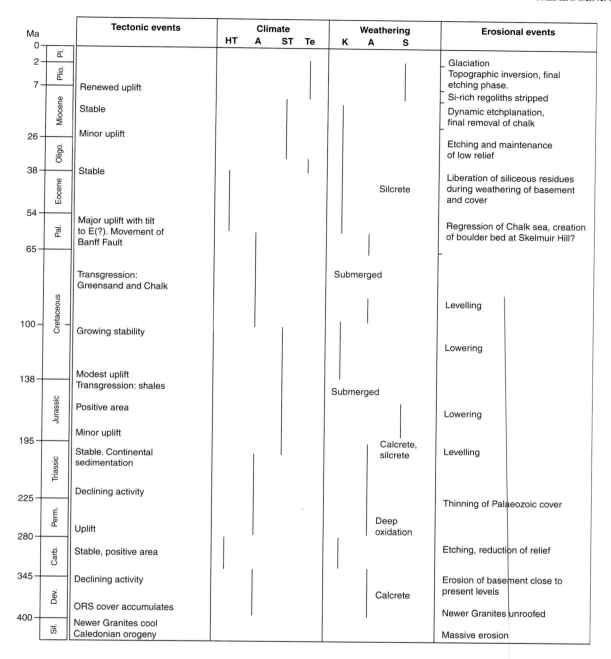

Figure 17   Summary of the main geomorphological events in the shaping of north-east Scotland (after Hall, 1987).

HT humid tropical;   A arid;   ST subtropical;   Te temperate;   K kaolinitic;   S sandy

PALAEOGENE AND NEOGENE

Tectonic activity

The onset of igneous activity in western Scotland in the Palaeogene at about 63 Ma was accompanied by widespread tectonic activity (Pearson et al., 1996). The Grampian Highlands were uplifted by at least 1 km and tilted eastwards towards the North Sea. Thick sand sequences derived from the raised block accumulated in the Moray Firth basin (Figure 2). Magmatism was at an end by about 52 Ma and both uplift and erosion slowed. Sedimentation rates dropped in the North Sea, with deposition of mud replacing that of sand. Uplift was renewed in the late Oligocene and it probably continued through the Neogene. The scale of Neogene uplift in Scotland may have been underestimated. Estimates of burial depths of Mesozoic rocks indicate that up to 1.5 km of basin fill was removed from the Inner

Moray Firth during the 'Tertiary' (Thomson and Hillis, 1995), possibly reflecting Neogene uplift of the western North Sea and adjacent areas (Japsen, 1997).

In north-east Scotland, major uplift of the eastern Grampians occurred in the Palaeogene. A proto-Dee and Don river system during the Eocene fed sediment to the Gannet Fan in the western North Sea (Figure 18). Uplift of the Cairngorms and accelerated erosion led to the exhumation of sub-Devonian valley systems. Yet the position of a major Paleocene depocentre in the inner Moray Firth (Andrews et al., 1990) suggests that the coastal fringe of north-east Scotland may not have been greatly uplifted at this time. Cretaceous cover rocks may have survived in eastern areas into the Neogene, finally being removed by erosion as a result of later regional uplift.

The subsequent evolution of the drainage network is of interest here (Figure 19). Despite the proximity of the Moray Firth, it is clear that the major drainage routes in north-east Scotland ran west to east. The headwaters of the Dee and Don, prior to river capture, lay in the Cairngorms and the topographic trench that extends from the Cabrach to Insch marks another west–east drainage line. Significantly, the Windy Hills Gravels of Neogene to Early Pleistocene age were also transported by a proto-Deveron–Ythan river system

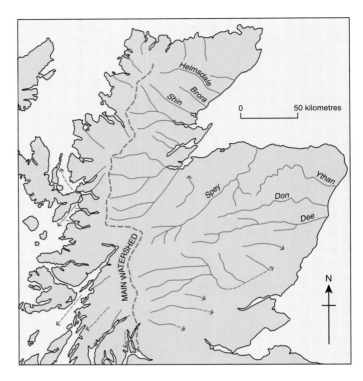

Figure 19 Neogene drainage patterns (after Hall, 1991).

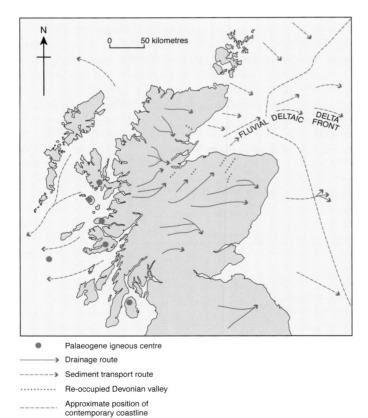

- ● Palaeogene igneous centre
- ⟶ Drainage route
- -----▸ Sediment transport route
- ············ Re-occupied Devonian valley
- ------- Approximate position of contemporary coastline

Figure 18 Palaeogene drainage and sediment transport (after Hall, 1991).

flowing eastwards towards the North Sea. River capture and glacial diversion disrupted this pattern of drainage to the advantage of rivers following the south-west to north-east Caledonian structural trend, some re-occupying Devonian valleys. Nevertheless, the ancient drainage pattern is still clear, and it clearly indicates tilting towards the North Sea and Neogene uplift of Buchan across the Banff Fault.

There is evidence of Neogene differential tectonic uplift and subsidence in the region. The existence of prominent scarps that appear to be unrelated to lithological boundaries, as around Bennachie and the Hill of Fare (Map 8), suggest relatively recent fault movement. Continued uplift along the Highland Boundary Fault also seems to have been necessary to raise the Mounth erosion surface 500 m or so above the floor of the Mearns. Block faulting seems widespread in the Elgin area (Hall, 1991) and such faulting may account for the marked elevation of Devonian sandstones and conglomerate of Hill of Fishrie (Map 3) and Windyheads Hill (Map 4). More general uplift of central Buchan is suggested by the elevation of the Buchan Ridge Gravels, which have been raised to around 150 m OD since deposition possibly at, or close to, contemporaneous sea level at some time in the Tertiary (Chapter 4).

Climate and weathering

Climates during the Paleocene and Eocene were warm and humid throughout north-west Europe and deep kaolinitic weathering profiles developed widely (Hall, 1991). Humid to subtropical conditions prevailed during the formation of organic deposits offshore to the north of Scotland in the late Oligocene (Evans et al., 1997). A sharp drop in temperature in the late Miocene (Buchardt, 1978), combined with an influx of less mature terrigenous debris derived from uplift of Fennoscandia and Scotland, led to increasing amounts of feldspar, illite and chlorite and decreasing quantities of kaolinite in North Sea sediments (Andrews et al., 1990). The mineralogical changes reflect a fundamental shift in weathering styles at this time from predominantly kaolinitic weathering profiles to less mature sandy weathering products (Chapter 4).

Landscape evolution

The relief of the area covered by this memoir is mainly lowland, dominated by a single, complex erosion surface, the 'Buchan Surface' (Hall, 1987; Figure 15), whose relative relief seldom exceeds 60 m OD. In detail, this subdued terrain resolves into a tiered landscape showing pervasive litho-structural control. An upper tier of isolated, low hills and broad interfluves developed on pelites and quartzites (with pockets of kaolinitic weathering) passes downslope into an extensive middle tier that includes open, saucer-like basins developed on deep sandy weathering covers (Chapter 4; Figure 22). Set into this middle tier are negative landforms developed on rocks of low resistance. These include the large, shallow basins of Maud and New Pitsligo, developed on norite and biotite granite respectively, and broad valleys such as that of the South Ugie Water, which follows a septum of biotite granite through the quartzite belt of central Buchan. The general correspondence between rock resistance to chemical weathering and topographic position reflects the dominant style of landscape evolution in this region during the Palaeogene and Neogene. In other words, deep weathering paved the way for subsequent erosion and heavily influenced the distribution of the main components of the preglacial topography.

The hilly terrain at the inland margin of the Buchan Surface reflects the interplay of differential weathering and tectonics. Uplift of blocks and erosion surfaces has created ridges and plateaux at different levels. The summit of Bennachie displays a fine series of tors. The tors have been only slightly modified by the passage of ice, with removal of blocks and the beginning of streamlining and lee-side plucking. They must predate at least the last ice sheet. Uplift has caused drainage incision. That this incision began prior to Quaternary glaciation is indicated by prominent benches that mark the preglacial valley floor along the middle reaches of many valleys in the eastern Grampians (Hall, 1991). Such benches occur along the Dee downstream of Banchory and above the Ythan gorge. Representatives of the intramontane basins of the eastern Grampians (Linton, 1951; Hall, 1991) occur along the inland margin of the district and include the basins of Feugh on the Dee, Alford on the Don, the Insch depression and the Knock basin on the Deveron. These basins have a long history of development,

perhaps extending in some cases back to the Devonian, and generally reflect the presence of rocks with low resistance to chemical weathering.

The oldest relief in the region, apart from localised sub-Devonian surfaces, is probably found in central Buchan (Figure 20). Here the juxtaposition of kaolinitic weathering profiles and the flint gravels of the Buchan Ridge defines an area of Neogene terrain (Chapter 4). The kaolinitic weathering appears to predate the cooling of climate that occurred in the late Miocene (Hall et al., 1989) and the constituents of the flint gravel indicate stripping of siliceous residues from pre-existing highly weathered landsurfaces. On the Buchan Ridge, the hilltops are close to the level of the sub-Cretaceous surface, as shown by the presence of Greensand and *remanié* deposits of flint. The flint gravels occur close to the highest tops in central Buchan. If it is accepted that the flint gravels are fluvial in origin then the headwaters of the rivers that deposited them have been lost to erosion (Chapter 4). The marked drop in elevation of the base of the Buchan Ridge Gravels, from 145 m OD at Whitestone Hill and 140 m OD at Skelmuir Hill to 70 m OD at Den of Boddam, 14 to 15 km to the east, suggests tilting towards the east. The age of the gravels is uncertain, but postdepositional uplift and tilting is likely to be late Neogene in age. Kaolinitic saprolites also occur to the north, within the outcrop of the Mormond Hill Quartzite. The inselberg of Mormond Hill is a very ancient feature and may have emerged from beneath Cretaceous cover rocks in the Neogene (Figure 22).

Elsewhere in the region, the widespread development of deep sandy weathering covers of late Miocene to early Pleistocene age indicates that all but the largest relief elements are of late Neogene age. Stripping of older kaolinitic regoliths and replacement by deep, but geochemically immature, weathering covers indicates regional erosion in response to continued uplift. Further etching out of lithological variations in the bedrock is demonstrated by the influence of geology on the distribution of sandy weathering patterns and by the landforms of differential weathering and erosion that comprise the meso-scale relief of the Buchan Surface.

CUMULATIVE GLACIAL EROSION DURING THE QUATERNARY

Compared with other areas of Scotland, the geomorphological impact of Quaternary glaciations on the lowlands of north-east Scotland has been modest. However, although most of the region is a zone of limited glacial erosion (Linton, 1963; Clayton, 1974) there is significant variation in the intensity of glacial erosion (Hall, 1986). Regional weathering zones indicate that deep weathering profiles are rare in areas affected by the relatively vigorous ice streams that occurred along the coastal fringe of the Moray Firth and North Sea. The area formerly covered by ice from the East Grampians (Figure 4) shows fewer signs of active glacial erosion and deep weathering is preserved widely. In the area between the Don and the Ythan valleys, there is clear

Figure 20 Summary of Cainozoic relief development in central Buchan (after Hall, 1987).

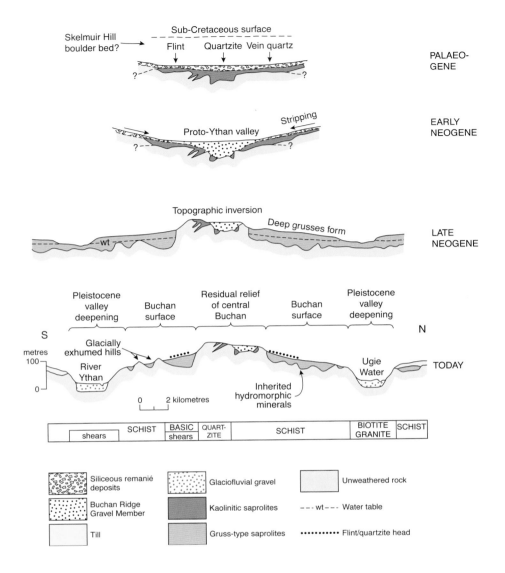

inverse relationship between the distribution of landforms of glacial erosion and of preglacial landscape remnants (Hall and Sugden, 1987; Sugden, 1989; Figure 21). By comparing morphology and sediment volumes along a transect from the Cairngorms to the central North Sea, it has been estimated that 16 to 42 m of weathered and fresh rock will have been removed from an area of relatively restricted glacial erosion (such as the lower Dee valley) during the Quaternary (Glasser and Hall, 1997). In central Buchan, the preservation of the Neogene Buchan Gravels Formation implies even more modest levels of glacial erosion.

The morphological contrasts described above reflect differences in glacier basal thermal regimes and rates of ice flow at different scales. The Moray Firth and North Sea ice streams were sourced well to the west of the district in areas with high snowfall and relatively elevated temperatures. These ice streams were warm-based and capable of scouring bedrock. In inland areas the survival of sandy weathering indicates a general lack of erosive capacity and it is clear that successive ice streams were generally cold-based and frozen to their beds. The local presence of ice-smoothed rock surfaces on hills, cols and in major valleys, however, shows that at these sites ice was at some time at its pressure-melting point and starting to slide and erode (Hall and Sugden, 1987). The contrast between the tor-studded plateau of Bennachie, where ice modification has been limited to removal of detached blocks, and the adjacent summit of Cairn William [NJ 656 168], with its striated pavements (Gould, 1997), is of interest here, as it shows that ice at this elevation (about 500 m OD) was only sliding in zones of convergent flow and on slopes that faced up-glacier.

The former glaciers played an important role in removing and reworking weathered materials. Weathering profiles are generally truncated and covered by a variable thickness of glacial deposits. In glacigenic deposits of the East Grampian Drift Group there is commonly a large proportion of material reworked from saprolites (FitzPatrick, 1963; Basham, 1974) and rafts of saprolite occur (Sugden, 1986). In soils developed on till, clay mineralogy is often found to be independent of drainage status and the clays are regarded as relict, being inherited from the underlying till (Wilson and Tait, 1977). In turn, the clay mineralogy of tills mirrors that of subjacent saprolites and it is clear that soil clays were also largely derived directly from former saprolites (Glentworth and Muir, 1963). Inherited material may also dominate the sand and gravel fraction of tills.

Abundant partially altered primary minerals occur in till below the depth of Holocene soil formation (Basham, 1968). Corestones are a conspicuous component of tills in areas down-ice from certain acid and basic igneous rocks (Wilson and Hinxman, 1890). In contrast, the amount of far-travelled material is substantially greater in glacial deposits of the Logie-Buchan and Banffshire Coast drift groups.

Figure 21 A comparison of the patterns of weathering and glacial erosion in eastern Aberdeenshire (after Hall and Sugden, 1987).

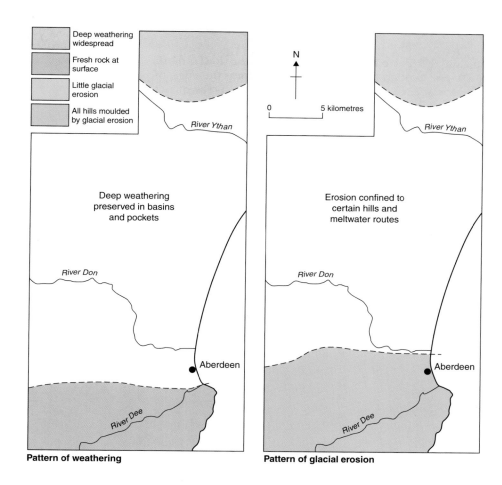

FOUR

Palaeogene and Neogene deposits, weathering and soil development

DEPOSITS

During the Palaeogene and Neogene, the lowlands of north-east Scotland and adjacent areas along the edge of the Moray Firth and North Sea basins were areas of net erosion (Chapter 3). Episodic uplift of the Inner Moray Firth Basin and adjacent land areas occurred at this time (Japsen, 1997) and consequently contemporaneous sediments offshore are largely confined to the region east of 1°W (Andrews et al., 1990). No Palaeogene (Paleocene and Oligocene) sediments are known on land in north-east Scotland, but Neogene (Miocene and Pliocene) deposits do occur in the form of isolated bodies of gravel belonging to the Buchan Gravels Formation (Figure 22). Furthermore, a characteristic feature of the lowlands of

north-east Scotland is the survival of deep weathering profiles (saprolites) that developed extensively during the late Cainozoic. Evidence of such chemical weathering is widely preserved in the current landscape and the saprolites preserved are potential resources of aggregate (Appendix 2) and are factors influencing ground stability, local groundwater resources and land use (Chapter 2).

The shoreline of north-east Scotland during the Neogene appears to have lain well to the east of the present coastline. Miocene littoral and sublittoral sands, containing glauconite, abundant shell debris and lignite, occur in the Moray Firth Basin in a north–south zone between 1° and 0°W (Andrews et al., 1990). Lower Pliocene and Miocene sediments are locally absent, implying mid-Pliocene uplift and erosion. Lignitic clays

Figure 22 Distribution of Miocene land surface and major topographical features (after Hall, 1985).

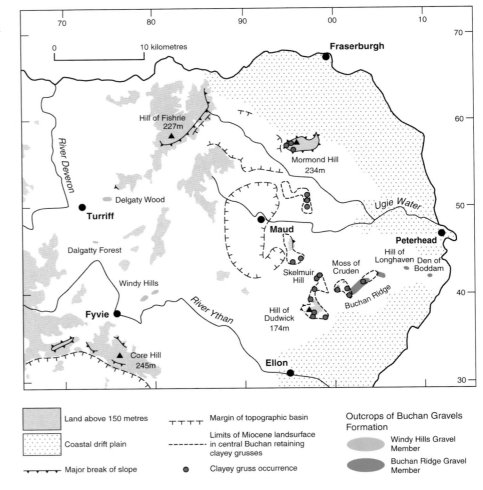

Land above 150 metres	Margin of topographic basin	Outcrops of Buchan Gravels Formation
Coastal drift plain	Limits of Miocene landsurface in central Buchan retaining clayey grusses	Windy Hills Gravel Member
Major break of slope	Clayey gruss occurrence	Buchan Ridge Gravel Member

of possible Pliocene age occur in BGS Borehole 81/19, 150 km north-east of Fraserburgh. Pollen of hickory is abundant, implying warm, generally frost-free conditions at this time. The abundance of well-ordered kaolinite invites comparison with the kaolinitic Buchan Gravels (Andrews et al., 1990).

The former, and perhaps continued, existence of small bodies of Upper Cretaceous, Palaeogene and Neogene rocks in the inner Moray Firth is indicated by the occurrence of chalk clasts in tills in north-east Buchan and Caithness, glacial rafts of Miocene clay at Leavad in Caithness (Crampton and Carruthers, 1914), and of Tertiary palynomorphs within the matrix of tills in eastern Buchan.

Studies of the clay mineralogy of sediments in the North Sea Basin show an increase during the late Miocene of illite, the appearance of chlorite and a corresponding reduction in kaolinite (Karllson et al., 1979; Berstad and Dypvik, 1982). These changes appear to correspond with transition to a cooler climate and a reduction in the intensity of chemical weathering. The ratios of clay minerals remain relatively stable through the Pliocene, despite fluctuations in climate. There are, however, some anomalous occurrences of kaolinite that may represent reworking of older material (Andrews et al., 1990). The overall change in clay mineralogy has been correlated with a switch towards the development of sandier, less mature weathering profiles under humid temperate climates on land in north-east Scotland (Hall, 1985).

Buchan Gravels Formation

Sediments assigned to the Buchan Gravels Formation occur in two distinct areas of Buchan as a discontinuous cover on ridges, hills and valley benches (Figure 22). The deposits crop out at elevations between 75 and 150 m above OD, between Den of Boddam [NK 115 416] and Delgaty [NJ 746 508]. The gravels in each area are lithologically distinct. In the west, quartzite and quartz-dominated gravels occur between Windy Hills and Turriff on Sheet 86E Turriff. In the east, flint dominated units are found on summits of the broad ridge ('Buchan Ridge') that extends from Hill of Dudwick [NJ 979 378] to Stirling Hill [NK 125 413] on Sheet 87W Ellon and Sheet 87E Peterhead, respectively.

Despite having been geologically investigated for more than 150 years and the subject of much recent research (Gordon and Sutherland, 1993 provide citations of the earliest studies), the precise age and origin of the gravels remain controversial. They have been assigned a variety of formal and informal names: Buchan Ridge gravels (McMillan and Aitken, 1981), Cruden flint gravels and Windy Hills-Turriff quartzite gravels (Kesel and Gemmell, 1981), Buchan Ridge Gravels and Windyhills Gravels (McMillan and Merritt, 1980; Merritt, 1981), Buchan Gravels Group (Hall, 1984, 1985), Buchan Gravels Formation (Hall, 1986), Buchan Ridge Formation (Hall, 1987), Buchan Gravels (Hall, 1987, 1991) and Buchan Ridge Gravel (Bridgland et al., 1997). The quartz-quartzite gravels are formally assigned here to the Windy Hills Gravel Member and the flint gravels to the Buchan Ridge Gravel Member.

WINDY HILLS GRAVEL MEMBER

In the type area, around Windy Hills [NJ 800 398], 12 km south-east of Turriff (Figure 22; Appendix Figure A1.17), white, coarse quartz and quartzite gravel underlies two flat-topped ridges trending north-east at surface elevations of between 115 and 125 m OD (Plate 4). Over 14 m of gravel occurs locally, interbedded with quartz- and mica-rich sand and resting on deeply weathered bedrock (Merritt, 1981). In the type sections of the Windy Hills Gravel Member [NJ 800 400] and BGS Borehole NJ73NE2, 700 m north-north-east of Windyhills [NJ 797 393], up to 11.3 m of gravel, interbedded with pale yellow, clayey pebbly sand, overlies deeply weathered and kaolinised Dalradian schistose pelite. Pebbles and cobbles of quartzite and vein quartz in the gravel are comparatively fresh and some carry chatter or percussion marks. In contrast, most pebbles of granite and metasedimentary rock are decomposed to kaolinitic sand, although fresh angular metamorphic clasts have been recently recovered from beneath kaolinised gravel in a trial pit close to the southern margin of the south-western spread of the member (Gemmell and Stove, 1999). Sparse flint pebbles are also present within the whitened gravel, as are very rare clasts of chert reported by Flett and Read (1921). The latter possibly originated as fossiliferous chert nodules in Cretaceous Greensand (information from N Trewin, Aberdeen University, 1981). Further detailed locational and sedimentological data from the gravel at Windy Hills is given in Appendix 1.

Quartz- and quartzite-rich gravels occur at two other localities near Turriff (Map 5). Exposure at Dalgatty Forest (Hospital Wood) [NJ 735 460] is poor, but a former pit, at a surface elevation of around 115 m OD, showed 3 m of quartzose gravel and white sand (Peacock et al., 1977). Flett and Read (1921) observed quartz- and quartzite-rich gravel and sand at elevations of 107 to 122 m OD in a number of pits in the Wood of Delgaty [NJ 744 508]. Characteristic features are an abundance of quartzite clasts, up to 30 cm in diameter, and flint representing up to 10 per cent of the coarse gravel clasts. Some flints contain fossils of Late Cretaceous age (Salter, 1857). Clasts of brown sandstone are also reported (Peacock et al., 1977).

Each of the deposits of the Windy Hills Gravel Member at the Windy Hills type site rests on a valleyside bench at an elevation of around 110 m OD. Transport from a westerly source is consistent with the distribution of the gravel bodies along a proto-Ythan–Deveron river system. A western provenance is also suggested by lithological similarities between the clasts and the Dalradian quartzites of Banffshire (Koppi, 1977; Kesel and Gemmell, 1981) and by the concentrations of ilmenorutile in the matrix, probably derived from the Younger Basic igneous masses (Hall, 1983). Recycling of clasts from Devonian conglomerates is likely, as the rounding of the pebbles and cobbles contrasts sharply with the rather angular nature of most of the quartz sand grains.

The member was derived from a terrain that had extensive development of kaolinitic saprolites, but also had significant areas of relatively fresh rock. Stripping of saprolites provided first-cycle quartz grains, mica sand and

a

b

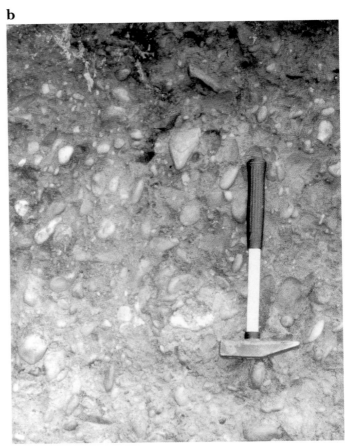

Plate 4   Buchan Gravels Formation at Windyhills.
a   Quartz-quartzite gravel of the Windy Hills Gravel Member at its type locality (P104101)
b   Cryoturbated till with vertically aligned pebbles capping the Windy Hills Gravel Member (P104102)
Scale: hammer is 32 cms long

kaolinitic silt. Erosion of fresh outcrops yielded clasts of granite and schist at Windy Hills and possibly Devonian sandstone at Delgaty. Some well-rounded pebbles and sand grains might have been recycled from older Tertiary gravel deposits. Following their deposition, the Windy Hills deposits were weathered throughout, with kaolinisation of clasts other than quartzite and also of the subjacent bedrock (Koppi, 1977; Koppi and FitzPatrick, 1980).

Broken sand grains in the Windy Hills Gravel Member may be the product of high-energy fluvial transport in the Neogene, or Quaternary glacial and glaciofluvial transport (Hall, 1983; Kesel and Gemmell, 1981; Gemmell and Auton, 2000). However, some of the grains may have been derived from the in situ decomposition of quartzo-feldspathic clasts within the deposit and others could be derived from glacigenic units within the Dalradian MacDuff Slate Formation. However, the member is regarded by most as being Neogene in age on account of its distinctive lithology and the extent of postdepositional chemical weathering (Appendix 1   Windyhills). It is locally overlain by till (Flett and Read, 1921) and must predate the cutting of the Ythan gorge, the floor of which is at least 75 m below the base of

the unit at Windy Hills. The position of the Windy Hills Gravel Member in a former course of a proto-Ythan–Deveron river system, contrasts with the ridge-top location of the Buchan Ridge Gravel Member and suggests that the Windy Hills Gravel Member is the younger of the two members of the Buchan Gravels Formation.

BUCHAN RIDGE GRAVEL MEMBER

The flint-dominated deposits of the Buchan Gravel Formation occur as isolated high-level masses in eastern Buchan (Figure 22; Maps 6 and 7). Exposure is generally poor and localised surface concentrations of flint suggest that other small gravel deposits may be concealed beneath Quaternary sediments.

MOSS OF CRUDEN   The largest mass of the Buchan Ridge Gravel Member underlies the ridge of the Moss of Cruden, extending from Moss of Auquharney [NK 018 399] to Hill of Aldie [NK 059 414] and includes small outliers on the flanks of Smallburn Hill [NK 016 405] (Appendix Figure A1.18). The size of the in situ gravel body has probably been over-estimated in the past, owing largely to the widespread development of younger flint-rich diamictons beyond the gravel outcrop. The diamictons have been recorded in an extensive network of trial pits, which have also shown that localised glacitectonic disturbance of Quaternary diamictons and gravels is widespread around Moreseat. Similar disturbance may also have affected parts of the gravel body in the Moss of Cruden type area.

Records from a BGS borehole (NK04SW3) suggest the gravel exceeds 25 m in thickness beneath the highest point of the Moss of Cruden (McMillan and Aitken, 1981) and subsurface data indicates that similar material infills at least two west-south-west-trending channels, one running along the Moss of Cruden ridge and the other transverse to the Hill of Aldie. Temporary excavations and boreholes on the Moss of Cruden show white, clay-bound coarse gravels with minor sand and silt units (Merritt, 1981). Gravel clasts are dominantly flint with quartzite and vein quartz; they are generally well-rounded pebbles and cobbles, many bearing numerous chatter marks. Some flints contain fossils of Cretaceous age (Jamieson et al., 1897) and a 75 × 30 cm block of Greensand was recovered from the gravel at Hill of Aldie [NK 062 410] by Kesel and Gemmell (1981).

The deposits originally contained less resistant clasts, probably mainly of granite, and these have decomposed to 'ghosts' of white sandy clayey silt. This decomposition extends throughout the known thickness of the deposit and into the underlying bedrock. Both sand and gravel units are bound, and in places, supported by white, kaolinitic sandy clayey silt. Flett and Read (1921) record clay pits on the Hill of Aldie, but it is unclear if these pits worked beds within the gravel or kaolinised bedrock.

Surface textures of sand grains from the Buchan Ridge Gravel Member at Moss of Cruden are similar to those from the Windy Hills Gravel Member, but grain breakage is more widespread (Hall, 1983). Rounded grains show thick precipitation surfaces, with silica coatings and fine euhedral overgrowths. Crescentic chocks are well developed indicating high-energy aqueous transport. Originally rounded grains are normally fragmented. However, interpretation of this latter feature is difficult, as the only samples from deep within the gravel body come from boreholes and breakage may be a drilling artifact.

As at Skelmuir Hill (see below), the base of the Buchan Ridge Gravel Member, in places, contains large nodules of flint and boulders of quartzite. The unworn flints resemble those in *remanié* deposits derived directly from solution of the chalk, such as the Eocene Tower Wood Gravels of south-west England (Hamblin, 1973). The gravel member rests on kaolinised granite and metasedimentary rock (McMillan and Aitken, 1981; Hall et al., 1989) and on highly weathered Lower Cretaceous Moreseat Sandstone (Hall and Jarvis, 1994). A kaolinitic silt, exceeding 6.5 m in thickness, occurs in a temporary pit (NK03NWP1) at East Backhill [NK 0089 3978] (Merritt, 1981). It was formerly recorded as either a unit within the member or deeply weathered granitic bedrock, but recent investigation has shown it to be a decomposed felsite dyke.

A ground probing radar (GPR) survey and resistivity soundings on the northern slope of the Moss of Cruden have confirmed that the feather edge of the Buchan Ridge Gravel Member rests on weathered Lower Cretaceous sandstone (Appendix 3 Shallow geophysics). Low-angle cross-stratification within the member dips towards the axis of the ridge. The resistivity and GPR data indicate that around 15 m of gravel lies beneath the crest of the ridge resting on granite weathered to clay. The gravels here fill a channel running parallel to the line of the ridge. Pitting and borehole data indicates that the gravel at Hill of Aldie fills another channel (McMillan and Aitken, 1981; Appendix Figure A1.18).

SKELMUIR HILL   A small deposit of yellowish brown, flint-bearing gravel was discovered at this locality [NJ 986 415] (Figure 22; Map 6) by Bridgland et al. (1997). The limits of the deposit are uncertain, but the total area of outcrop is unlikely to exceed 0.04 km². The gravel reaches a maximum altitude of 148 m OD and attains a thickness of at least 8 m, apparently resting against a relatively steep rock margin. The deposit has a number of distinctive features. It contains large numbers of boulders up to 0.8 m across. These include relatively angular clasts derived from the local psammites and quartzites, but also rounded boulders with percussion scars (Plate 5). The clasts include a greater proportion of less durable rock types than has been recognised at other sites. These 'ghost' clasts originally formed part of a clast-supported gravel containing pebbles and cobbles of brown flint. A basal boulder layer includes distinctive large, angular flint clasts, up to 0.25 m across, with flaked surfaces resulting from impact. The Skelmuir deposit appears to represent a basal facies of the Buchan Ridge Gravel Member resting within a steep-sided channel or against a cliff (Bridgland et al., 2000).

HILL OF DUDWICK   A small body of flint gravel rests within a depression at an elevation of about 170 m OD to the north of the summit of the Hill of Dudwick [NJ 979 378] (Kesel and Gemmell, 1981; Figure 22; Map 6). Up to 5 m of structureless clay-bound gravel is reported, and it is clear that at least the upper part of the deposit has been disturbed by glacial action. Surface concentrations of flint and quartzite suggest that another small gravel mass may be concealed in the ground to the south of Whitestones Hill [NJ 978 391].

HILL OF LONGHAVEN   Flett and Read (1921) reported flint gravels above the 300 foot (90 m) contour in the vicinity of Newton and Mount Pleasant, presumably beneath the summit of the Hill of Longhaven [NK 085 420] (Figure 22; Map 7).

DEN OF BODDAM   The flint gravels in the vicinity of the Den of Boddam [NK 114 414] partly fill a deep channel cut into the Peterhead Granite (Figure 22; Map 7). The deposit has a base at around 70 m OD, considerably below other occurrences, and it has been suggested that the gravel has been periglacially or glacially reworked (Hall, 1993a). However, recent pitting at Den of Boddam has shown that only the upper part of the gravel body has been thus disturbed, whereas the presence of intact weathered clasts and traces of bedding suggest that the rest of the deposit is in situ. The gravel consists of rounded cobbles and pebbles of flint and quartzite, up to 25 cm across, in a kaolinitic matrix. Kaolinised clasts of granite, probably Peterhead Granite, are common. The deposit has been considerably disturbed by the activities of late Neolithic flint miners (Bridgland et al., 1997; Bridgland and Saville, 2000).

Origin of the Buchan Ridge Gravel Member

The origins of this member remain controversial in terms of the provenance of the gravel constituents, depositional environment and age.

The main constituents of the gravel are flint, quartzite, vein-quartz and kaolinitic silt and sand. It has long been held that the flints are derived from a former Chalk cover (Wilson, 1886). The discovery of little-worn flints at the base of the gravels at Skelmuir Hill (Bridgland et al., 1997, 2000) and Moss of Cruden (Hall, 1993a) confirms that Chalk once covered this area. Furthermore, the recent discovery of Lower Cretaceous strata preserved beneath the gravels at Moss of Cruden suggests that a succession of Cretaceous sedimentary rocks formerly existed in this area. The quartzite clasts bear mineralogical similarities to

a

b

Plate 5   Flint clasts recovered from the base of the Buchan Ridge Gravel Member in a trial pit on the Moss of Cruden [NK 0253 4024]. Little-worn flints such as these were either derived directly from a former cover of Chalk in the area or from a *remanié* flint deposit. This evidence suggests that the present terrain lies close to the level of the sub-Cenomanian surface. (a P104122, b P104121).

metaquartzitic rocks in north-east Scotland (Kesel and Gemmell, 1981), but may have been recycled from ORS conglomerates (Hall, 1993a). Careful observations of 'ghost' clast structures at Skelmuir Hill (Bridgland et al., 1997, 2000) suggest that the amount of original less-durable material in the gravels may have been significantly under-estimated by other workers, including Hall (1982). This indicates erosion of the local granite and metamorphic bedrock and incorporation into the basal facies of the gravels.

The kaolinitic matrix of the Buchan Ridge Gravel was partially derived from highly weathered saprolites, because discrete beds of kaolinitic silt occur within the gravels and the clay mineralogy shows little variation with depth in the gravels (Hall, 1982). However, the gravels were deeply weathered after deposition, and kaolinisation of nonquartzitic clasts has occurred throughout the deposit at Moss of Cruden and into the underlying bedrock. Secondary alteration is less thorough at Skelmuir Hill, but the former weathering profile may have been considerably truncated there. The clay-bound nature of the beds of gravel is largely a result of secondary infill of voids by clays derived from the weathering of feldspathic clasts and sand, followed by collapse and deformation (McMillan and Merritt, 1980; Merritt and McMillan, 1982; Bridgland et al., 1997). The member is apparently undisturbed at depth, but the individual bodies have been affected to different degrees by glacial disturbance, deformation and thrusting during the Pleistocene.

Marine, fluvial and glacial origins have been suggested for the Buchan Ridge Gravel Member. Evidence adduced for deposition as beach gravel includes:

- abundance of well-rounded and chatter-marked flint clasts (Flett and Read, 1921; Koppi and FitzPatrick, 1980; Bridgland et al., 1997)
- the apparent original open-work character of parts of the deposit (McMillan and Merritt, 1980; Merritt and McMillan, 1982)
- the presence of large, flaked flint blocks at the base of the Skelmuir deposit (Bridgland et al., 1997, 2000)

- the coarse calibre of parts of the gravel body suggesting that it formed in a high-energy beach environment, rather than a river environment, because the surrounding land surface was probably of low relief (Chapter 3)

Difficulties for the last interpretation include the presence of well-rounded and chatter-marked clasts. These are similar to those occurring within the Windy Hills Gravel, which most agree is of fluviatile origin. The beds of kaolinitic silt and clay are more difficult to account for within beach gravel than a fluvial sequence. Evidence in favour of fluviatile deposition includes:

- the occurrence of each of the gravel bodies within channels cut in bedrock
- the association of kaolinitic gravels with a deeply weathered land surface is characteristic of other Palaeogene, Neogene and early Pleistocene river gravels in north-west Europe, including south-west England (Hamblin, 1973) and the Low Countries (van den Broek and van der Waals, 1967; Friis, 1976)

Deposition in a glacial or glaciofluvial environment is supported by the presence of beds of matrix-supported gravel and of broken, previously rounded quartz grains. However, trial pits have shown that the gravels contain a range of relatively undisturbed sedimentary structures and glacial disturbance is likely to have been restricted in its effects. Transport by meltwater is more difficult to refute if the widespread breakage of quartz grains at depth within the Buchan Ridge Gravel Member is not a result of the drilling process.

Despite much investigation, the age of the member remains uncertain. The gravels are clearly post-Cretaceous in age, as they contain flint, and they predate the Late Pleistocene, as they are overlain by till (Whittington et al., 1998). The degree of postdepositional alteration is striking and certainly far exceeds that shown by the felsite-rich Leys Gravel of Middle Pleistocene age found at Kirkhill (Hall and Jarvis, 1993a). The constituents of the Buchan Ridge Gravel Member were derived in part from a highly weathered land surface and the kaolinitic style of weathering has been related to warm climates that prevailed before the Pliocene (Hall, 1985; Chapter 3). However, absolute dating of the weathering of bedrock and of the gravels is lacking and kaolinitic fluvial deposits are known from Europe to span the whole of the Palaeogene, Neogene and the earliest Pleistocene. Significant landscape modification has occurred since deposition, with probable uplift and topographic inversion (Figure 20), but the timing of these events is unknown.

If the Buchan Ridge Gravel Member is fluvial, rather than glaciofluvial in origin, then the upper parts of the former river catchment have been destroyed by erosion. Furthermore, the energy required for boulder transport is likely to have come from steep river gradients resulting from regional tectonic uplift and tilting towards the North Sea (Chapter 3). Uplift such as this brought about deposition of kaolinitic sands in the inner Moray Firth both in the Eocene and Pliocene (Andrews et al., 1990). If the gravels are marine in origin then deposition occurred during a period when sea level fluctuated over 75 m and all contemporaneous fine-grained sediments have been destroyed by erosion, which is perhaps unreasonable. Most would agree that the wide elevation range and localised distribution of deposits of the Buchan Ridge Gravel Member suggest that the individual bodies were deposited over a considerable time period. There is a distinct possibility that beach deposits formed during the post-Cretaceous marine regression were later fluvially reworked.

Channel-fill deposits at the mouth of the River Spey

A borehole survey of sand and gravel resources around Garmouth, at the mouth of the Spey (Aitken et al., 1979) revealed the existence of a deep rock-cut channel extending below present sea level. The channel is filled with medium- and fine-grained greenish grey quartzose sands. Thin seams of pale greenish grey, sandy silty clay are also present, containing illite with kaolinite; 20 m or more of glaciofluvial gravel cover the deposits. The channel is interpreted as a possible preglacial channel of the River Spey. An enigmatic kaolinitic 'channel-fill' deposit was also found at depth in a borehole sited on the 'Mosstodloch' glaciofluvial terrace on the Muir of Stynie [NJ 3284 6152] (Aitken et al., 1979).

DEEP WEATHERING AND SOIL DEVELOPMENT

An important feature of the landscape of the lowlands of north-east Scotland is the extensive development of deeply weathered rock (Wright, 1997). Although deep weathering is known from other parts of Scotland, it tends to be relatively restricted in its depth and distribution. In Buchan, in particular, it is common to find extensive areas with few, if any fresh rock outcrops. Boreholes show the weathering commonly extends to depths of 10 to 20 m and may exceed 50 m. The characteristics of these saprolites (in situ weathered rock) and weathering profiles are now quite well known. Investigation of the deep weathering is important for understanding the long-term evolution of the landscape, the variable impact of glacial erosion and the origins of soils (Chapter 3). The deep weathering is also of importance in engineering geology, for the weathering reduces rock mass strength in a complex manner, varying in depth and degree of alteration over short distances (Chapter 2). Finally, the disaggregation that accompanies weathering allows certain types of weathered granular rock to become a source of aggregate (Appendix 2).

Distribution

Deep weathering is common throughout the lowlands of north-east Scotland, but varies in its frequency and depth (FitzPatrick, 1963; Hall, 1986). Information on weathering sites comes from exposures, both long-term and temporary, and from boreholes. Commonly, weathered rock has been wrongly identified as drift in borehole logs. A point distribution map of weathering sites (Figure 23) is useful in showing the widespread distribution of the 500 or so sites known in north-east Scotland. However, data on the degree of rock weathering are unevenly distributed, and it is likely that many areas that lack fresh outcrops and with a thin drift cover also conceal extensive zones of weathered rock. One such area lies around the headwaters of the River Ythan, where smooth, valley-floor slopes developed across pelite are mantled by periglacial slope deposits (Galloway, 1958), but may also hide pockets of deep weathering. In addition, a point distribution map may give a false impression of the importance of weathering in an area, as small pockets of weathered rock commonly occur in areas of dominantly fresh rock. Nonetheless, it is clear that rock type is a basic control over the distribution and degree of weathering.

Deep weathering is most commonly developed on biotite granites, basic igneous rocks and in feldspathic psammite bands within quartzites (Hall, 1986). It affects a wide range of metamorphic rocks, but is comparatively uncommon on pelites, although data on weathering in areas underlain by slate is sparse. Deep chemical weathering also has not been noted widely on the Devonian sandstones and conglomerates (Hall, 1986), although geophysical surveys suggest weak alteration on the Turriff outlier down to 10 m in places (Ashcroft and Wilson, 1976). Decomposed conglomerates occur in the Elgin district (Peacock et al., 1968) and between Stonehaven and Inverbervie (Auton et al., 1990).

Depths of weathering are recorded in deep quarries and in boreholes. Weathering to depths of more than 5 m is common in eastern Buchan: for example in the quartzites at the western edge of Mormond Hill [NK 950 568], in quartz-mica psammite at Northseat [NJ 930 408], near Auchnagatt, and in granites at Cairngall, Longside [NK 053 471] and at Hill of Longhaven [NK 084 423]. Weathering to a depth of over 60 m is recorded in a borehole south-west of the Hill of Dudwick [NJ 979 378] and in this part of the Buchan, depths of weathering exceed 20 m in numerous boreholes (Hall, 1985). A similar depth is recorded from a fracture zone in the Peterhead Granite (Edmond and Graham, 1977). Leaching of Permo–Triassic sandstone reaches depths of more than 50 m near Lossiemouth (Peacock et al., 1968). Seismic profiles indicate alteration to depths of 50 m

Figure 23 Deep weathering sites in north-east Scotland (after Hall, 1986).

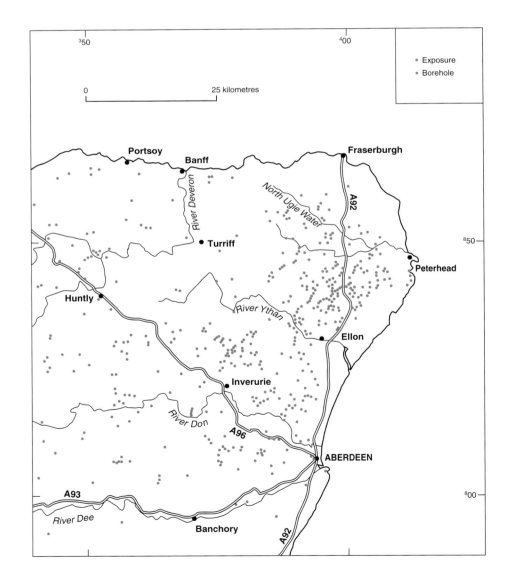

on the sheared margins of the Arnage and Insch basic masses (Leslie, 1984). Weathering depths of 10 to 20 m or more are known from boreholes from the Huntly, Knock and Maud basic masses, the Skene and Peterhead granites and schists near Aberdeen, but these depths tend to be exceptional and normally correspond to fracture zones (Hall, 1983, 1986). Similar depths of weathering are expected to occur along the northern margin of the Insch depression and beneath the floor of the Drumblade depression (Ashworth, 1975).

Many sections and boreholes reveal rapid lateral variations in weathering depths. In sections at Northseat (Hall, 1986; Map 6), Cairngall (Hall, 1983), Kirkhill (Hall, 1983; Map 7) and Ythanbank (Hall and Sugden, 1987; Map 6), zones of weathering lie adjacent to fresh outcrops, yet extend more than 15 m below the surface. Very closely spaced boreholes at Cuttlehill [NJ 499 473], near Ruthven, and Lumphart [NJ 769 269], near Old-meldrum, demonstrate variations in depth of more than 20 m over distances of less than 100 m. (Hall, 1983, fig. 5.3.ii). Boreholes in the Knock basin in the area east of Ruthven show local development of 20 to 30 m of weathered basic igneous rock, but again weathering depths are highly variable (Hall, 1983, fig. 12.6.iv). These

variations are associated with fractured or hydrothermally altered rocks and with sequences of rock of widely different resistance. Rocks of more homogeneous composition tend to show more gradual changes in weathering depths. In general, however, lateral variations in weathering depths of over 5 m can be expected over short distances in most areas.

Given the irregularity of many rockhead profiles, stripping of saprolite should have resulted in slopes with upstanding rock knobs and ribs. In fact, slopes generally show a marked homogeneity in weathering patterns. Lowland areas north of the Don valley show gentle convexo-concave slopes with few rock outcrops. In section, smooth slopes are developed across thin drift covers resting on rocks at various stages of alteration. Bosses of fresh rock have protected adjacent pockets of weathering from erosion, yet these 'risers' commonly fail to have any topographic expression. This discordance seems to be largely a result of glacial and periglacial activity. Although glacial erosion has been generally moderate to low in its intensity, it has been sufficient to remove small, upstanding rock forms. Frost shattering has reduced rock knobs and solifluction has contributed to the smoothing of slopes (Hall, 1986).

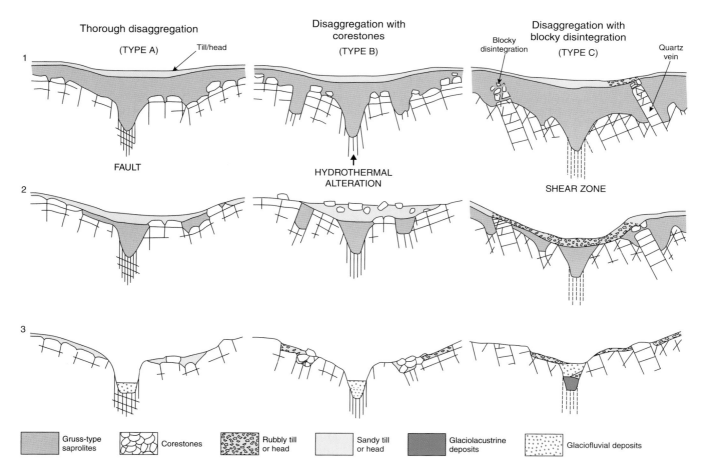

Figure 24 Slope development and the progressive removal of different types of weathering profile (after Hall, 1983).

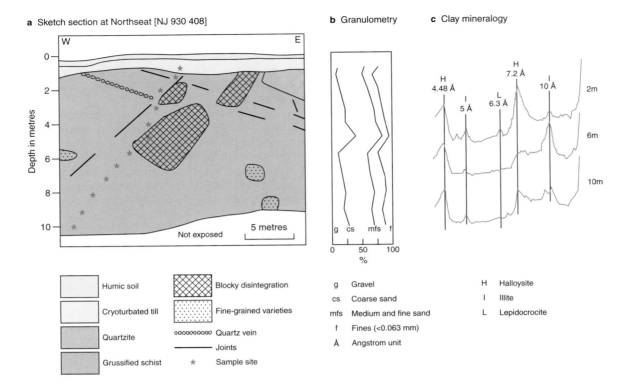

a Sketch section at Northseat [NJ 930 408]   **b** Granulometry   **c** Clay mineralogy

Humic soil

Cryoturbated till

Quartzite

Grussified schist

Blocky disintegration

Fine-grained varieties

Quartz vein

Joints

Sample site

g   Gravel

cs   Coarse sand

mfs Medium and fine sand

f   Fines (<0.063 mm)

Å   Angstrom unit

H   Halloysite

I   Illite

L   Lepidocrocite

Figure 25   Vertical changes in a gruss weathering profile in quartz-biotite psammite, Northseat [NJ 930 408] (after Hall, 1986).

Characteristics of the weathering profiles

Three basic types of weathering profile have been identified in north-east Scotland (Hall, 1983; Figure 24). The first, type A, is characterised by thorough disaggregation, with a gradual downward increase in coherence to hard rock. These profiles are typically associated with coarse-grained granites and other homogeneous or closely fractured rock types (Plate 6). A good example is seen at Hill of Dens quarry (Map 7) on the Hill of Longhaven (Hall, 1993c). The second type of profile (B) shows an upper zone of thorough disaggregation and a lower zone of corestone development, where kernels of fresh rock are isolated by penetrative weathering along joints (Plate 7). Profiles with corestones are typical of most widely jointed acid and basic igneous rocks. Good examples were formerly exposed in quarries to the south-east of New Pitsligo. Corestones occur widely on the Knock, Huntly and Insch basic masses and to the east of the Strichen granite. The third type of profile (C) is characteristic of weathered metamorphic rocks. At the base of the profile the metasedimentary rocks break up into angular blocks separated by thin seams of decomposed rock. The blocks become progressively reduced in size up-profile and are surrounded by weathered rock. The heterogeneous composition of many metamorphic rocks in the district means that zones of blocky disintegration persist even in the upper zones of weathering profiles.

Borehole records indicate that the transition from fresh to hard rock at the base of weathering profiles is variable in character. Around one third of boreholes examined show a well-developed basal surface of weathering. This abrupt change from weathered to fresh rock is most common where weathering is shallow. However, around half of the boreholes show that fresh rock is overlain by alternating bands of decomposed, weakened, shattered and fresh rock. Such transition zones may persist over depths of 15 m or more. In about one in five boreholes the transition from weathered to fresh, more competent rock is gradational over several metres. Overlying glacial and periglacial deposits generally truncate the weathering profiles, but exposures commonly show pockets of weathering protected by adjacent bands of hard rock.

In many deep exposures, such as Northseat [NJ 930 408] (Hall, 1986; Figure 25; Map 6), there are few signs of soil horizon development within the weathering profiles. Colour, grain size and clay mineralogy all change gradually down the weathering profile. Downward translocation of clay is indicated by the presence of clay coatings along joint boundaries and around blocks. However, in more altered saprolites, near-surface zones of kaolinisation and reddening are replaced at depth by less weathered materials. The lateral variations in the degree of weathering of the Moreseat Sandstone on the Moss of Cruden have been interpreted as reflecting the differential truncation of a deep weathering profile (Hall and Jarvis, 1994). Reddening (rubefaction) is associated with saprolites developed on the Peterhead Granite where it probably results from

Plate 6 Weathered megacrystic Crathes Granite in Littletown quarry [NJ 6988 0969], west of Dunecht (P104104). Scale: hammer is 40 cms long.

the presence of iron minerals in the granite. Reddened quartzitic saprolites occur at several sites including Sunnyside [NJ 983 372], Drinnies Wood [NJ 973 497] and Howe of Dens [NJ 975 805].

At the base of weathering profiles, staining by manganese or manganese-iron oxides is widespread, especially on rocks rich in mafic minerals. The presence of manganese dioxide, in particular, has been interpreted as an indicator of present or former hydromorphic conditions at, or beneath, the water table (Koppi, 1977). This observation is significant as many weathering profiles that contain these minerals lie well above contemporary water tables, for example at Northseat. The relationship between weathering

profiles and water tables is poorly understood, although normally the basal surface of weathering provides a permeability contrast, which corresponds with the level of the water table. On parts of the Buchan Ridge, where weathering depths exceed 15 m, water is not met for 10 m or more below the surface. Several deep quarries in partly weathered rock, such as Cairngall and Hill of Dens, have floors above the local water table. Here it appears that the free-drainage offered by the disaggregated rock allows a lowered water table. However, weathering is known to penetrate deep below the water table in some boreholes indicating that weathering processes must occur locally in the deeper zones of groundwater circulation.

Plate 7 Corestones developed in weathered gabbroic rock near Maud [NJ 8897 5053]. (P104105). Scale: spade is 0.9 m long.

Weathering patterns

At scales of 1 to 10 km², rockhead profiles show three main features:

- alternating low risers and depressions
- linear zones of deep alteration
- scarp-foot weathering zones

Each of these distinctive weathering patterns may have topographic expression.

In the lower Ythan valley, topography appears to reflect the progressive stripping of saprolites from a gently undulating rockhead surface (Hall, 1986, fig. 4). The intensity of glacial erosion increases towards the axis of the valley, where ice-moulded bedrock surfaces are locally developed (Hall and Sugden, 1987). The formation of the Barra basin, one of a series of small basins along the lower Don valley (Hall, 1983, fig. 12.5.ii), has involved the removal of weathered rock by a combination of fluvial and glacial processes. The lowest part of the basin floor corresponds with a zone of intense shearing and is underlain by deeply weathered rock. In contrast, the adjacent mass of Hill of Barra (Map 8) is composed of dislocated, but relatively unsheared, basic cumulates and is comparatively little weathered (Ashcroft and Munro, 1978).

Zones of deep linear alteration exist at a variety of scales. Trenches up to 300 m wide in the weathering front at Crichie [NJ 970 440] (Map 6) and Minnes [NJ 944 237] (Map 9) appear to relate to localised fracturing. Both have been partly excavated by meltwater. Boreholes for a bypass site investigation south-west and west of Keith, to the west of the district, showed that weathering beneath the floor of Strath Isla penetrated to depths exceeding 30 m in schists. Adjacent metalimestones showed cavities, presumably resulting from solution, down to depths of 20 m (Ove Arup and Partners, personal communication, 1993). These linear zones of alteration generally reflect fracturing or shearing (Leslie, 1984).

Scarp-foot weathering zones occur widely. The acceleration of weathering at the foot of water-gathering slopes has led to the development of deep weathering zones transverse to the scarp-foot, itself generally located on a geological discontinuity. These zones may be found at various stages of development dependent on the present drainage (Figure 26).

The key controls on the distribution of weathering at the small scale are rock type, structure and topography. At the regional scale (greater than 10 km²), the varied intensity of glacial erosion is an important factor. Identification of regional weathering zones in north-east Scotland (Hall, 1986; Figure 27) suggests that weathering is most sparsely distributed in areas of more vigorous ice flow. The most common occurrences of fresh rock typically occur in areas that were eroded by the coastal ice streams and by ice flowing down the Dee valley. This contrast is particularly apparent on a transect northwards from Aberdeen to the Buchan Ridge, where there is a close inverse correlation between the depth and frequency of weathering and the occurrence of ice-moulded bed forms (Hall and Sugden, 1987; Figure 21).

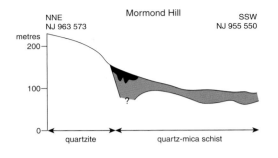

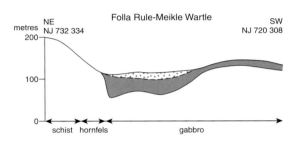

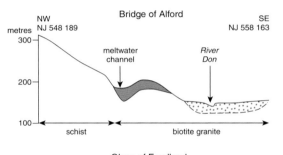

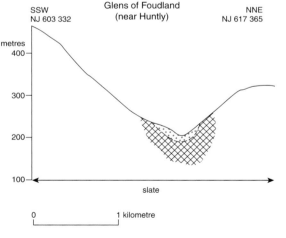

Figure 26 Scarp-foot weathering zones and their progressive exploitation (after Hall, 1986).

Figure 27 Regional weathering patterns (after Hall, 1986).

ZONE 1   Deep and continuous saprolites with few fresh outcrops. Saprolite thicknesses generally are at least 3 m and commonly exceed 10 m. Fresh outcrops are confined to highly resistant rocks

ZONE 2   Thinner and discontinuous saprolites with fresh outcrops. Pockets of deep alteration remain but saprolites are generally less than 3 m thick. Fresh outcrops are common, especially on steep slopes and along valley floors

ZONE 3   Fresh rocks with numerous pockets of weathering. Fresh rocks underlie most of the area. Weathering generally is found only in ice-lee locations on interfluves and along valleys, and is common only in tributary valleys and basins

ZONE 4   Fresh rocks with rare pockets of weathering. Fresh rock predominates and local pockets of weathering usually relate to fracture zones. A distinction is made between areas of gentle slopes and/or rocks of low to moderate resistance in which weathering was probably extensive prior to glaciation (Zone 4a) and areas of steep slopes and/or resistant rocks in which weathering was probably thinly or sporadically developed prior to glaciation (Zone 4b)

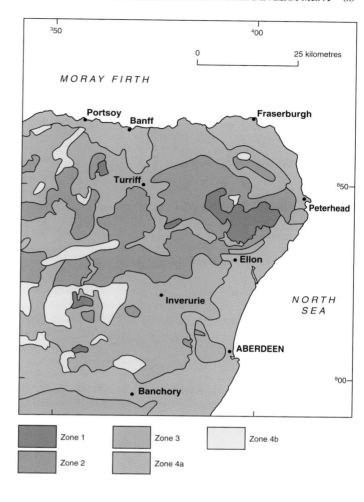

These contrasts probably reflect differences in the basal thermal regimes of the ice streams. The coastal areas and the Dee valley were traversed by warm-based ice, capable of significant erosion. In contrast, inland areas were covered by ice that was usually cold-based and frozen to its bed, allowing only limited erosion.

Saprolite characteristics

Saprolites in the lowlands of north-east Scotland vary widely in grain size, geochemistry and clay mineralogy. Grain size within weathering profiles generally does not change systematically with depth and near-surface samples commonly have a similar texture to those at depth (Hall, 1983). Analysis of grain size for 78 exposures of saprolite shows that grain size characteristics vary according to rock type and degree of chemical alteration. It is also significant that grain size characteristics vary widely within a section. At the initial stages of weathering, granitic, basic igneous and metasedimentary weathering profiles form distinct granulometric populations (Figure 28) and the resultant saprolites possess different mechanical properties. On granites, initial disintegration forms granular grit, with median grain sizes above 1000 μ and very limited development of fines. This style of disaggregation is well displayed on the coarse Peterhead Granite at Hill of Dens quarry on the Hill of Longhaven (Figure 28; Map 7). At that site, total fines (silt and clay) content is usually less than 5 per cent. The granulometry of such immature granite saprolites closely reflects the dimensions of the minerals in the fresh granite, with fine- to medium-grained granites breaking down into saprolites with median grain size of 400 to 500 μ and coarse-grained granites providing median grain sizes of 1200 to 1500 μ (Hall, 1983). On medium- to coarse-grained gabbro and norite, initial breakdown also produces

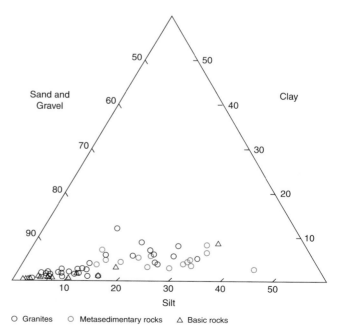

Figure 28   Granulometry of weathered rocks.

granular sands (Figure 28). Fines contents are low (generally less than 10%), and median clay content (0.7%) is below that of the weathered granites (2.8%). Quartz psammites produce varied grain size characteristics that reflect the diverse lithology of the weathered, less resistant rock bands within the host rocks and the advanced degree of alteration at these sites. Other metasedimentary rocks produce saprolites with elevated fines (27.4%) and clay (4.5%) contents (Figure 28). In comparison with overlying soils developed on the saprolites, the soils are greatly enriched in fines and deficient in coarse sand.

The grain size characteristics of saprolites developed in the initial stages of breakdown on different rock types indicates a particular style of disaggregation (Figure 24). The initial breakdown of granitic and gabbroic rocks produces a polymineralic granular gruss (or sand). This process has been termed 'granular disintegration' (Basham, 1974). Metasedimentary rocks first break into angular blocks, a process termed 'blocky disintegration' (Basham, 1974). Alteration then penetrates along the fractures and produces thin seams of decomposition containing significant amounts of fines. These contrasts may reflect the ways in which residual stresses are released from the rock during weathering and unloading. In finer grained metasedimentary rocks, the opening of fractures involved in blocky disintegration largely relieves residual stress. In igneous rocks, stress retained after jointing is relieved by the propagation of micro-cracks, possibly after the modest expansion of certain minerals, notably biotite, in the initial stages of weathering.

With increasing alteration, the granulometric differences between rock types become less clear, although metasedimentary saprolites continue to have a relatively high fines content. The size of the clay fraction ($< 2\,\mu$) is a useful indicator of the degree of chemical alteration. Advanced alteration gives saprolites on all rock types with a clay content of more than 6 per cent, reaching 20 to 25 per cent on metasedimentary rocks.

Weathering types and age

Two distinct types of saprolite or weathered rock mantle are identified in north-east Scotland (Hall, 1985).

A more evolved weathering type, *clayey gruss*, shows an elevated clay content (>6%), with clay mineralogy dominated by kaolinite and illite, and with small amounts of hematite. Within the district, this degree of weathering is known from only a small number of sites, mainly on the high ground of central Buchan (Hall, 1985; Figure 22). It is represented by kaolinisation of clasts within, and bedrock beneath, the deposits of the Buchan Gravels Formation. It also occurs within the Mormond Hill Quartzite. Kaolinised granite containing a hematite/layer-silicate clay mineral, macaulayite, is found at the foot of Bennachie [NJ 693 245] (Wilson et al., 1981, 1984; Koppi, 1977). Kaolinistion affects the pelites beneath the quartzite gravels at Windy Hills (Figure 22) and silicate clasts within the gravels themselves (Koppi and Fitz-Patrick, 1980; Hall et al., 1989). Koppi (1977) also describes a highly weathered feldspar-biotite psammite from Clashindarroch Forest [NJ 435 305], just west of the boundary of the district. Comparison with the mineralogy of North Sea sediments suggests that highly kaolinitic weathering mantles formed prior to the Pliocene in north-east Scotland under warm humid conditions (Hall, 1985).

The vast majority of the saprolites in the district fall within the *gruss* weathering type. The saprolites are dominantly sandy, with limited development of fines, and the clay mineralogy is closely controlled by rock type. Granitic saprolites contain kaolinite-mica clay mineral assemblages (Hall et al., 1989). Basic igneous saprolites contain a wide range of clay minerals. For example, weathering of the 'Insch Gabbro' has left feldspar and hornblende largely unaffected. Pyroxene alters initially to iron oxides and later to vermiculite, while biotite weathers to hydrobiotite and vermiculite and, locally, to kaolinite and gibbsite (Basham, 1974). Acid metamorphic rocks give kaolinite and mica clays, but an increasing content of primary ferromagnesian minerals leads to a reduction in kaolinite and an increase in smectite content (Hall et al., 1989). Altered metalimestones are dominated by smectite, partly inherited from the parent rock (Wilson et al., 1968). Grusses are thought to have formed in north-east Scotland under the humid temperate environments of the Pliocene and warmer periods of the Pleistocene (Hall, 1985).

FIVE

Quaternary Period

GLOBAL RECORD OF CLIMATE CHANGE

In the last two decades significant advances have been made in the understanding of Quaternary environmental change in Scotland. This is largely because the timing and pace of climatic change in Scotland can be viewed now as part of the broader pattern affecting the whole North Atlantic region (Boulton et al., 1991). Evidence of global and regional events is found in deep-sea sediment cores, in cores of ice through the Greenland ice sheet and from extended sequences of interbedded loess and organic deposits from continental Europe. The fluctuations in climate appear to be driven by minor orbitally controlled variations in solar radiation that are amplified through complex interactions between the atmosphere, oceans, ice sheets and global tectonics.

The onset of glaciation in the Northern Hemisphere probably began in the late Miocene with a significant build-up of ice over southern Greenland, although progressive intensification did not begin until 3.5 to 3 Ma when that ice sheet expanded into northern Greenland (Maslin et al., 1998). This suggests that the Scottish climate had begun to deteriorate long before the beginning of the Quaternary Period. The Quaternary is presently defined as beginning at about 1.77 Ma (Shackleton et al., 1990), but studies of ice-rafted debris in ocean floor sediments in the North Atlantic and elsewhere have shown that significant climatic deterioration occurred some 600 ka earlier when the mid-latitude continents of the Northern Hemisphere first became glaciated (Shackleton et al., 1984; Ruddiman and Raymo, 1988). Many now agree that the beginning of the period should be placed either at 2.4 Ma, as accepted here (Figure 6), or at 2.6 Ma (Funnell, 1995) in order to take into account the above evidence and recent advances in micropalaeontology, magnetostratigraphy and climatostratigraphy (Mauz, 1998). Although no direct evidence of glaciation at this time has been found on the Scottish mainland or neighbouring offshore shelves, the relatively minor decline in temperature needed for glaciers to develop, suggests that they did so, at least in the western Highlands.

Deep ocean record

The frequency, rapidity and intensity of climatic change are a key feature of the Quaternary. Studies of the isotope geochemistry and micropalaeontology of deep ocean-floor sediments have revealed that climatic conditions have fluctuated continuously throughout the period and that climate systems have switched rapidly between *interglacial* and *glacial* modes (Figure 29). At least 50 significant 'cold–warm–cold' oscillations have been recognised. Many theories have been put forward to explain the initiation of Northern Hemisphere glaciation (Maslin et al., 1998), most involving changes in atmospheric composition (such as caused by increased volcanism) or in total solar radiation. The initial cooling that began during the Neogene has been attributed to the slow changes in the global configuration of the continents as a consequence of sea-floor spreading. These include the emergence of the Panama Isthmus and the deepening of the Bering Straits, both of which had pronounced affects on the ocean circulation patterns (Raymo, 1994). The cooling has also been attributed to the uplift of high mountain ranges such as the Tibetan Himalayas and the Sierra Nevadan and Coloradan mountains of North America, causing perturbations of the upper atmosphere and subsequent climatic changes (Ruddiman and Kutzbach, 1991). Uplift of the Himalayas also may have resulted in a massive increase in chemical weathering during the late Cainozoic, leading to increased sedimentation of calcium carbonate and atmospheric depletion of carbon dioxide. Global cooling, the inverse of the 'greenhouse effect' would thus occur (Raymo and Ruddiman, 1992; Raymo, 1994). However, none of these mechanisms can wholly explain the rapid intensification of glaciation observed in the deep ocean record at about 2.7 to 2.5 Ma (Maslin et al., 1998).

The driving force of climatic change during the Quaternary appears to be the long-term cyclical variation in solar energy. Cyclical changes in solar insolation appear to be associated with the Earth's orbital periodicities, rather than on long-term changes in energy output of the Sun, although the latter is difficult to disprove (Shackleton and Opdyke, 1973). There are three significant orbital ('*Milankovich*') cycles, caused by:

i   precession ('wobble') of the planet's axis
ii   the tilt of the Earth's axis
iii the eccentricity of the Earth's orbit (Imbrie et al., 1984)

The periodicities of these cycles are roughly 23 ka, 41 ka and 100 ka, respectively. The climatic fluctuations that occurred during each major glacial–interglacial cycle have been attributed to the first two cycles (Imbrie and Imbrie, 1979), but now it is generally agreed that they alone could not have produced the magnitude or rapidity of the documented changes. Maslin et al. (1995) suggested that the increase in the amplitude of orbital obliquity cycles deduced from the deep ocean record from 3.2 Ma onwards may have increased the seasonality of the Northern Hemisphere, thus initiating the long term cooling trend. The subsequent sharp rise in the amplitude of precession between 2.8 and 2.55 Ma may have forced the rapid intensification of glaciations that began then.

Figure 29
Oxygen isotope
curve (Pacific Site
849) representing
ice volume change
over the past 1.2
million years and
a curve
predicting
summer solar
insolation at 65°N
(after Raymo,
1997).

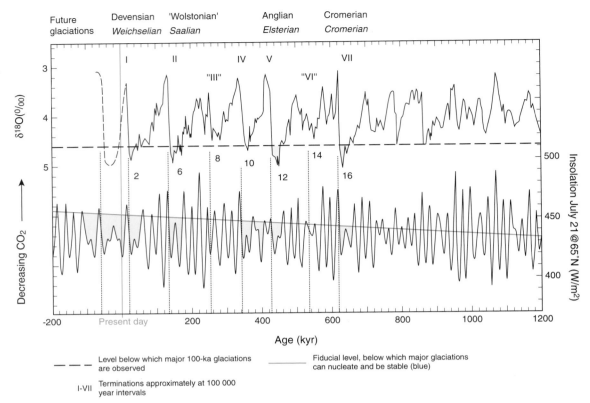

The primary Milankovich fluctuations of solar insolation must have been amplified substantially by additional factors involving physical, biological and chemical interactions and 'feedback loops' between the atmosphere, oceans and ice sheets. For example, complex changes in the surface and deep water circulation patterns of the oceans, and the concentrations of carbon dioxide and other 'greenhouse' gases in the atmosphere all play a crucial role (Broecker and Denton, 1990). Of particular importance to the British Isles are changes in the position of the Gulf Stream. This northward-flowing current of warm surface water is compensated by the return southwards of cold, dense water at depth. Sudden changes in this circulation pattern, the so-called 'North Atlantic conveyor', may have had a major impact on climate (Skinner and Porter, 1995; Figure 30). For example, large volumes of fresh water released during rapid warming events may have temporarily 'switched off' the conveyor leading to cooling in north-west Europe (Lagerklint and Wright, 1999).

The 'SPECMAP' deep ocean oxygen isotope record (Figure 6) indicates that during the early Quaternary, up until about 760 ka, each dominant warm–cold cycle lasted about 40 ka (Ruddiman et al., 1989). Ice caps in Greenland, Alaska, Iceland and Scandinavia expanded to the coast and glaciers probably developed in the western Scottish Highlands at high elevations (Clapperton, 1997). Iceberg dropstones found in sediment cores taken offshore from north-west Europe provide indirect evidence of these early glaciations (Holmes, 1997; Sejrup et al., 2000). Following a major change at about 760 ka, the build-up of much larger ice

sheets in the Northern Hemisphere occurred, and to date there have been seven major glacial–interglacial cycles (Figure 29). Each 'glacial' episode lasted between 80 and 120 ka and was followed abruptly by an 'interglacial' lasting 10 to 15 ka. The rapid deglaciations are defined as 'Terminations I–VII' (Broecker, 1984; Figure 29). The glacial periods included long, cold intervals, termed *stadials*, and less cold, and even warm, intervals lasting for a few thousand years termed *interstadials*. Most terrestrial evidence preserved in Scotland relates to the last glacial–interglacial cycle (Holocene, Devensian and Ipswichian stages), but older deposits are preserved locally (Figure 7).

The SPECMAP timescale of Imbrie et al. (1984) has been the cornerstone of Quaternary palaeoclimate studies. It assumes that the climatic cycles observed in deep ocean sediment records were driven by Milankovitch orbital parameters. Although this assumption has been challenged (Schrag, 2000), the validity of the timescale has been confirmed independently by Raymo (1997). She hypothesises that the general intensification of the major glaciations during the Quaternary results from long-term reduction of atmospheric carbon dioxide levels by tectonic processes (as shown by the gently inclined straight line on Figure 29). This gradual weakening of the 'greenhouse effect' has caused cold periods to become colder, allowing ice sheets to expand in regions that were previously too warm. Raymo's (1997) model shows that major glaciations (coloured areas on the insolation curve of Figure 29) only occurred when both 'obliquity' and 'eccentricity' Milankovitch cycles reinforced themselves, causing long intervals of low summer insolation in northern latitudes. Furthermore, her model

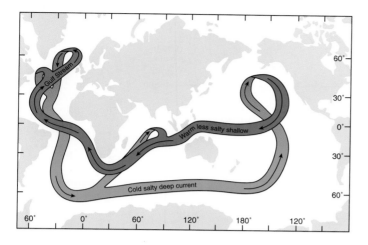

Figure 30   Major thermohaline circulation cells that make up the global conveyor system (after Skinner and Porter, 1995). The cells are driven by exchange of heat and moisture between the atmosphere and ocean.

(Figure 29) predicts that the present Holocene interglacial is nearly over and that an intense 'double' glaciation is likely to follow, terminating about 64 ka from now.

Greenland ice core record

New, high resolution evidence derived from cores taken through the Greenland ice sheet near its summit (Figure 31), demonstrate that the Earth's climate has been much more variable during the past 250 ka than previously thought (Dansgaard et al., 1993). Dramatic climatic changes have occurred, both on the millennial scale and over just a few years. Some 24 interstadial intervals have been identified in the last (Devensian) glacial stage alone, compared with the five or six that were recognised previously from the pollen record and formalised in the north-west European and British stratigraphy (Mitchell et al., 1973; Figure 7). Each of the 24 interstadials within the Devensian started with abrupt warming and typically cooled over a period of 1 to 3 ka, ending in a cold event (Broecker, 1994). These periods, known as *Dansgaard–Oeschger* (*D-O*) *cycles*, are sometimes grouped together within longer, *Bond cycles*, during which temperatures also declined gradually (Bond et al., 1993; Figure 32). It is important to note that even the milder interstadials in Greenland had a climate in which average temperatures were five to six degrees colder than on the summit of the Greenland ice sheet today (Dansgaard et al., 1993). Intervals when glaciers were probably present in Scotland during the past 25 ka are indicated in Figure 31).

Many Bond cycles appear to culminate in a *Heinrich event* (Figure 32). Originally it was concluded that temperatures rose substantially within a decade or so following these events (Bond and Lotti, 1995), but there is growing evidence that they are actually the result of extremely rapid warming (Lagerklint and Wright, 1999). Heinrich events, as originally defined, were massive discharges ('armadas') of icebergs from the Laurentide ice sheet, recognised in deep ocean cores in the North Atlantic by the presence of ice-rafted debris (IRD) at intervals in the sediment sequence (Heinrich, 1988; Figure 31). These horizons were thought to result from the sudden mass wastage, via iceberg calving, of the Laurentide ice sheet every 11 ka or so, possibly as a result of it having grown too large to sustain itself and resulting in catastrophic collapse. Such 'binge-purge' cycles are thought by some to have played a pivotal role in driving world climate variability (Broecker, 1994; Alley and MacAyeal, 1994). More recently, Heinrich events have been identified in North Pacific cores (Hicock et al., 1999) and ice sheets in Iceland, Britain and Scandinavia are now known to have contributed IRD during Heinrich events (Frontal et al., 1995; Sejrup et al., 2000). It therefore seems unreasonable that all of the ice sheets 'binged and purged' in harmony unless there was some independent forcing mechanism (Bond and Lotti, 1995; Kotilainen and Shackleton, 1995). Furthermore, it appears that there was a synchronous deposition of ice-rafted layers in the Nordic seas and North Atlantic (Dowdeswell et al., 1999) and glacial events in north-west Europe may actually have driven events across the Atlantic, not the other way around (Grousset et al., 2000). One suggestion is that Heinrich events are triggered every 6100 years as a result of direct changes in solar radiation (Lehman, 1996), their effects being greatest during the longer cold periods when glaciers, ice sheets and tidewater glacier margins were most extensive.

Heinrich events probably led to 'armadas' of icebergs in the seas around the British Isles. The release of large volumes of glacial meltwater and icebergs from North America possibly 'switched off' the 'North Atlantic Conveyor' for a while leading to cooling of the north-east Atlantic and north-west Europe (Lagerklint and Wright, 2000). Minor re-advances of ice streams flowing through the major firths of the west and east coasts of Scotland are possibly due to such events (Clapperton, 1997). The Heinrich events of the North Atlantic occurred at about 10 ka BP (H0), about 14.2 ka BP (H1), 21.4 ka BP (H2), 26.7 ka BP (H3), 34.8 ka BP (H4) and 47.2 ka BP (H5) (Chapman et al., 2000; Figure 31).

The future challenge is to determine to what extent the various fluctuations in climate have left a recognisable signature in the geological record of sediments, landforms, and fossil floras and faunas (Alverson and Oldfield, 2000 and papers cited therein). The offshore stratigraphical record obtained over the last two decades has already significantly augmented the relatively incomplete terrestrial record, particularly for the Early and Middle Quaternary. Onshore, recent integrated studies at the few sites at which organic materials are well preserved, involving sedimentology, the analysis of pollen, beetle remains and marine molluscs, and the use of a range of dating techniques, have also provided much new information on past Quaternary environments.

OFFSHORE STRATIGRAPHICAL RECORD

An understanding of Scotland's evolving landscape requires knowledge of how the position of the coastline

Figure 31 Greenland (GRIP Summit) oxygen isotope record (after Dansgaard et al.,1993) and predicted former ice cover in Scotland (after Clapperton,1997). The GRIP timescale was determined by counting annual ice layers back to 14.5 ka, and beyond by estimation based on ice-flow modelling. Peaks in ice rafted detritus (Heinrich events H0-H6) are placed after Bond et al. (1993).

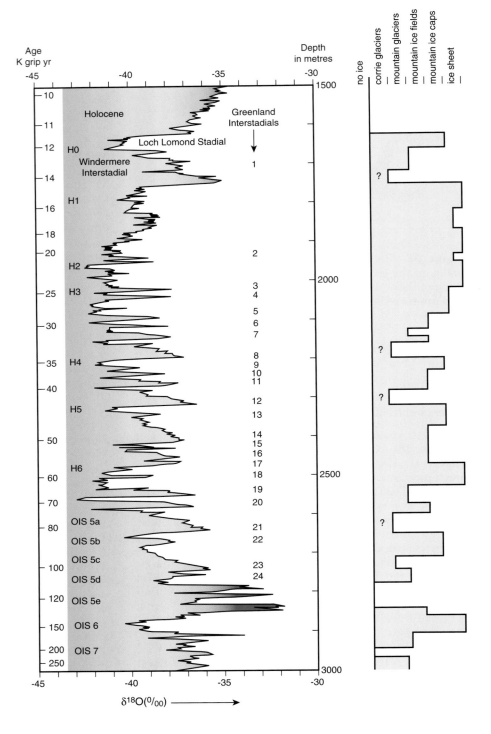

has varied throughout the Quaternary in response to changes in climate and relative sea level. The current position of the coastline is only a transient feature. For example, most of the southern North Sea was low-lying land like the Netherlands until the first major glaciation. Moraines are preserved at the shelf-edge north-west of mainland Scotland, more than 60 km west of the nearest coastline (Holmes et al., 1993; Stoker et al., 1993). Conversely, marine deposits are preserved more than 10 km inland from the modern coastline at the head of the Moray Firth and at least 6 km inland near Elgin. For the following account, the modern coastline is taken as the boundary separating onshore and offshore Quaternary sedimentary deposits.

Compared with the fragmented onshore Quaternary sedimentary record, that offshore is considerably more complete, especially towards the shelf-edge where large submarine fans formed during major glaciations (Table 1).

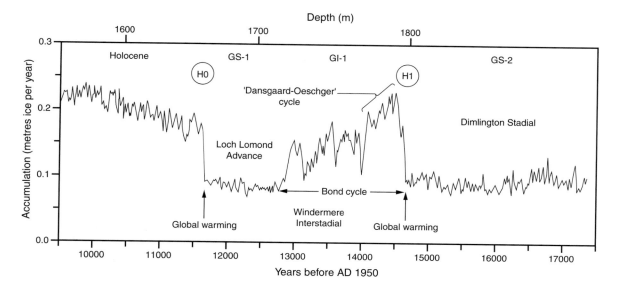

Figure 32 Proxy climate record of the last termination based on a graph of Late-glacial snow accumulation on the Greenland ice sheet summit (after Alley et al., 1993; Clapperton, 1997).

GS-1 Greenland Stadial 1; GI-1 Greenland Interstadial 1; GS-2 Greenland Stadial 2 (after Walker et al., 1999). H0 and H1 are approximate positions of Heinrich events so-named.

The offshore sequence is generally thicker, more extensive and can be correlated regionally using seismostratigraphical methods (Stoker et al, 1985; Holmes, 1997). At least five major glacial episodes have been recognised in the North Sea basin, within a sequence that is dominanted by deltaic, low salinity cold-water marine and glaciomarine conditions (Sutherland, 1984a). The thickest and most complete sequence (more than 500 m) is preserved in the Central Graben of the North Sea, which has subsided tectonically throughout the Quaternary. At least ten till units are present within the Norwegian Channel with six or seven extending to the distal shelf break (Sejrup et al., 2000). An unconformity occurs at the base of the Quaternary and the magnitude of the hiatus increases away from the graben towards north-east Scotland (Andrews et al., 1990; Johnson et al., 1993; Gatliff et al., 1994). In the Moray Firth, most Quaternary sediments rest on Paleocene, Cretaceous, and Jurassic strata, with successively older formations cropping out as the modern shoreline is approached (Andrews et al., 1990, fig. 2). This style of basin–tectonic subsidence was initiated in the mid-Miocene, and was coeval with Miocene, Pliocene and Quaternary uplift of the mainland, which resulted in increased sediment supply to offshore areas (Jordt et al., 1998; Chapter 3). The truncation of Quaternary, Neogene and older sediments towards the margins of the North Sea basin is consistent with a hypothesis involving cycles of subglacial and marine erosion removing terrestrial sediments and contributing to long-term isostatic uplift of the mainland and near-shore areas (Japsen, 1998).

The following account is based largely on Holmes (1997) and the BGS regional reports covering the Moray Firth (Andrews et al., 1990) and the central North Sea (Gatliff et al., 1994). Details of map coverage are given in *Information*

Sources. Interpretations are based mainly on seismic reflection data calibrated by borehole logs. Regionally mappable seismostratigraphical units have been established as formations and members where there is sufficient lithostratigraphical, biostratigraphical or chronostratigraphical control (Figures 33; 34), but difficulties in correlation are common. Moreover, the Moray Firth basin was investigated early in the offshore programme when seismic acquisition technology and many procedures and dating techniques were in their infancy. Hence the available data is not as high quality or detailed as that obtained more recently (Figure 35). A summary of important events is given in Table 1.

The north-west European chronostatigraphical classification has been adopted offshore because it is more complete than the British one, especially in the Lower and Middle Pleistocene (Figure 33). The oxygen isotope stages (OIS) of Shackleton and Opdyke (1973) provide a link between the two schemes, but there is presently vigorous controversy concerning their correlation. The boundary between the Lower and Middle Pleistocene is taken here at the Brunhes–Matuyama boundary of the palaeomagnetic record of Funnell (1995; Figure 33).

Early Quaternary (2.44 to 0.78 Ma)

The 'non-glacial' Early Quaternary (2.44 to 1.2 Ma)

This part of the record is contained within the Aberdeen Ground Formation (Figures 33; 34), the oldest and thickest formation mapped offshore (Holmes, 1977; Stoker et al., 1985). The greater part of the formation off northeast Scotland comprises prodeltaic silts, clays and fine-grained sands deposited in a shallow sea (less than 50 m deep) that lay adjacent to extensive deltas occupying most

Figure 33
Quaternary
stratigraphy of
the central
North Sea
(from Gatliff et
al., 1994).

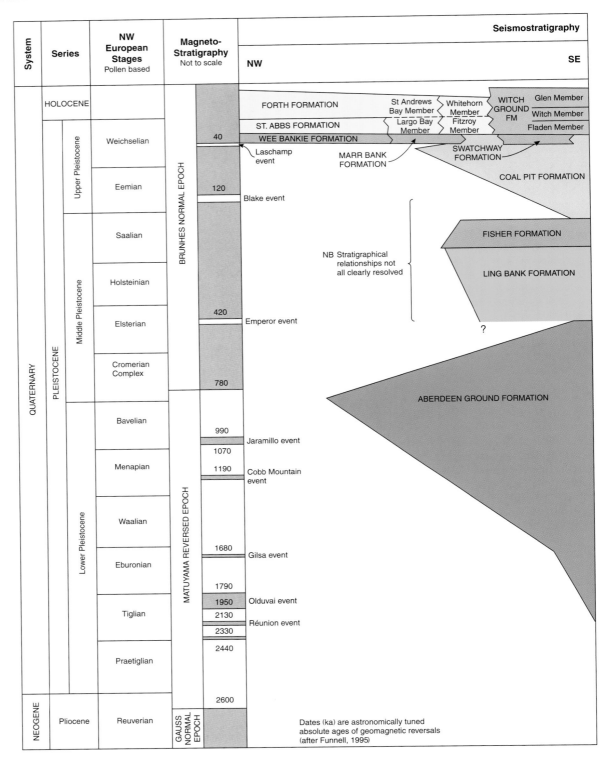

of the southern North Sea basin (Stoker and Bent, 1985; Jeffrey and Long, 1989; Cameron et al., 1992). The deltas were formed mainly by precursors of the Rhine and Thames and by a major river draining the area now largely occupied by the Baltic Sea.

The unconformable base of the Aberdeen Ground Formation is represented in the central North Sea by the 'crenulate reflector' (Holmes, 1977, 1997). Locally it lies 50 m above the Gauss–Matuyama palaeomagnetic boundary (Figure 33; Table 1). To the north and west of

Figure 34
General stratigraphical relationships of Quaternary formations in the central North Sea north of 56°N (from Gatliff et al., 1994).

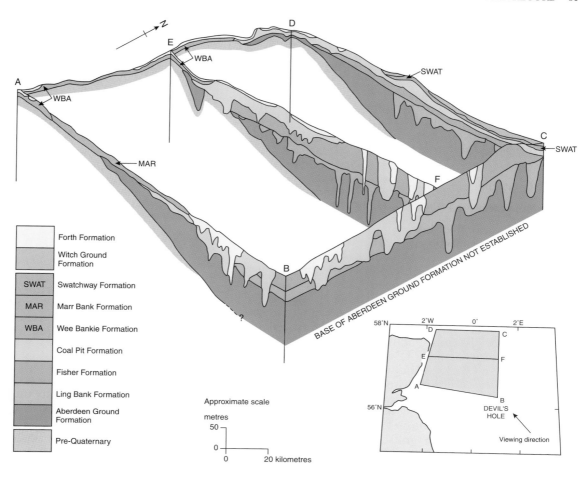

Forth Formation

Witch Ground Formation

SWAT   Swatchway Formation

MAR   Marr Bank Formation

WBA   Wee Bankie Formation

Coal Pit Formation

Fisher Formation

Ling Bank Formation

Aberdeen Ground Formation

Pre-Quaternary

Approximate scale

metres

50

0

0   20 kilometres

58°N 01°E, the unconformity is interpreted from 3-D seismic evidence to take the form of fluvial channels. These features merge downslope into furrowed surfaces like those known to have been produced by grounded icebergs or sea-ice keels (Holmes, 1997). This evidence for ice scour within the Aberdeen Ground Formation is reinforced by the occurrence of periodic ice-rafted debris in sequences of equivalent age on the Atlantic margins off north-west Scotland (Stoker et al., 1994). Hence, the Scottish Highlands were probably glaciated to some extent during the colder stages of the Early Quaternary before 1.2 Ma.

Early Quaternary glaciations (1.2 to 0.78 Ma)

In the northern North Sea, the Shackleton Formation (Johnson et al., 1993) includes an erosion surface that separates sand-rich sediments below from mud-rich sediments above. The incoming of mud-rich sediments is reflected by a change of acoustic facies that is tentatively attributed by Johnson et al. (1993) to the first impact of regional glaciation on the depositional environment of the northern North Sea. The facies change is numerically undated, but on the basis of seismostratigraphical correlation, it predates the 1.07 to 0.99 Ma Jaramillo Subchron (Figure 33), which has been reported from the

overlying Mariner Formation (Stoker et al., 1983; compare with Skinner et al., 1986).

There is more robust evidence in the Troll region of the Norwegian Channel for the 'Fedje Glaciation', which postdates the 1.19 Ma Cobb Mountain Event within the Matuyama Reversed Polarity Chron (Funnell, 1995; Table 1). On the basis of the micropalaeontology, Sr-isotopes, palaeomagnetism and amino-acid geochronology, this glaciation has been assigned an age close to 1.1 Myr (Sejrup et al., 1995). The base of the Mariner Formation is defined regionally on seismic reflection profiles by an irregular erosion surface, which is locally overlain by a diamicton (Johnson et al., 1993). The presence of the diamicton, the geometry of the reflector and other evidence have been put forward as evidence of shelf glaciation extending from Norway across the northern North Sea to areas west of Shetland (Holmes, 1997). However, there is no secure basis for such wide-reaching correlations and the early shelf glaciations may not even have affected offshore areas as far south as the Moray Firth.

The earliest evidence for glaciation of the sea bed offshore from north-east Scotland occurs in the Fladen area of the central North Sea where a 10 m-thick unit of diamicton lies just above the Jaramillo Subchron within the Aberdeen Ground Formation. As the diamicton is overlain by sediment containing microfossils with an interglacial aspect provisionally correlated with the Leerdam Netherlands Pollen Stage, Sejrup et al. (1987) place the glacial event within the Bavelian Complex, between 800 and 900 ka. It has not been established whether the ice that deposited the diamicton flowed from Scotland or Scandinavia, although Sutherland and Gordon (1993) have argued that the latter is more likely. Nevertheless, the Scottish mountains were most probably extensively glaciated during that event.

Middle Quaternary (0.78 Ma to 130 ka)

Cromerian glaciations (OIS 14, 16 and 18)

The oldest known glacial deposits laid down offshore by ice flowing from the Scottish Highlands have been found in boreholes in the Forth Approaches and the Moray Firth (Stoker and Bent, 1985; Bent, 1986). There, towards the top of the Aberdeen Ground Formation, glaciomarine sediments laid down by a grounded ice sheet occur in the west, with the facies becoming increasingly more distal to the east. The deposits lie immediately above the Brunhes–Matuyama palaeomagnetic boundary (Figure 33; Table 1) suggesting that the glaciation occurred in OIS 18, within the Cromerian Complex.

The top of the Aberdeen Ground Formation is cut by a succession of isolated, anastomosing and locally stacked channels, some of which may have been formed as subglacial or ice-marginal 'tunnel' valleys. The channels, together with the oldest units of sediment contained within them (Ling Bank Formation), have been correlated on the basis of regional seismostratigraphy with the widespread Elsterian glaciation of the north-west European mainland (Stoker et al., 1985; Cameron et al., 1987). However, several lines of evidence indicate that the basal sediments of the Ling Bank Formation at its type locality were laid down during a late-Cromerian interglacial (Penney, 1990; Ansari, 1992 Knudsen and Sejrup, 1993; Sejrup and Knudsen, 1993) and that they are overlain by arctic glaciomarine sediments thought to be Elsterian in age, on the basis of amino-acid dating of shells (Sejrup and Knudsen, 1993). The channelled surfaces at the top of the Aberdeen Ground Formation may well have formed in more than one glacial cycle, including the severe Cromerian glaciation of OIS 16 (Holmes, 1997). It follows that the first major interruption of the growth of deltas across the southern North Sea may have occurred in the Cromerian and not the Elsterian as commonly believed (Table 1).

Elsterian glaciation (OIS 12) and Holsteinian interglacial deposits (OIS 11 and 9).

Sediments near the top of the Ling Bank Formation at its type locality have been correlated with 'type' Holsteinian sections in Denmark and Germany (Knudsen and Sejrup, 1993). However, it is still widely believed that marine Holsteinian deposits, in general, directly overlie the major erosion surface at the top of the Aberdeen Ground Formation (Figure 34). There was, therefore, a widespread

Olive green clayey calcareous soft mud with arctic mollusc shells

Red-brown, grey-brown and pink-grey soft mud

Brown to grey-brown muddy sand, sandy clay and diamicton

Grey pebble free mud and rhythmites

Local gravelly mud

Local olive-grey sandy diamicton

Grey pebbly diamicton

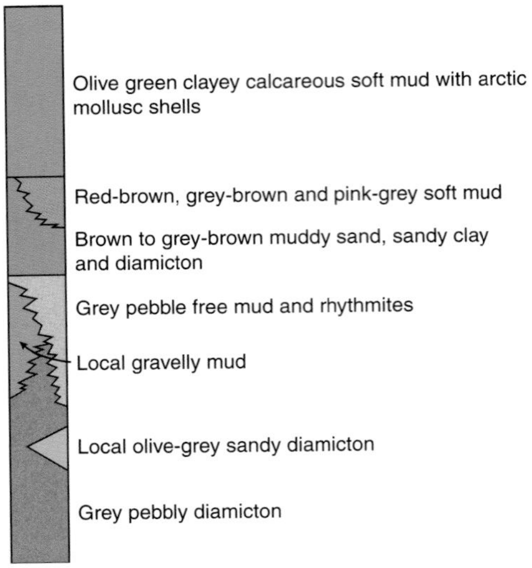

Figure 35 Generalised sequence of Quaternary deposits in the Moray Firth (from BGS Moray–Buchan Sheet 57°N–04°W). Ages uncertain, possibly spanning Middle Devensian to early Flandrian.

glaciation during the Elsterian stage, which involved ice flowing from Scandinavia into the southern North Sea (Cameron et al., 1992) and extending to the continental shelf edge north-west of Scotland (Stoker et al., 1994). The timing of the glaciation is significant because the palaeogeography of the North Sea basin and hence the climate of the surrounding lands subsequently changed profoundly. Furthermore, Scotland and the adjoining continental shelves would have been glaciated during that event, which is generally correlated with the Anglian stage of the conventional British chronostratigraphy (Table 1).

Saalian glaciation(s) (OIS 6 to 8)

Evidence for Saalian glaciations off north-east Scotland occurs in the Fisher and Coal Pit formations. The predominantly arctic glaciomarine Fisher Formation rests on a major, gently undulating unconformity that cuts across (onlaps) the Ling Bank and Aberdeen Ground formations (Figures 33, 34). The unconformity results from a marine transgression and the overlying Fisher Formation is thought to be no older than OIS 7 (Jensen and Knudsen, 1988; Holmes, 1997). A till has been identified within the Fisher Formation (Sejrup et al., 1987) and contemporaneous subglacial erosion is thought to have occurred to the north of 58.6°N in the Moray Firth (Bent, 1986). The top of the Fisher Formation is defined by another regional unconformity, but unlike the one at its base, this one is typically crenulate (Figure 34) and is believed to be the result of the Saalian glaciation (Gatliff et al., 1994). The channels are mainly infilled with glaciomarine deposits belonging to the overlying Coal Pit Formation (Sejrup et al., 1987), although some fragmentary beds deposited in warmer waters have been identified within the formation in the northern and central North Sea (Gatliff et al., 1994).

The glaciomarine basal deposits of the Coal Pit Formation are succeeded by marine sediments containing foraminiferal assemblages typical of the Eemian–Ipswichian Interglacial (Cameron et al., 1987). These in turn underlie a horizon that has been correlated with the palaeomagnetic Blake Event at the OIS 5e/5d boundary (Stoker et al., 1985).

The evidence cited above suggests that there was a regional glaciation during OIS 6 of the Saalian and possibly an earlier, more limited Saalian glaciation in the north. The glaciation of OIS 6 is thought to have affected most of Scotland, because ice flowing from the mainland affected the northern North Sea at least as far south as 56°N (Sutherland and Gordon, 1993) and it reached the shelf break to the north-west of Scotland (Skinner et al., 1986; Stevenson, 1991; Holmes, 1997). Sejrup et al. (2000) conclude that ice streams occupied the Norwegian Channel during every glacial stage between OIS 12 and 6, each one representing a regional glaciation.

Late Quaternary (130 ka to present day)

Eemian Interglacial (OIS 5e)

A record of the Eemian (Ipswichian) is contained within the central part of the Coal Pit Formation, which fills channels that were probably eroded during, or immediately after, the Saalian glaciation in the Moray Firth and central North Sea (Andrews et al., 1990; Gatliff et al., 1994; Figure 34). Dinoflagellate cyst assemblages, like those found at present in the North Sea, have been identified in interbedded bioturbated sand and stiff, dark grey, shelly, pebbly clay with wood fragments in the type borehole (Stoker et al., 1985). The final occurrence of *Elphidium ustulatum* (Todd) has been noted; this is a foraminiferid that became extinct in the North Sea by the early Weichselian (Gregory and Bridge, 1979). In southern Norway, the Eemian is now thought to have been succeeded by a prolonged period of interstadial conditions with restricted mountain glaciation (Sejrup et al., 2000).

Early Weichselian glaciation (OIS 4)

No secure evidence of early Weichselian–Devensian glacial deposits has been reported off north-east Scotland, but an ice sheet is believed to have reached the shelf break to the north-west of the Scottish mainland where it formed submarine end-moraines (Stoker, 1988, 1990; Stewart and Stoker, 1990). Sejrup et al. (2000) conclude that an ice stream occupied the Norwegian Channel after 80 ka, suggesting that a regional glaciation occurred equivalent to the Karmoy Glaciation established in south-west Fennoscandia (Figure 43). Micromorphological studies of sediments from several boreholes in the central North Sea also suggest regional glaciation at that time involving coalescing Scottish and Scandinavian ice sheets (Carr, 1998).

Middle Weichselian (OIS 3)

The Coal Pit Formation in the central North Sea includes a sequence of shelly glaciomarine clays that have been placed tentatively in OIS 3 on palaeomagnetic evidence (Stoker et al., 1985) indicating that this area was free of glacier ice during this stage. In the northern North Sea and on the West Shetland Shelf a widespread surface of marine erosion has been correlated with the OIS 4/3 boundary. It is overlain by the mainly arctic marine Cape Shore Formation, which is securely placed in the Middle Weichselian on several lines of evidence (Johnson et al., 1993; Skinner et al., 1986; Sejrup et al., 1994; Holmes, 1997). Thus the seas to the north and west of Scotland were also free of glacier ice during this stage and perhaps much of the Scottish mainland.

Indirect evidence of former near-shore environments along the southern coast of the Moray Firth is provided onshore by glacial rafts that were transported by ice during the Late Devensian. At Clava, near Inverness, rafts of high-boreal to low-arctic shallow marine mud originally deposited in Loch Ness, then a fjord, are probably early to middle Devensian in age (Merritt, 1992). The deposits correlate with those of the Bø Interstadial in Norway on amino-acid dating evidence (Figure 43). Rafts of broadly similar age have been located at the Boyne Limestone Quarry, King Edward and Gardenstown sites (Appendix 1).

Late Weichselian/Late Devensian glaciation (OIS 2)

The offshore record of Upper Weichselian deposits is of particular importance in establishing models of the main Late Devensian ice sheet in northern Britain, but it is complicated and, as explained below, controversial (Figure 3). BGS mapping suggests that an ice sheet extended to the continental shelf break, and beyond, to the north and west of Scotland (Holmes, 1991; Stoker and Holmes, 1991; Stoker et al., 1993; Figure 41). The resultant sediments are correlated on the basis of regional seismostratigraphy with Late Weichselian (OIS 2) deposits in the northern North Sea (Johnson et al., 1993). A radiocarbon date of about 22.5 ka BP from glaciomarine deposits within the limit of glaciation on the outer shelf to the west of St Kilda (Selby, 1989) appears to be consistent with the Late Devensian glacial maximum predating 18 ka BP (Figure 41).

The important units off north-east Scotland are, from west to east, the Wee Bankie, Marr Bank, Swatchway and Coal Pit formations (Figures 33, 34). The Wee Bankie Formation (Stoker et al., 1985) lies directly off the east coast of Scotland. It has a sheet-like geometry with an uneven, ridged top and comprises up to 40 m of stiff, matrix-dominated diamicton with some interbeds of sand, pebbly sand and silty clay. The formation is generally thought to have been laid down beneath the main Late Devensian ice sheet, and its mapped eastern boundary (Figure 44) has been used in many reconstructions of that ice sheet (e.g. Sutherland, 1984a; Boulton et al., 1985; Boulton et al., 1991), although some have envisaged more extensive glaciation at this time (e.g. Ehlers and Wingfield, 1991).

The Wee Bankie Formation is replaced eastwards by the Marr Bank Formation, commonly at a low, eastward-facing scarp interpreted as a former ice-contact slope, but the two formations probably interdigitate locally (Stoker et al., 1985). The Marr Bank Formation consists mostly of sands and muddy sands of Scottish provenance with a sparse microfauna indicative of shallow, high boreal to arctic waters. It forms a sheet-like deposit up to 25 m thick resting on an extensive surface of marine planation dipping north-eastwards (Holmes, 1977). As it is traced eastwards, the basal reflector of the Marr Bank Formation becomes acoustically indistinguishable from the upper part of the adjacent Coal Pit Formation, and the two formations probably pass laterally into one another locally (Gatliff et al., 1994).

In the central North Sea, the lower part of the Swatchway Formation, at Borehole 77/2 (Figure 44), is formed mainly of glaciomarine sediments from which AMS radiocarbon dates of 22.7, 20.9 and 19.7 ka BP have been obtained on in situ mollusc and benthic foraminiferids (Sejrup et al., 1994). This evidence suggests that the area was free of grounded ice during that period. However, a diamicton underlying the glaciomarine deposits in that borehole is interpreted as a till (Sejrup et al., 1994). It contains reworked arctic benthic foraminiferids that have provided a maximum AMS radiocarbon age of 42.3 ka BP. The diamicton rests on cold marine deposits assigned to the Ålesund Interstadial of north-west Fennoscandia and it is concluded by Sejrup et al. (1994) that the till was laid down between 28 ka BP and 22 ka BP during an initial, maximal stage of the Late Devensian glaciation. This conclusion hinges on the identification of the diamicton as a till and that the foraminiferids are unlikely to be in situ. Although Sejrup et al. (1994) describe deformation structures in the diamicton, they do not discriminate between subglacial deformation and disturbance by sea-ice. However, recent micromorphological evidence has revealed brittle shear at the base of the diamicton suggesting that it is indeed a subglacial, deforming bed till (Carr, 1998). Furthermore, Sejrup et al. (2000) and Carr (1999) both conclude that the central North Sea was also glaciated in the previous Skjonghelleren Glaciation of Fennoscandia, between about 50 and 40 ka BP (Figure 43).

There are important implications to these findings, discussed below, which indicate that the Scottish and Scandinavian ice sheets reached their maximum extent in the Late Devensian prior to about 22 ka BP and that they very probably coalesced (Figure 41). At least the uppermost part of the Wee Bankie Formation postdates this early phase. Following retreat to an unknown position at about 20 ka BP, during the Hamnsund Interstadial of Norway (Valen et al., 1996), the Scottish ice sheet probably then re-advanced to the eastern boundaries of the 'Bosies Bank Moraine' (Bent, 1986; Figure 44). This event probably equates with the Tampen Glaciation of Norway, when ice re-advanced onto the shelf and an ice stream reoccupied the Norwegian Channel (Sejrup et al., 2000; Figure 43).

Deglaciation, the Late-glacial period and Holocene (OIS 2 and 1)

Much of the northern North Sea north of 58°N had been deglaciated by 16 to 14 ka BP, and was either subaerially exposed or inundated by a very shallow sea (Peacock, 1995). This is compatible with a maximum age of about 14.1 ka BP for the onset of glaciomarine sedimentation following retreat of ice from the Witch Ground area

(Figure 44; Sejrup et al., 1994) and at about 15 ka BP in the Norwegian Channel (Sejrup et al., 1995). The onset of deglaciation on the Hebridean Shelf has been dated to about 15.2 ka BP, predating the onset of warming in the North Atlantic (Peacock et al., 1992; Austin and Kroon, 1996). Shells within glaciomarine sediments occurring onshore near Peterhead have yielded ages of about 14.3 and 14.9 ka BP (Appendix 1 St Fergus).

Radiocarbon dates from *Portlandia arctica* Gray, a high arctic marine bivalve, indicate that polar water, and probably seasonal sea-ice, remained in the northern North Sea until at least 13.2 to 13.1 ka BP when warmer waters arrived. Glaciomarine and estuarine silts and clays of the Errol Formation accumulated along the coasts (Peacock, 1999), while the muddy St Abbs Formation was laid down in this polar sea off the eastern coast of Scotland (Stoker et al., 1985; Figure 34). Warm North Atlantic waters did not reach the north-east Atlantic and western Scotland until about 13 ka BP, when conditions changed from high arctic to boreal possibly in less than 50 years (Kroon et al., 1987; Peacock and Harkness, 1990).

A record of the Windermere Interstadial is probably contained within the Swatchway Formation, which occurs to the north-east of Buchan (Figure 34). It comprises shelly muds and sands with a mixed northern temperate to arctic microfauna (Stoker et al., 1985; Harland, 1988). It occurs more certainly in the Largo Bay Member of the Forth Formation, which is more widespread off north-east Scotland (Stoker et al., 1985; Figure 34). It includes up to 30 m of silty muds that become coarser grained and pebbly upwards with concomitant decreasing faunal diversity. The trends probably reflect lowering sea level and cooling seas towards the onset of the Loch Lomond Stadial. The overlying St Andrews Member was laid down as coastal sand bars in a very shallow sea during the subsequent Loch Lomond Stadial. High arctic marine fauna returned during the stadial, during which nearshore marine summer temperatures were approaching 10° below present levels (Graham et al., 1990; Peacock, 1996).

Most sea-bed sediments of Holocene age have been mapped lithologically rather than lithostratigraphically. The return of warm North Atlantic Drift waters to the Scottish seas occurred within a few decades just prior to 10 100 ka BP (Peacock and Harkness, 1990). At first sea temperatures were 2 to 3° lower than those of the present day, but a warming occurred at about 9600 BP.

ONSHORE STRATIGRAPHICAL RECORD

Despite over a century of research, the record of events prior to deglaciation of the main Late Devensian ice sheet on land is less well understood than offshore. This is mainly because of the fragmentary nature of terrestrial deposits, the scarcity of organic material that is suitable for dating by radiocarbon methods and the relative unreliability of age determining methods beyond the range of radiocarbon dating, such as thermoluminescence (TL) techniques and amino-acid geochronology. Although more information is normally available at onshore sites compared to boreholes offshore, the absence of seismostratigraphy and palaeomagnetic chronology makes any correlation between those sites very difficult indeed. Inevitably there is controversy and the chapter concludes with a review of the published ice-sheet models for north-east Scotland. A summary of the important events is given in (Table 1).

The heirachy and description of named lithostratigraphical units are given in Chapter 8, and correlations between these units and the Oxygen Isotope Stages are presented in Table 7. Many correlations between units are based on a 'count from the top' basis, which means that any reinterpretation of the status of the uppermost unit has consequences for others lower in the sequence.

Early Quaternary (2.44 to 0.78 Ma)

No Early Quaternary deposits have been identified in north-east Scotland, but mollusc shell fragments in the Kippet Hills Sand and Gravel Formation of the Logie-Buchan Drift Group (Chapter 8) have most probably been derived from the Aberdeen Ground Formation lying offshore. The Early Pleistocene age assigned to the shells on faunal grounds by Jamieson (1882a) has been confirmed by amino-acid analysis (Appendix 1 Kippet Hills).

Middle Quaternary (0.78 Ma to 130 ka)

Anglian glaciation (OIS 12) (=Elsterian)

The oldest known deposits in the district have been found at Kirkhill (Connell et al., 1982; Appendix 1). The basal fluvial or glaciofluvial sands and gravels at this site contain erratics derived from the west, as does the basal till (Leys Till Formation) occurring nearby. The two units were originally assigned to the Anglian glaciation on the basis of minimum ages for the overlying glacial and interglacial deposits using a simple chronostratigraphical model (Connell and Hall, 1984a; Hall and Connell, 1991). However, they are now assigned tentatively to OIS 8 on the basis of new luminescence dates on the Corse Gelifluctate Bed higher in the sequence (Appendix 1; Table 7).

Glacial erratics of Norwegian origin are relatively common within glacigenic deposits in north-east Scotland (Read et al., 1923; Bremner, 1939; Hall and Connell, 1991). Although none of these clasts occurs in a till assigned to OIS 12, it is possible that they were transported to north-east Scotland during that stage (Ehlers, 1988).

Hoxnian interglacial (OIS 11) (=Holsteinian)

The oceanic climate record of the North Atlantic suggests that there have been only two periods in the last 600 ka that were as warm, or warmer, than the Holocene, namely the Hoxnian (OIS 11) and Ipswichian (OIS 5e) interglacials (Ruddiman and McIntyre, 1976). Other interglacials were temperate (e.g. OIS 9 and 7), but probably not as warm. The sands overlying the Kirkhill Palaeosol Bed (Lower Buried Soil) at Kirkhill (Appendix 1 Site 7) contain detrital organic matter derived from the erosion of soil and vegetation formed in an interglacial climate, presumed from its stratigraphical position to be Hoxnian (Connell, 1984a; Lowe, 1984). The presence of pollen grains of pine, alder

and lime appear to corroborate this presumption. However, while an OIS 11 age cannot be confirmed, the absence of significant unconformities within the Kirkhill sequence indicate that the soil and overlying organic sediment (Corse Gelifluctate Bed) may date from OIS 7, as indicated by luminescence dates reported by Duller et al. (1995). The Kirkhill Palaeosol Bed itself was originally thought to be of interglacial origin (Connell et al., 1982), but has also been compared to a cold-water gley soil formed in an interstadial climate (Connell and Romans, 1984). However, the strong development of the truncated Ea horizon, together with the presence of the mineral proto-imogite/allophane in the Bs horizon (Appendix 1 Site 7 Kirkhill and Leys), suggests the interglacial interpretation is correct. The nearest other site with a possible Hoxnian record is Dalcharn, near Inverness, where the presence of significant quantities of holly pollen implies that the palaeosol there did form in a particularly warm interglacial (Walker et al., 1992). However, there too, subsequently obtained luminescence dates suggest a much younger age (Duller et al. (1995).

'Wolstonian' glaciation(s) (OIS 6–10) (=Warthe/Saal/Drenthe/Fuhne)

Two distinct phases of 'Saalian' glaciation are thought to have affected continental north-west Europe during this period (Ehlers et al., 1991) and three regional glaciations affected the North Sea basin (Sejrup et al., 2000), but no equivalent terrestrial glacial deposits have been shown unequivocally to occur in Scotland. Several tills in north-east Scotland may date to this period, but arguments for their 'Wolstonian' age rely mainly on indirect evidence. This includes stratigraphical position, whether the tills are judged to have been weathered in an interglacial climate (as against incorporating materials weathered previously) and whether that weathering is judged to have occurred in the Ipswichian (OIS 5e). The Rottenhill Till (Kirkhill Lower Till) has been assigned to the 'Wolstonian' (Hall, 1984; Hall and Connell, 1991; Appendix 1) and luminescence dates from Leys Quarry nearby apparently confirm an OIS 6 age for the unit. Several other tills have been correlated with the Rottenhill Till (Table 7), including the Red Burn Till at Teindland (Hall et al., 1995a), the Camp Fauld Till (Corse of Balloch Till) on the 'Buchan Ridge' (Whittington et al., 1993), the supposedly weathered 'Kings Cross' till at Aberdeen (Synge, 1963), the Craig of Boyne Till at Boyne Bay (Connell and Hall, 1984b; Peacock and Merritt, 2000), and the Bellscamphie Till near Ellon (Hall and Jarvis, 1995). Apart from the last two named tills, the others were laid down by ice that flowed out of the Moray Firth and south-eastwards across Buchan towards Aberdeen (see ice sheet models below). A 'Wolstonian' age is also likely for the Benholm Clay Formation occurring to the south of Inverbervie (Auton et al., 2000; Appendix 1 Site 26 Burn of Benholm).

Late Quaternary (130 ka to the present day)

Ipswichian Interglacial (OIS 5e) (=Eemian)

Evidence from England and continental north-west Europe indicates that at least the early part of the last interglacial was a little warmer than the present day (Aalbersberg and Litt, 1998). Teindland and Kirkhill represent two of the four sites in Scotland at which organic deposits or palaeosols have been assigned to the Ipswichian (Sutherland and Gordon, 1993). Lowe (1984) disagreed with Fitz-Patrick (1965) and Edwards et al. (1976) that the Teindland Palaeosol Bed (Teindland Buried Soil) could not be attributed to the interglacial with any confidence, but more recent work there, and at Red Burn nearby, largely confirms the earlier conclusions (Hall et al., 1995a; Appendix 1 Site 1 Teindland). For example, the soil contains pollen indicative of deposition towards the end of an interglacial with the progressive replacement of hazel-alder woodland, first by pine woodland, then by heathland and, ultimately, sparse grassland. Luminescence dating suggests an Ipswichian age.

The truncated Backfolds Palaeosol Bed (Kirkhill Upper Buried Soil) is now more confidently assigned to the Ipswichian, following the acquisition of new luminescence dates on sediments underlying it at Leys (Duller et al., 1995). Other supposed interglacial deposits at Tipperty and Balmedie (Bremner, 1938, 1943), north of Aberdeen, and at King's Cross (Synge, 1963) in Aberdeen, have not been confirmed (Peacock, 1980; Edwards and Connell, 1981). Another site at Errollston, Cruden Bay (Appendix 1 Site 17 Errollston Clay Pit), includes pre-Quaternary palynomorphs and clearly contains much reworked material (Peacock, 1984b; Connell et al., 1985). Reworked shell fragments of Ipswichian age are possibly present at Kippet Hills (Appendix 1 Site 16).

Early Devensian (OIS 5a-d) (=Early Weichselian)

The north-west European pollen and beetle record indicates that the last interglacial was followed by two short, cold stadials (the Herning 5d and Rederstall 5b) separated by cool and wet interstadials (the Brörup 5c and Odderade 5a) (Aalbersberg and Litt, 1998; Figures 6, 7). Organic deposits of one or more of these interstadial periods appear to be more common in Scotland than those of the Ipswichian, possibly because the conditions were more conducive to the accumulation of peat. The most complete record in the region of an early Devensian interstadial, probably 5c, comes from the Allt Odhar site, near Inverness (Walker et al., 1992). Here, the pollen and beetle evidence indicates deteriorating climatic conditions towards the end of a cool interstadial as reflected in the replacement of birch woodland and willow scrub with grassland and heath, and then by open communities of grass and sedges. The record of the latter stages of this interstadial, followed by a colder phase of tundra environment, may be represented at both the Camp Fauld and Crossbrae sites in Buchan (Whittington et al., 1998; Table 7; Appendix 1 Sites 5 and 14). The Burn of Benholm Peat Bed at the Benholm site, south of Inverbervie (Appendix 1 Site 26), is also now correlated tentatively on pollen evidence with OIS 5a or 5c (Auton et al., 2000).

Evidence of major glaciation was formerly reported during OIS 5d and 5b in the north-west Fennoscandian record (Baumann et al., 1995; Mangerud et al., 1996),

but it now seems that only limited glaciation occurred (Sejrup et al., 2000; Figure 43).

Early Devensian glaciation (OIS 4)

It has been suggested for some time (Sutherland, 1981) that much of Scotland was glaciated during this very cold stage, but no unequivocal evidence has been presented (Bowen et al., 1986; Worsley, 1991). Nevertheless, judging from the record of events in the North Sea basin and north-west Fennoscandia (Figure 43), a regional glaciation occurred during OIS 4 (Sejrup et al., 2000), and much of Scotland is likely to have been glaciated. Several sites in the district probably include a record of this period in the form of periglacial phenomena, for example Corsend Gelifluctate Bed (Gelifluctate IV) at Kirkhill. The Woodside Diamicton (Teindland Till) at Teindland was probably laid down during an Early Devensian glaciation (Hall et al., 1995a), but similar claims for the Hythie Till (Kirkhill Upper Till) (Hall and Jarvis, 1993a), and the Pitlurg Till near Ellon (Hall and Jarvis, 1995) are less secure; the units are assigned here tentatively to OIS 2 (Table 7).

Sutherland (1984c) concluded that the clayey deposits containing beds of cold-water, interstadial-type marine molluscs beneath till at King Edward were in situ (Appendix 1 Site 4). The mid-Devensian amino-acid and radiocarbon ages obtained from the shells (Miller et al., 1987) therefore indicated mid-Devensian marine inundation. For this to have occurred during a period of low global sea level (Pirazzoli, 1993), there must have been substantial glacio-isostatic depression following an Early Devensian glaciation. However, the shelly deposits at King Edward almost certainly have been glacially reworked and form part of the Whitehills Glacigenic Formation (Peacock and Merritt, 1997). Sutherland (1981) used a similar argument in connection with the Clava Shelly Formation at Inverness (Fletcher et al., 1996). However, despite these shelly deposits also being glacial rafts, their derivation from Loch Ness, then a fjord, does seem to imply Early Devensian glaciation of at least the western Highlands (Merritt, 1992).

Middle Devensian (OIS 3)

Periglacial and glacial environments prevailed across much of continental north-west Europe during OIS 3 and glaciers probably existed in the western Highlands for most of the time. There were relatively warm periods between 50 to 41 ka and 37 to 36 ka on the Continent (Huijzer and Vandenberghe, 1998), the former being correlated with the Upton Warren Interglacial of the British chronostratigraphy (Figures 6; 7). There is also evidence of two interstadials at roughly equivalent times in the Scandinavian record (Figure 43). The younger of the two, the 'Sourlie Interstadial' is apparently represented by organic deposits beneath till in the lowlands around Glasgow, where reindeer, woolly rhinoceros and mammoth roamed in a tundra-like environment (Jardine et al., 1988; Sutherland and Gordon, 1993). Reindeer bones found in caves near Inchnadamph, Ross-shire, also date from the Middle Devensian (Lawson, 1984). However, there are no known representative organic deposits in north-east Scotland, although the district was probably free of ice. The Crossbrae Peat (Appendix 1 Site 5) was originally thought to date from between 22 and 26.5 ka BP, but it is now correlated with OIS 5a or 5c (Whittington et al., 1998).

Sand within the glaciofluvial Byth Gravel at the Howe of Bythe Quarry (Appendix 1 Site 6) has yielded luminescence ages of about 45 and about 37 ka, implying the presence of glacier ice in the vicinity (Hall et al., 1995b). This Middle Devensian glaciation would correlate with the Skjonghelleren glaciation of Norway (Figure 43), when ice probably crossed the North Sea basin (Carr, 1998; Sejrup et al., 2000).

Amino-acid ratios and radiocarbon dates on shells within rafts and tills in the Whitehills Glacigenic Formation (Chapter 8) suggest that the deposits are derived from cold-water marine muds of Middle Devensian age. More specifically, the ratios correlate with the Bö Interstadial of Norway (Figure 43), for which ages from 40 to 80 ka have been proposed by Miller et al. (1983) and the higher estimates are favoured by Sejrup et al. (2000). The rafts of the Clava Shelly Clay near Inverness are also thought to have been originally deposited during that interstadial (Merritt, 1992b). Oddly, there appear to be no correlatives of the younger Ålesund Interstadial of the Norwegian sequence in north-east Scotland (Peacock and Merritt, 1997).

Dimlington Stadial of the Late Devensian (28 ka to 13 ka BP) (OIS 2)

Although most of the drift deposits in the district were laid down during this period, the sequence of events that occurred is not fully understood. Several conflicting models have been published (Figure 3), and the controversy is discussed more fully below, where a more detailed history is given. It is the view of the authors that the district was overwhelmed entirely by ice in the Late Devensian, approximately coeval with the mainland ice sheet reaching the continental shelf break to the north-west of Scotland (Figure 41). There is growing evidence from Scandinavia, the North Sea and elsewhere that the ice reached its maximum extent early in the Late Devensian, between about 28 ka and 22 ka BP, not at about 18 ka BP, as was widely believed until recently (e.g. Bowen et al., 1986). The Scottish and Scandinavian ice sheets coalesced during this early phase (Carr, 1999; Sejrup et al., 2000). It was probably also at this time that ice of the Moray Firth ice stream was forced to flow south-eastwards across Buchan to lay down at least part of the Whitehills Glacigenic Formation and its correlatives (Table 7). There probably followed a period of glacial retreat that lasted from about 21 to 18 ka BP, roughly equivalent to the Hamnsund Interstadial of Norway (Valen et al., 1995; Figure 43). A major re-advance then probably occurred between 18 and 15 ka BP equivalent to the Tampen Re-advance of Norway (Sejrup et al., 1994; Figure 43). This would have mostly involved the coastal ice streams. The Moray Firth ice stream had retreated from the north-east coast of Buchan by about 15 ka BP,

and the whole district would have been ice-free by the beginning of the Windermere Interstadial at 13 ka BP.

Throughout this memoir, the Scottish ice sheet that formed between 28 and 13 ka is referred to as the 'Main Late Devensian' in order to to distinquish it from that of the Loch Lomond Stadial, also part of the Late Devensian. The term 'Dimlington Stadial ice sheet' is not used because there are now strong arguments for redefining the time span of the Dimlington Stadial to 18–13 ka BP (Sejrup et al., 1994).

Windermere ('Lateglacial') Interstadial (13 ka to 11 ka BP) (OIS 2)

A sudden amelioration of the climate occurred at about 13 ka BP when the oceanic polar front in the North Atlantic migrated northwards (Ruddiman and McIntyre, 1973, 1981; Ruddiman et al., 1976) and temperate waters returned off the western coasts of the British Isles (Peacock and Harkness, 1990). Atmospheric temperatures rose within decades to close to present values (Bishop and Coope, 1977; Atkinson et al., 1987). It is almost certain that no ice survived the interstadial in north-east Scotland, although glaciers might not have disappeared completely in the central and western Highlands (Sissons, 1974a; Sutherland, 1984a; Sutherland and Gordon, 1993; Clapperton, 1997). Masses of ice buried within sediments from all phases of the Main Late Devensian glaciation (if not before) commonly melted out relatively slowly resulting in kettleholes (Peacock in Harkness and Wilson, 1979).

Organic sequences preserved in some kettleholes in the region provide a record of environmental change during the Late-glacial period (Donner, 1957; Vasari and Vasari, 1968; Vasari, 1977; Gordon, 1993; Appendix 1 Mill of Dyce, Rothens, Loch of Park). Elsewhere, beds of peat preserved beneath solifluction deposits and landslips provide important evidence (Godwin and Willis, 1959; Clapperton and Sugden, 1977; Hall, 1984; Connell and Hall, 1987; Aitken, 1991; Appendix 1 Site 25 Glenbervie). Radiocarbon dates are given in Table 8. The freshly deglaciated ground was first colonised by a pioneer vegetation of open habitat species followed by the immigration of crowberry heath, juniper and dwarf varieties of birch and willow. Eventually open birch woodland developed with juniper and isolated stands of Scots pine locally. A stepwise climatic deterioration occurred throughout the Windermere Interstadial (Lowe et al., 1999; Mayle et al., 1999) and it is likely that glaciers had already started to build up in the Western Highlands before more sustained cooling began at about 11.2 ka BP.

Relative sea level in north-east Scotland probably had fallen to below its present level by the beginning of the interstadial and it continued to fall, especially in the west of the district, where glacio-isostatic rebound was greater.

Loch Lomond Stadial (11 ka to 10 ka BP) (OIS 2)

The North Atlantic Polar Front migrated rapidly southwards to the latitude of northern Portugal at about 11 ka BP (Bard et al., 1987), and the climate of the British Isles reverted temporarily to arctic conditions.

Although no glaciers are thought to have developed in the district they did so in the corries of the Cairngorms and the Mounth, and substantial ice caps formed over Rannoch Moor and in the North-west Highlands (Sutherland and Gordon, 1993).

A tundra environment existed in north-east Scotland during the stadial. Soils that had developed during the Windermere Interstadial were largely destroyed by periglacial processes such as solifluction and frost churning (geliturbation) (FitzPatrick, 1956, 1958, 1969, 1972, 1976, 1987; Galloway, 1961a–c, Connell and Hall, 1987). Vegetation was quickly replaced by open-habitat plant communities tolerant of the Arctic conditions and unstable soil (Gunson, 1975). Some fossil frost polygon networks observed in the district (Chapter 7) might have been formed during the stadial (Clapperton and Sugden, 1977), but most probably formed earlier during the retreat of the Main Late Devensian ice sheet (Gemmell and Ralston, 1984, 1985; Armstrong and Paterson, 1985; Connell and Hall, 1987; Appendix 1 Ugie valley). Fluvial and debris-flow activity would have been enhanced during the stadial, especially during springtime snow-melts (Ballantyne and Harris, 1994; Maizels and Aitken, 1991). Slopes would have been particularly prone to land-slipping (Appendix 1 Glenbervie) and head deposits formed widely across the district. Much of the coarse-grained alluvium in minor valleys across the district would also have been laid down then.

Holocene (10 ka BP to the present day) (OIS 1)

At about 10 ka BP the global climate changed rapidly, possibly within an average human lifetime, to its present interglacial mode, marking the beginning of the Holocene epoch. The oceanic polar front in the North Atlantic rapidly shifted northwards and the warm Gulf Stream current became established, providing an ameliorating influence on the climate of Scotland. At first the glaciers in the western Highlands might have re-advanced locally as a result of increased snowfall (Benn et al., 1992), but rapid disintegration soon followed. The widespread occurrence of bare, unstable soils at the beginning of the Holocene apparently led to a period of intense fluvial erosion and deposition with enhanced debris flow activity on mountain sides and extensive formation of landslips as the ground thawed (McEwen, 1997). Soils gradually became more stable following the establishment of vegetation: firstly shrubs and scrub communities were established during the early part of the Windermere Interstadial, and was replaced later by woodland (Durno, 1956, 1957; Vasari and Vasari, 1968; Edwards, 1978). A distinct phase of juniper dominance was replaced by birch woodland, followed by the arrival of hazel, elm and shortly after, by oak (Appendix 1 Loch of Park, Nether Daugh). Pine forest became established in the East Grampians by the mid-Holocene whereas birch-hazel forest was predominant in the east and across the coastal lowlands (Vasari and Vasari, 1968; Gunson, 1975; Birks, 1977).

Climatic deterioration may have begun soon after the beginning of the Holocene and it has continued in a stepwise fashion. Distinct layers of pine stumps preserved quite widely in upland blanket peat deposits in the East Grampians indicate that pine forest locally succumbed to the spread of Sphagnum moss between 6.7 ka and 5 ka BP (Pears, 1968, 1970). Colder and wetter climatic conditions caused this change, rather than the influence of people, but human impact is apparent in the pollen and sediment records before 5 ka BP (Edwards, 1978, 1979b; Edwards and Rowntree, 1980). It can also be seen in the late Holocene pollen and Coleopteran record obtained from a silted-up ox-bow lake on the floodplain of the River Don at Nether Daugh, near Kintore (Appendix 1).

Although the imprint of glaciation remains dominant, postglacial processes have superimposed subtle, but distinctive, modifications on the landscape. Steep hillsides have been modified by landslips, soil-creep and debris flows, whereas valley floors have been sculptured by rivers forming spreads of alluvium and river-terrace deposits. In general, braided rivers with gravelly beds would have been replaced by the single-thread, and locally meandering, rivers of the present day by the mid-Holocene (McEwen, 1997). The Spey is an exception in that it still maintains a braided, gravelly floodplain in its lower reaches (Lewin and Weir, 1977).

Sea level was below that of the present day during the early Holocene, but the Main Postglacial Transgression resulted in the formation of raised beaches and marine inundation of the lower reaches of valleys. The latter led to the deposition of estuarine silts, fine-grained sands and clays, locally on terrestrial peats, tree stumps and fluvial muds (Chapter 7). The transgression reached its maximum after 6.1 ka BP in the Ythan valley (Smith et al., 1999; Figure 48) and between 6.3 ka and 5.7 ka BP in the Philorth valley, between Fraserburgh and Peterhead (Smith et al., 1982; Appendix 1 Site 9). The subsequent regression, which has been attributed to a general lowering of global sea level following expansion of the Antarctic ice cap (Goodwin, 1998), was accompanied by renewed terrestrial sedimentation in the valley mouths. Horizons of fine-grained sand identified in both estuarine and near-shore terrestrial deposits of the lower Ythan and Philorth valleys were laid down at about 7200 BP by a tsunami caused by a major submarine slide on the Norwegian continental slope (Dawson et al., 1988; Long et al., 1989; Appendix 1 Philorth).

PREVIOUS MODELS OF GLACIATION

The systematic study of glacial deposits and landforms in north-east Scotland began in the mid-19th century with the pioneering work of Thomas Jamieson. During the following 150 years of painstaking research there have been numerous attempts to unravel the complicated record of glacial events. Although the glacial reconstructions that have been published are generally based on sound evidence, they were inevitably influenced by the contemporary knowledge of glacial processes and events. The conflicts that have arisen result largely from the limited stratigraphical record of events in the region and the general absence of materials that can be reliably dated, a situation that still exists today.

The work of Jamieson

In a series of papers stretching over a period of fifty years, Jamieson developed modern concepts of Quaternary glaciation from previous theories of the great biblical submergence and erosion by floating ice (Jamieson, 1858, 1859, 1860a, b, 1862, 1865, 1866, 1874, 1882a, b, 1906, 1910). He published a wealth of stratigraphical detail establishing sequences of glacial events and reconstructed former flow lines from the distribution of erratics, striae, and orientations of *roches moutonnées*. He was one of the first to realise that the glaciation of Scotland had involved the formation of a major ice cap like that of Greenland rather than Alpine-style valley glaciers.

In Jamieson's final model (Figure 36) the oldest glacial event involved ice from the Moray Firth laying down 'shelly indigo boulder clay' in the Ellon district. The deflection was interpreted to have been caused by Scandinavian ice impinging on the eastern coast of Buchan. The indigo diamicton was largely stripped away by ice flowing from inland during a subsequent event, which laid down the locally derived, non-shelly 'Lower Grey Boulder Clay' of Aberdeenshire. That till was subsequently overlain by the 'Red Clay Series' along the coast to the north of Aberdeen. These vivid reddish brown deposits with Old Red Sandstone erratics had been deposited by ice flowing

northwards from Strathmore, but deflected onshore by Scandinavian ice in the North Sea. Jamieson noted that the Red Clay Series intermingled with, and thus was contemporaneous with, the 'Dark Blue Boulder Clay' in the Peterhead area. The latter till contained many large erratics derived from Mesozoic rocks in the Moray Firth. Local ice had retreated far to the west during this incursion, although the Lower Grey Boulder Clay was regarded as broadly similar in age to the red and dark blue coastal tills. There was local marine incursion following the decay of the coastal ice streams leading to the deposition of clays. The final event involved an expansion of ice from inland. During this 'Aberdeen Re-advance', a glacier descended the valley of the River Dee to reach the coast, where it scoured away the Red Clay Series deposits locally.

The work of Bremner and Read

The mantle of Jamieson was taken over by Alexander Bremner in the early 20th century. He also published a wealth of detail on the Quaternary of north-east Scotland, paying particular attention to the pattern of supposed ice-marginal glacial meltwater channels formed during glacial retreat (Bremner, 1912, 1915, 1916, 1917, 1919, 1920a, b, 1921, 1922, 1928, 1931, 1934a, b, 1936, 1938, 1939, 1942, 1943). Bremner developed the triple glaciation model, although he disagreed

Figure 36 Patterns of ice flow deduced by Jamieson (1906).

a Deposition of Indigo Till
b Deposition of Lower Grey Boulder Clay
c Deposition of Dark Blue Boulder Clay and Red Clay Series
d Aberdeen Re-advance

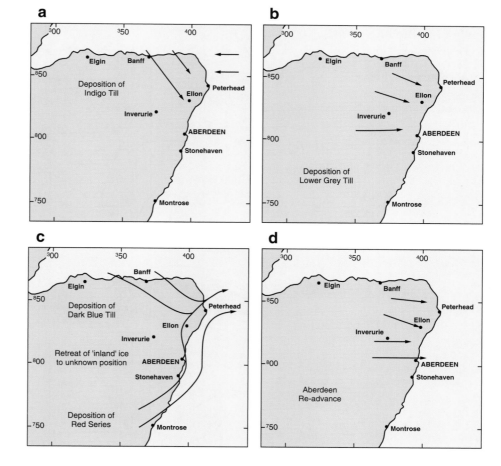

with Jamieson about the sequence of events and many details. Unlike Jamieson, Bremner claimed to have found evidence confirming that the main glacial episodes were separated by warm, nonglacial periods.

In Bremner's final model (Figure 37), the first glaciation involved ice sourced in Sutherland, Ross and Cromarty and the Great Glen flowing into the region from the west and north-west. It picked up Mesozoic rocks and shells from the Moray Firth and laid down dark blue shelly till along the northern coast. The Moray Firth ice stream coalesced with Scandinavian ice off Buchan and the combined ice masses flowed southwards leaving a train of boulders of Peterhead Granite and sparse erratics of rhomb-porphyry and larvikite. Peat preserved in the vicinity of the Burn of Benholm site (Appendix 1) supposedly formed in the subsequent interglacial period.

Ice traversed Buchan from the south-west and south in Bremner's second glaciation. He used evidence cited by Read (1923) to show that it flowed northwards into the Moray Firth along the northern coast. Ice from Strathmore deposited red tills along the eastern coast and laminated clays were deposited in freshwater lakes ponded against the landward margin of that ice stream during its retreat.

Tills resulting from Bremner's third glaciation were supposedly difficult to identify owing to the glacial reworking of older deposits. Patterns of ice flow were determined mainly from the orientation of supposed ice-marginal drainage channels and other morphological evidence. During this glaciation, the 'Aberdeen Re-advance' was supposedly contemporaneous with another advance of Moray Firth ice that affected the northern coast as far east as Fraserburgh. Widespread glaciofluvial deposits and morainic deposits in the lower Dee valley were attributed to the Aberdeen Re-advance and another spread at Dinnet, in the Tarland Basin, was attributed to a subsequent re-advance.

Bremner disagreed with H H Read regarding the evidence for the number of glacial episodes that had affected the northern coast. However, he acknowledged the careful observations that Read (1923) had made during the detailed geological survey of Banffshire, in particular Read's wide knowledge of the source rocks from which the glacial erratics were derived. Read deduced that the earliest ice-movement to affect the Banffshire coast came from the west and north out of the Moray Firth, conditioned by the presence of Scandinavian ice at the mouth of the firth. Glaciolacustrine deposits of his 'Coastal Series' formed in lakes following the partial withdrawal of the Moray Firth ice stream. Inland ice then 'crept' northwards towards the coast, redistributing erratics, but depositing little recognisable till.

Figure 37 Patterns of ice flow deduced by Bremner (1943).

a First ice sheet
b Second ice sheet
c Aberdeen Re-advance

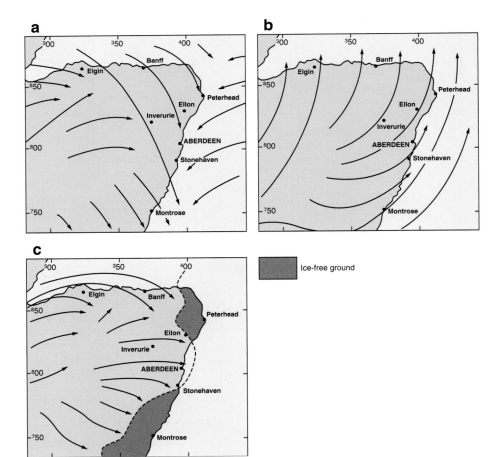

Ice-free ground

Figure 38   Patterns of ice flow deduced by Synge (1956) (after Clapperton and Sugden, 1977).

a   Greater Highland Glaciation
b   Moray Firth-Strathmore Glaciation
c   Aberdeen Re-advance
d   Dinnet Re-advance

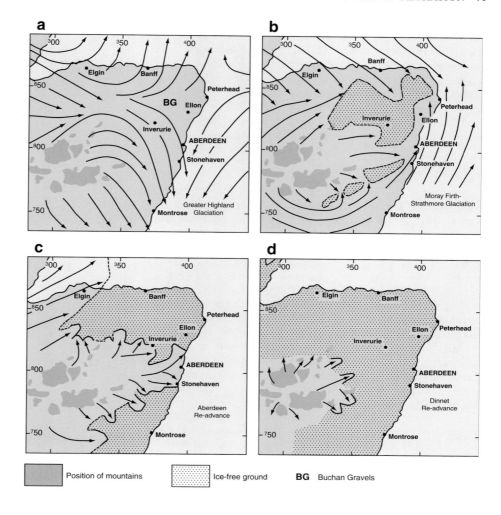

Position of mountains   Ice-free ground   **BG** Buchan Gravels

Work in the post-war period

Detailed, accurate stratigraphical work in the Aberdeen area was continued by Simpson (1948, 1955). He demonstrated that the moraines and meltwater channels attributed by Bremner (1938, 1943) to the Aberdeen and Dinnet re-advances of his third glaciation were simply produced during periods of stagnation (still-stands) that interrupted the decay of contemporaneous coastal and inland ice masses. Simpson (1955) support-ed Jamieson (1865, 1906) in that he believed the dark shelly tills of the northern coast were contemporaneous with the red tills of eastern Aberdeenshire, an interpre-tation disputed earlier by Bremner (1943).

Charlesworth (1956) and Synge (1956) both concluded that a substantial portion of the Buchan Plateau had not been glaciated during the last glaciation. This resulted in the concept of 'Morainless Buchan'. They supported Jamieson (1865, 1906) believing that ice from the Moray Firth and Strathmore had coalesced in the vicinity of Peterhead while there was no ice immediately inland. Charlesworth concluded that many valleys became blocked by ice at the coast causing a series of proglacial lakes to form. These lakes were supposedly interconnect-ed by glacial spillways. The concept of ice-free enclaves

during the Devensian and earlier glaciations was support-ed by the recognition of widespread saprolites, weathered tills and periglacial phenomena (FitzPatrick, 1958, 1972; Galloway, 1961a, b, c).

Synge (1956, 1963) recognised two major glaciations in his model (Figure 38), with two re-advance stages occur-ring during the decay of the final ice sheet. Ice first moved into Buchan from the east or north-east, leaving Scandina-vian erratics along the coast. There followed a 'Greater Highland Glaciation' (Wolstonian or Anglian) in which the Scandinavian ice caused Scottish ice to bifurcate over Buchan, thus preserving the Buchan Gravels in an enclave of minimal glacial erosion. Extensive weathering of bedrock supposedly occurred in the region during the succeeding interglacial period. Ice reoccupied the coasts during the second (Devensian) glaciation, but a substan-tial area of central Buchan purportedly remained ice free. The decay of the Strathmore ice lobe was accompanied by the deposition of estuarine red clays. Two short cold inter-vals lead to re-advances of Grampian ice at Aberdeen and Dinnet, the latter being correlated incorrectly with the Loch Lomond Re-advance of western Scotland.

Sissons (1965, 1967) placed Synge's Greater Highland Glaciation in the Devensian and correlated the Aberdeen and Dinnet events with his Aberdeen–Lammermuir and Perth re-advances, respectively (Figure 39). He later abandoned the concept of the first re-advance and agreed that there was no irrefutable evidence for the latter at Perth (Paterson, 1974; Sissons, 1974b, c). Peacock et al. (1968) proposed that another 'oscillation' occurred in the Elgin area and that earlier in the deglaciation, ice-dammed lakes developed agaist the Moray Firth ice stream along the Banffshire coast.

Late 1960s to the early 1980s

The evidence for multiple glaciations and active, phased retreat of the last ice sheet became seriously questioned during this period of research. This stemmed from the general recognition that complex depositional sequences could result from rapidly changing processes and environments within single glacial events. Most drift deposits in the region were assigned to the last glaciation (Murdoch, 1975; McLean, 1977; Clapperton and Sugden, 1975, 1977), although doubts remained about its timing and the main directions of ice flow (Sissons, 1981).

The apparent spatial continuity of the pattern of glacial landforms and deposits, particularly of glaciofluvial origin, led Clapperton and Sugden (1975, 1977) to conclude that the entire area of north-east Scotland had been covered by the Late Devensian ice sheet (Figure 40). They concluded that the ice sheet was largely wet-based, and it was composed of three distinct regional ice streams with a triple junction zone located over central Buchan. The ice sheet decayed comparatively quickly leading to widespread stagnation in many lowland sites and no major re-advances were recognised. Central, northern and eastern Buchan became free of ice 'early' in the deglaciation, allowing time for proglacial lakes to form against the more persistent coastal ice streams and for previously deposited sediments and saprolites to become periglacially modified.

It was widely held that the whole of Scotland together with the northern and central North Sea basin was glaciated during the Late Devensian (e.g. Sissons, 1976; Boulton et al., 1977; Price, 1983).

Early 1980 to 2002

The discovery that stiff, stony lodgement tills were absent beneath most of the northern North Sea basin had a profound influence on modelling work during the 1980s for it was generally assumed that this indicated that Scandinavian ice had not crossed the North Sea during the Late Devensian glaciation (e.g. Cameron et al., 1987). This led to Sutherland's (1984a) reconstruction of an independent British ice sheet of restricted size, in which ice terminated at the eastern margin of the Wee Bankie Formation off the eastern Scottish coast (Figure 3d). Sutherland (1984a) also cited evidence from the Crossbrae and King Edward sites (Appendix 1) to argue that the northern coast of Banffshire and Buchan to the east of Portsoy had not been glaciated during the Late Devensian. The nonglaciated enclave he proposed differed from that envisaged by Synge (Figure 38) and enclaves of various extent have appeared

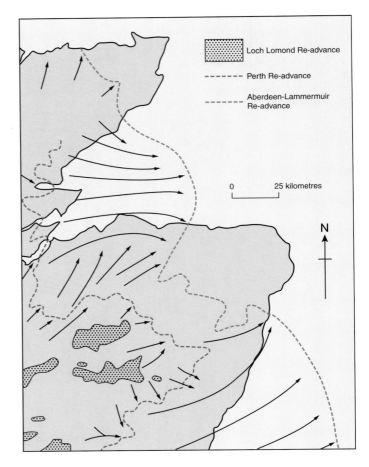

Figure 39 Pattern of glacial re-advances following the Main Late Devensian maximum, when Scottish and Scandinavian ice sheets coalesced, as deduced by Sissons (1967).

on subsequent reconstructions of the north-eastern sector of the last British ice sheet (e.g. those shown in Figure 3 and Bowen et al., 1986; Sejrup et al., 1987; Nesje and Sejrup, 1988; Lambeck, 1993, 1995), some long after Sutherland's evidence had been seriously questioned (Peacock, 1985; Hall and Bent, 1990; Hall, 1997). Ehlers and Wingfield (1991) also recognised an unglaciated enclave in Buchan, but suggested Scottish ice expanded farther east than the Wee Bankie Moraine, coalescing in places with Scandinavian ice (Figure 3f).

Over the past twenty years, field-related research into the glaciation of north-east Scotland has been undertaken predominantly by the present authors and contributors. Modified versions of the single stage Clapperton and Sugden model for the last glaciation have been adopted in recent publications and maps of the Geological Survey on north-east Scotland over this period (e.g. Merritt, 1981; BGS, 1992). However, although stratigraphical evidence now appears to confirm that the 'inland', Moray Firth and Strathmore ice streams were confluent, both Hall (1983, 1984) and Hall and Connell (1991) reverted to a two-stage model similar to that originally proposed by Jamieson (1906).

This was because there was conflicting geochronometric dating evidence for the age of the 'inland' tills in Buchan.

The whole of Scotland was ice-covered during the first phase of Hall's 1983 model, but ice from Strathmore reached only as far as Stonehaven. Ice flowing down the Dee valley and Insch depression (Wilson and Hinxman, 1890) crossed the coast between Aberdeen and the Ythan, depositing basic igneous erratics at Nigg Bay (Appendix 1). In Buchan, the pebbles of the distinctive Buchan Gravels from Whitestones Hill and of erratics from the quartz-biotite norite of the Maud intrusion indicates flow towards the north-east and east (Wilson, 1886). 'Inland' ice may have been confluent with Moray Firth ice along the valley of the North Ugie Water. Ice flowed north-eastwards across Banffshire to join the Moray Firth ice stream at the coast, west of the mouth of the Deveron (Read, 1923).

The East Grampian ice sheet retreated to an unknown position in the west during the second phase. Strathmore ice flowed northwards until it met the more vigorous Moray Firth ice stream, which diverted 'inland' ice east-wards into the lower Spey valley (Read, 1923) and pushed across northern parts of Buchan.

New information from offshore allowed the model to be refined (Hall and Bent, 1990; Hall, 1997). The maximum limit of the main Late Devensian ice sheet was placed at the Bosies' Bank Moraine (Figure 44), where a tide-water margin was identified some 25 km north-east of Buchan. East Grampian ice had already retreated before the maximum offshore extent was attained, allowing incursion of Moray Firth ice across northern Buchan. This timing of maximum expansion was not known, but deglaciation was well advanced by 15.3 ka BP, when ice had retreated from the Buchan coast (Appendix 1 St Fergus). Glaciomarine sedimentation was also occurring to the west of Bosies' Bank at the retreating ice margin at this time (Bent, 1986).

In the last decade or so it has been demonstrated that low gradient ice sheets crossing wet, deformable beds, especially in marine areas, are unlikely to form thick stony lodgement tills. Instead, fine-grained diamictons (deformation/deforming bed tills) are produced that can be mistaken easily for undeformed glaciomarine deposits. Additionally, some diamictons are too thin to be detected on seismic records. This knowledge, together with new AMS ^{14}C dates on individual molluscs and hand-picked foraminiferids, led Sejrup et al. (1994) to return to the view that the British and Scandinavian ice sheets had coalesced in the northern North Sea basin during the early part of the Late Devensian, between 28 and about 22 ka BP. This agreed with the revised interpretation of the Dimlington type locality in Yorkshire (Eyles et al., 1994). The central part of the northern North Sea was deglaciated by 20 ka BP, followed by a re-advance in the eastern North Sea between 18.5 and 15.1 ka BP. A similar re-advance probably affected north-east Scotland during this period, when ice possibly reached the limits that Sejrup et al. (1987) used in their earlier reconstruction of the main Late Devensian maximum (Figure 41). Peacock (1997) argued that the glacitectonic disturbance of the St Fergus Silts (Appendix 1) could have resulted from a

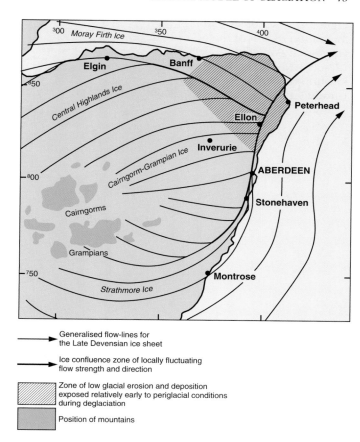

Legend:
Generalised flow-lines for the Late Devensian ice sheet

Ice confluence zone of locally fluctuating flow strength and direction

Zone of low glacial erosion and deposition exposed relatively early to periglacial conditions during deglaciation

Position of mountains

Figure 40 Pattern of ice flow deduced by Clapperton and Sugden (1977).

subsequent re-advance of the Moray Firth ice stream between 15 and 14 ka BP.

Recent micromorphological examination of sediment thin sections from boreholes has enabled Carr (1998) to demonstrate that ice crossed the North Sea basin on three occasions during the Devensian–Weichselian. Sejrup et al. (2000) conclude that it did so during the Karmoy, Skjonghelleren and main Late Weichselian glaciations of Fennoscandia (Figure 43).

The glacial record in Banffshire has been re-appraised by Peacock and Merritt (2000). Ice first flowed south-east-wards from the Moray Firth during the first phase of the Late Devensian glaciation, as originally deduced by Read (1923). The Moray Firth ice stream then withdrew from the coast, allowing ice from the East Grampians to creep northwards until being steered eastwards by the more powerful coastal ice-stream. The inland ice subsequently withdrew, allowing glacial lakes to form against the Moray Firth ice stream, which continued to remain active and made forays southwards from time to time (Figure 42).

PRESENT MODEL OF GLACIATION

The glacial reconstructions described above are clearly conflicting, and some of their aspects are certainly flawed,

Figure 41 Reconstruction of the maximum extent of the Main Late Devensian–Weichselian ice sheets (28–22 ka BP) showing possible location of former ice streams (modified from Sejrup et al., 2000).

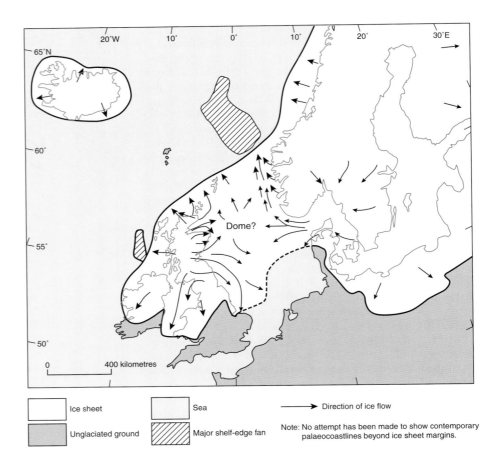

but they all contain elements that remain valid. It is salutary that the stratigraphical approach of Jamieson has perhaps best survived the test of time. The sequences he described have generally been confirmed in recent site excavations and most of his deductions about the origins of deposits presented in his final paper are sound. Although there is still doubt about the time span involved, the basic tripartite sequence he established in the Ellon area still holds, and his evidence supporting the Aberdeen Re-advance has not been rejected beyond doubt.

Bremner's final model was not so firmly based on stratigraphy, but his evidence for incursion of Scandinavian ice is compelling. Although some of the meltwater channels that he thought were formed at the retreating margin of the last ice sheet contain pre-Late Devensian deposits, and hence are older features, most indicate the pattern of ice retreat that he deduced. Conversely, the evidence does not support Clapperton and Sugden's (1977) model of a regionally integrated, contemporaneous network of channels forming beneath a warm-based ice sheet. Nevertheless, the continuity of channel development is strong evidence in support of full ice cover during the last glaciation of the region. The pattern of ice movement in Clapperton and Sugden's model is in broad agreement with available stratigraphical evidence, although it has to be modified in eastern Buchan to take into account the evidence of west to east glacial streamlining (Map 6), rather than north-west to south-east. Furthermore, their flow lines for the Strathmore ice stream

do not account for the significant onshore carriage of Permo–Triassic lithologies into eastern Buchan.

Extent of glaciation

Until recently, the tills occurring in central Buchan could only be loosely dated as Devensian. This allowed repeated claims, discussed above, that parts of the area remained ice-free during the Main Late Devensian glaciation (Figure 3c, d, e). The evidence for so-called 'Moraineless Buchan' was equivocal, but the discovery in 1980 of the Crossbrae Peat with its interstadial pollen record soon become central to the debate (Appendix 1 Site 5). The peat appeared to rest on till deposited by ice flowing from the west and it was overlain by gelifluction deposits, but not by till. Furthermore, radiocarbon dates implied that the peat formed early in the Late Devensian. The site seemed to provide crucial evidence that 'inland series' tills in the area predated the Late Devensian and that central Buchan had indeed not been glaciated during that period (Sutherland, 1984a; Connell and Hall, 1987). However, on re-examination in 1992, the peat was found to rest on bedrock and to be overlain by coarse gravels of possible glaciofluvial origin (Whittington et al., 1998). Furthermore, new dating and palaeoenvironmental evidence from the peat indicate that it was formed most probably during the Early Devensian in OIS 5c or 5a (Table 7). The coarse gravel unit is probably, but not unequivocally Late Devensian in age.

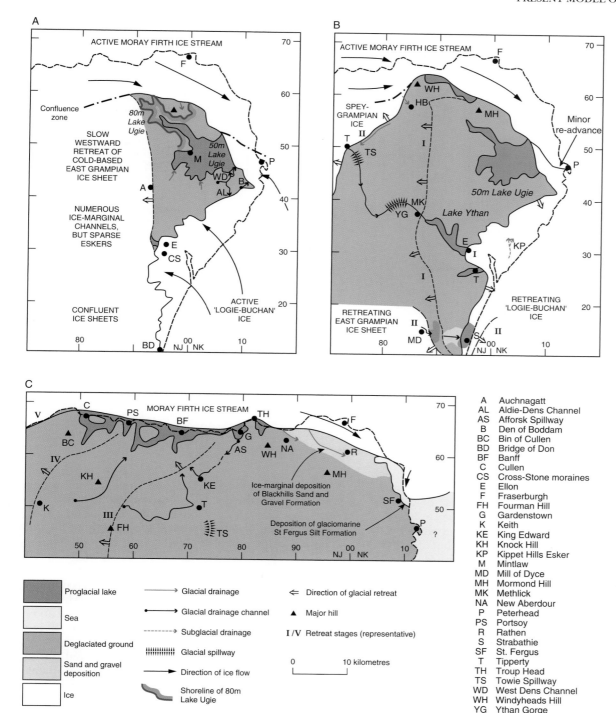

Figure 42   Tentative reconstructions of former proglacial lakes in north-east Scotland.

a   Creation of '50 m' Lake Ugie shortly after the maximum of the second major expansion of the Main Late Devensian ice sheet (after 18 ka BP) and following the earlier ponding of the '80 m' lake. Widespread glacial over-riding of glaciolacustrine deposits occured east of Lake Ugie.
b   Parting of East Grampian and 'Logie-Buchan' ice with the formation of Lake Ythan.
c   Diachronous ponding along the Banffshire coast and glaciomarine incursion around St Fergus at about 15 ka BP

Figure 43 Devensian–Weichselian events in Britain and south-west Fennoscandia (after Peacock and Merritt, 1997; Sejrup et al., 2000). Norwegian data from Baumann et al. (1995) and Mangerud et al. (1981). D/L amino-acid ratios corrected according to Miller and Mangerud (1985). British data from Baker et al. (1995), Bowen (1989), Gordon et al. (1989), Lawson and Atkinson (1995) and Miller et al. (1987).

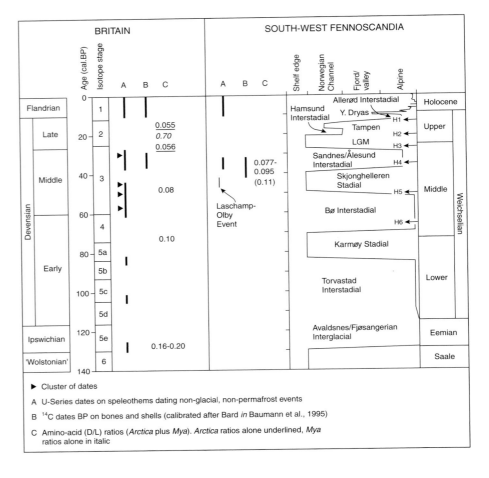

Evidence from Crossbrae can no longer be used confidently to support the argument that parts, or all, of Buchan escaped glaciation in the Late Devensian. Recent mapping confirms that the entire area is crossed by a system of glacial meltwater channels that formed diachronously during the general westward retreat of ice (Chapter 7). Several lines of reasoning suggest that this was the Main Late Devensian ice sheet.

1 Time-dependent amino-acid epimerisation data on reworked shells in the Whitehills Glacigenic Formation at Boyne Limestone Quarry, Gardenstown and King Edward Quarry (Appendix 1   Sites 2, 3, 4) suggest that this unit and overlying tills can be attributed either to a Middle Devensian (OIS 3) glacial event equivalent to the Skjonghelleren glaciation of Norway (Figure 43), or to the Late Devensian (Peacock and Merritt, 1997).

2 Full ice cover in Buchan during the Late Devensian is implied by the projection of ice surface profiles from terminal moraines of probably Late Devensian age in the Outer Moray Firth (Hall and Bent, 1990) and by evidence that the Scottish and Scandinavian ice sheets coalesced in the northern North Sea between 28 and 22 ka BP (Sejrup et al., 1994, 2000).

3 The existence of raised glaciomarine silts and a raised beach at St Fergus (Appendix 1   Site 11), radiocarbon dated to 14 to 15 ka BP (Hall and Jarvis, 1989; Peacock, 1997), implies the previous removal of a thick ice cover, because considerable isostatic depression would be required for such sediments to be laid down during a period of low global sea level.

4 Glaciofluvial gravels beneath till of westerly origin at the Howe of Byth (Appendix 1   Site 6) have yielded luminescence dates of 45 ± 4 ka and 37 ± 4 ka BP (Hall et al., 1995b), although doubts have been expressed about the validity of these dates (Peacock and Merritt, 1997).

It is concluded that all of Buchan was glaciated during the Late Devensian. Furthermore, apart from nunataks in the north-west Highlands and Inner Hebrides (Ballantyne et al., 1998), it is unlikely that ice-free areas existed anywhere in Scotland during the Main Late Devensian glaciation, including Caithness and the islands of Orkney and Lewis (Hall and Whittington, 1989; Hall, 1995). A return to earlier models (e.g. Sissons, 1976; Boulton et al., 1977, 1985) of extensive ice sheet glaciation in the Late Devensian is suggested, with Scottish ice terminating at, or close to, the continental shelf to the north and west of Scotland and covering most of the northern North Sea basin (Hall and Bent, 1990; Sejrup et al., 1998, 2000). This view is supported by the recent reconstruction of the last ice sheet in north-west Scotland (Ballantyne et al., 1998), based on the detailed mapping of glacial trimlines, and assuming low gradient profiles for ice streams crossing deformable sediments.

Multistage models of the Main Late Devensian glaciation

The detailed record of climate change derived from the Greenland ice sheet (Figure 31) together with the discovery of various horizons of ice-rafted debris (Heinrich events) in North Atlantic cores, has led to increased scrutiny of the evidence for Late Devensian glaciation in Britain. It had been widely accepted that the last major glaciation of the Northern Hemisphere peaked at about 18 ka BP (Climap, 1976), and evidence from the type locality of the Dimlington Stadial in Eastern Yorkshire appeared to support a glacial maximum between 18 and 16 ka BP (Rose, 1985). However, subsequent Late Devensian amino-acid ratios on shells from the supposed pre-Ipswichian Basement Till at Dimlington indicate that the maximum extent of the main Late Devensian glaciation in eastern England occurred before 18.4 ka BP (Eyles et al., 1994). Several surges of coastal ice followed. As explained above, Sejrup et al. (1994, 2000) have presented evidence that the maximum Late Devensian glaciation also occurred before 18 ka BP in the northern North Sea. Further support for an 'older' Late Devensian maximum is provided by a ^{14}C date of 22.5 ka BP for glaciomarine deposits associated with the maximum position of the independent Outer Hebrides ice cap near the edge of the continental shelf (Selby, 1989). Similarly, Peacock (1997) argued for a re-advance of Scottish ice into the western North Sea between 15 and 14 ka BP.

Supporting evidence from Ireland and the northern Irish Sea basin also suggests that maximum glaciation occurred early in the Late Devensian (by about 22 ka BP), followed by stillstands and at least one re-advance that has been age-constrained by AMS radiocarbon dating on foraminiferids (McCabe, 1996; McCabe and Clark, 1998). Following a period of ice retreat between 16.7 and 14.7 ka BP (Cooley Point Interstadial), ice re-advanced, reaching its maximum position at about 14 ka BP (Killard Point Stadial), which is roughly coincident with Heinrich Event 1 (McCabe et al., 1998).

Arguably, the north-west Fennoscandian sequence of Sejrup et al. (2000; Figure 43) provides the best yardstick against which events in north-east Scotland can be judged. However, there are outstanding difficulties requiring further work and clarification. For example, Olsen (1997) concludes that Scandinavian ice advanced on to the continental shelf four times between 40 and 14.5 ka BP, with ice advances peaking at about 40, 30–29, 24–21 and 17–14.5 ka BP. Furthermore, radiocarbon ages on bone fragments at the type locality of the Hamnsund Interstadial suggest that it occurred at about 24.5 ka BP and not at about 20 ka (Mangerud et al., 1996). There are also difficulties with the radiocarbon dates for the Swatchway and Witch Ground formations in the Fladen area and the correlation of these units with the Wee Bankie and Tampen formations (Carr, 1998). In essence, ice might have been present in the Witch Ground area until almost 13 ka BP. It is speculated here that this ice could be the remains of a previously more extensive

ice dome (Figure 41). The dome had possibly expanded in size during the 'Dimlington Advance' event (18 to 13 ka BP) causing the deflection of ice from Strathmore into the lower Ythan valley to lay down the deposits of the Logie-Buchan Drift Group. The speculative ice dome shown on Figure 41 is based on evidence of a more or less radial pattern of channels (incisions) of Late Weichselian age in the Witch Ground area (Ehlers and Wingfield, 1991).

Another consequence of accepting a more widespread glaciation between 28 and 22 ka BP is that there must have been considerably more glacio-isostatic depression of the land in north-east Scotland, and hence higher relative sea levels, than determined by Lambeck (1993b) from numerical, geophysically based modelling.

Likely sequence of events during the Late Devensian in north-east Scotland

Early stage of the Main Late Devensian glaciation (28 to 22 ka BP)

The district was probably overwhelmed entirely by ice in the early part of this period (Hall and Bent, 1990; Peacock and Merritt, 1997; Whittington et al., 1998; Figure 41). Initially, ice would have flowed from centres in the western and north-western Highlands, followed by the establishment of an independent ice cap over the Cairngorms and eastern Grampian Highlands. This probably began prior to the Late Devensian. A major ice stream occupied the Moray Firth where it would have flowed relatively swiftly across the underlying soft, deformable, muddy sea bed and Mesozoic sedimentary rocks. Another Highland-sourced terrestrial ice stream flowed north-eastwards along Strathmore towards the North Sea and a further one flowed down the Spey Valley towards the Moray Firth (Sutherland and Gordon, 1993). The interaction between the relatively fast-moving and probably warm-based ice streams and the relatively sluggish, and probably largely cold-based, East Grampians ice sheet, has profoundly influenced the distribution of drift deposits in the district, leading locally to complicated sequences (Peacock and Merritt, 2000; Figure 4). Each ice stream produced distinctive suites of deposits, forming the basis of the five lithostratigraphical groups described in Chapter 8.

Many of the terrestrial deposits of this early phase of glaciation would have been eroded or reworked during subsequent phases. Evidence is preserved that indicates that the Moray Firth ice stream invaded the coastal lowlands of Moray, Banffshire and Buchan to lay down the dark grey shelly tills and rafts of Mesozoic strata, the main components of the Whitehills Glacigenic Formation. The ice possibly crossed the Buchan Plateau towards Aberdeen during this event (Hall and Jarvis, 1995), but if not, it did so in a pre-Late Devensian glaciation, laying down 'indigo' tills around Ellon (Table 7). On all occasions, the deflection of Moray Firth ice across Buchan was probably caused by the presence of Scandinavian ice in the central North Sea.

It is clear that following the deposition of the Whitehills Glacigenic Formation at Boyne Bay, Gardenstown and up the Deveron valley, the power of the Moray Firth ice stream waned sufficiently to allow ice sourced inland to flow eastwards towards the coast and to lay down the Old Hythe, Crovie, Hythie, Byth, Sandford Bay and laterally equivalent tills (Table 7). Although there is no apparent unconformity in the sequence, there was possibly a significant hiatus accompanying the change in direction of flow, perhaps an interstadial event coincident with this glacial reorganisation.

A mid-Late Devensian interstadial? (22 to 18 ka BP)

Judging from the north-west Fennoscandian record (Sejrup et al., 2000; Figure 43), it is likely that there was substantial ice-sheet recession in Scotland at about 20 ka BP during a cold, dry interstadial period. The initial retreat of the East Grampian ice sheet from central Buchan had probably occurred by then, with the development of ice-marginal lakes between it and the coastal ice streams (Figure 42). Deposits of the Ugie Clay Formation were laid down in these lakes. At about this time, the lower reaches of the valley of the River Deveron north of Turriff were blocked by the Moray Firth ice stream causing meltwaters to flow south and east, first via the Towie spillway, near Fyvie, then into the valley of the River Ythan (Figure 42).

Later stages of the Main Late Devensian glaciation (18 ka to 13 ka BP)

There is growing evidence elsewhere that a significant build up of ice occurred during this period, but there is no unambiguous record of any substantial re-advance of the East Grampian ice sheet. There is evidence that the Moray Firth ice stream expanded across Lake Ugie, because a thick sequence of laminated clay (Ugie Silt Formation) is capped by blue-grey deformation tills (Essie Till Formation) inland of Peterhead (McMillan and Aitken, 1981; Chapter 8). The Logie-Buchan ice also advanced over glaciolacustrine deposits (Appendix 1 Site 12 Sandford Bay), and the moraines at Cross Stone, near Ellon, indicate a late re-advance onshore from the North Sea (Map 6). The position of the Cross Stone moraine and the composition of the Logie-Buchan Drift Group deposits associated with it, suggest that Scandinavian ice was present in the North Sea basin, causing ice derived from Strathmore to flow northwards, then westwards and north-westwards penetrating onshore. However, the central North Sea, at least, is thought to have been ice free at this time (Sejrup et al., 1994). If correct, the Moray Firth and Strathmore ice streams possibly advanced to the position of the Bosie's Bank and Wee Bankie moraines (Figure 44),

Figure 44 Tentative reconstruction of ice margins at the maximum stage of the second major expansion of the Main Late Devensian ice sheet (after Hall and Bent, 1990 and Sejrup et al., 1987). This stage is correlated with the maximum of the 'Dimlington Advance', 18.5–15.1 ka BP (Sejrup et al., 1994).

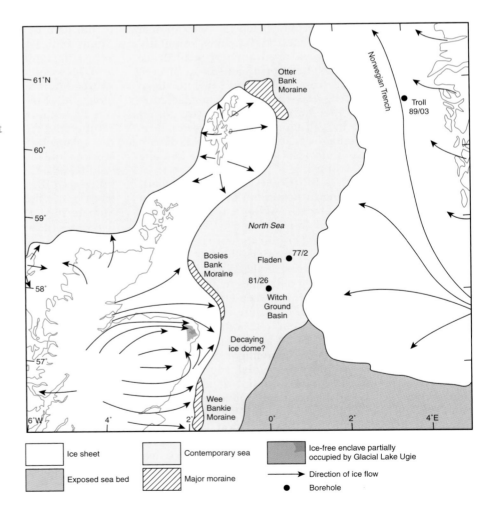

equivalent to the Tampen Re-advance of north-west Fennoscandia (Sejrup et al., 1994).

The Moray Firth ice stream possibly stood at the entrance of the Moray Firth at about 15 ka BP, where its activity disturbed the deposits of the St Fergus Silt Formation (Appendix 1 Site 11). There was substantial ponding of meltwaters between the slowly retreating East Grampians ice sheet and the Moray Firth ice stream. The glaciolacustrine Kirk Burn Silt Formation was deposited in these lakes along the Banffshire coast and its hinterland (Figure 42). Local glacial readjustments led to a minor re-advance of the East Grampians ice sheet towards the Banffshire coast following the retreat of the Moray Firth ice stream (Peacock and Merritt, 2000). A subsequent oscillation of the Moray Firth ice stream affected the area around Elgin, possibly at about 14 ka BP, and a fourth one formed a prominent push moraine at Ardersier, near Inverness, possibly at about 13 ka BP (Merritt et al., 1995). This event has since been correlated tentatively with the Killard Point Stadial at about 14 ka BP (McCabe et al., 1998). Deglaciation of the whole district was complete 13 ka (Clapperton and Sugden, 1977).

In the Aberdeen area, substantial ponding occurred between the slowly retreating East Grampian ice sheet and ice offshore (Thomas and Connell, 1985). The lakes gradually expanded southwards as the two bodies of ice 'un-zipped' in that direction (Simpson, 1954; Aitken, 1991; Appendix 1 Strabathie, Mill of Dyce).

On mainland Scotland, the eustatic lowering of sea level caused by the abstraction of water to form the continental ice sheets was more than offset by the isostatic depression of the land. Relative sea level was probably high during the periods of deglaciation, making the coastal ice streams particularly vulnerable to rapid retreat by the process of iceberg calving (Stewart, 1991). The resulting glaciomarine deposits and shorelines formed during deglaciation have been subsequently raised above sea level (Chapter 7). The St Fergus Silt Formation, dated to about 14 to 15 ka BP, was laid down when sea level stood at least as high as 12 m OD (Figure 48). The Spynie Clay Formation, associated with raised beaches lying at about 30 m above OD near Elgin, has yielded an arctic fauna and is comparable with the Errol Clay Formation of the Firth of Tay (Chapter 8). It was laid down in arctic conditions preceding the Windermere Interstadial (Peacock, 1999).

SIX

Quaternary deposits

This chapter provides a systematic description of the drift deposits depicted on the eleven sheets covering the district (Figure 1). The deposits are classified here following the current BGS morpho-lithogenetic scheme (Figure 5). Names of units vary from sheet to sheet, but some of the most common synonyms are given in Table 3 for comparison. Some deposits on the more recently published maps have also been classified lithostratigraphically and descriptions of these units are given in Chapter 8. Further descriptive accounts may be found in the older memoirs covering specific sheets, and in various sand and gravel assessment reports (*Information sources*). Geomorphological features are described in Chapter 7. The engineering properties of glacigenic deposits are described in Chapter 2.

A set of generalised maps showing the distribution of drift groups and a selection of deposits, landforms and localities mentioned in the text is provided at the back of this publication (Maps 1 to 11).

GLACIAL DEPOSITS

TILL

Tills are the most widely distributed of the glacial deposits. They crop out over much of the district and also occur beneath younger superficial deposits. Tills are composed mainly of *diamictons,* materials that are characterised by a lack of sorting (in the geological sense). They are mostly matrix supported, dense, cohesive, commonly unstratified and comprise a mixture of rock fragments, gravel, sand, silt and clay (Table 6). Not all tills are diamictons and not all diamictons are tills. For example, many diamictons were formed as cohesive debris flows and mudslides, not necessarily in a glacial environment: others were formed by periglacial processes involving repeated freezing and thawing. The mapped tills probably include some of these nonglacial diamictons because it is not practical to delineate them separately.

Tills consist of ice-transported material laid down *subglacially* at the base of active glaciers and ice sheets, *proglacially* and *supraglacially* at the margins of retreating ice sheets, and *paraglacially* soon after glaciation when sediment deposited earlier is remobilised. Deposits formed in the proglacial, supraglacial and paraglacial environments commonly occur at the surface and comprise a metre or so of heterogeneous, very poorly sorted, crudely stratified, gravelly diamicton that is commonly intercalated with gravel, silty sand, silt and clay. These sediments accumulated at ice-sheet margins, mainly as debris flows that were modified and redeposited by ephemeral meltwater streams and sheet wash. They are generally permeable and

include large boulders, up to several metres in diameter. In contrast, tills formed in the subglacial environment are relatively homogeneous (massive) and impermeable. In general, it is not practical to map out the various types of till separately. Locally, where proglacial, supraglacial and paraglacial deposits are several metres or more thick and form constructional mounds, they have been mapped out as 'hummocky glacial deposits' (see below).

The subglacially formed diamictons include lodgement tills and deformation (deforming-bed) tills, although the two types may be difficult to distinguish, resulting as they do from a continuum of processes. There is commonly a sharp boundary between deformation till and underlying units of pervasively sheared and deformed sediment or weathered bedrock. The amount of deformation in such units, which are now commonly referred to as *glacitectonites* (Table 6), generally decreases downwards. Some sandy deformation tills are weakly stratified on a scale of a few millimetres to a few centimetres and the stratification is formed of laterally impersistent laminae and wisps of pale, very fine-grained sand and silt. The presence of these laminae may indicate that some resedimentation in running water has occurred at the ice–till interface during melt-out, in which case the deposit may be classified as a *melt-out till* (Boulton, 1970; Shaw, 1979). However, the stratification can also result partially, if not wholly, from subglacial shearing and associated comminution of granular material (Boulton, 1979; Boulton et al., 1974). True melt-out and ablation tills are rare because they have poor preservation potential (Paul and Eyles, 1990).

Tills vary considerably in thickness and lithology across the district where they tend to underlie poorly drained ground of low relief and smooth slopes (Plate 8). The tills of the district are described in terms of the five lithostratigraphical groups to which they are assigned (Figure 4). The groups relate to the major bodies of ice responsible for depositing the sediments (Chapter 8). However, the groups have been recognised only on the more recently published 1:50 000 maps. Diamictons belonging to different groups interdigitate locally, especially towards the coasts. The following account describes tills that are most likely to be encountered at the surface and in shallow excavations. They are predominantly Late Devensian in age, although much of the material forming them has probably been reworked from older glacial deposits and weathered bedrock. Information on the sparse, but stratigraphically important occurrences of older till is contained within Chapter 8 and Appendix 1.

Central Grampian Drift Group

This group of deposits, which mainly lies to the west of the River Spey (Figure 4), includes some of the thickest tills in

Table 6 Commonly occurring types of till and related sediments.

Classification	Genesis	Typical lithology	Common attributes
Lodgement till	Formed beneath actively moving glaciers as a result of frictional retardation of debris particles and debris-rich ice masses against the glacier bed (Boulton and Deynoux, 1981)	Extremely stiff, very stony, sandy, clayey diamicton with matrix support. Little stratification, but commonly with platy structure. Clasts typically less well dispersed than in deformation tills	Small boulders with bevelled and striated surfaces. Subhorizontal fissures becoming more pronounced upwards. Concavo-convex discontinuities and shear planes lined with silt, clay or silty fine-grained sand. Fissure fillings commonly ferruginous due to passage of groundwater
Deformation till (Deforming-bed till)	Formed by the disaggregation and homogenisation of sediments and weak rocks in the subglacial 'deforming layer' (Boulton, 1987; Hart and Boulton, 1991; Benn and Evans, 1996, 1998)	As above, but more variable depending on nature of parent material. Clayey sediments commonly yield unstratified silty clays with well-dispersed pebbles. Far-travelled lithologies may be sparse	As above, but with deformed inclusions and laminae of sand and decomposed rock that range in size from a few centimetres to large glacial rafts many tens of metres across. Sharp, planar, basal contacts with underlying penetrative glacitectonites
Glacitectonite	Subglacially sheared and deformed sediment or weathered bedrock (Benn and Evans, 1996, 1998)	Materials where primary features are replaced by tectonic lamination ('penetrative' glacitectonite)	Lamination parallel to planar base of overlying deformation till. Strain decreasing downwards. Gradational basal contacts with non-penetrative glacitectonites below
		Materials retaining some original sedimentary structures or igneous/metamorphic fabric ('non penetrative' glacitectonite)	Various extensional structures e.g. boudins, low-angle shears, conjugate microfaults, brecciation and folding. Strain decreasing downwards
Flow-till complex	Formed as cohesive debris flows at the ice margin, amongst decaying ice or paraglacially from remobilised glacial deposits (Boulton, 1968; Boulton and Paul, 1976)	Friable, sandy, matrix-supported diamicton interbedded with pebbly silty sand, clast-supported diamicton, and gravel and thinly laminated silt and clay	Individual beds generally less than 50 cm thick (typically 5 to 20 cm), laterally impersistent with gradational contacts. 'Fold noses' may bound individual flows

the district. Sequences of glacigenic deposits with individual till units up to 15 m or so in thickness are common, and similar to successions described in detail nearer to Inverness by Fletcher et al. (1996). Tills occurring at the surface to the north of a line roughly between Rothes and Portsoy are predominantly reddish brown to brown sandy lodgement tills derived from sandstones and conglomerates of the Old Red Sandstone Supergroup cropping out to the west (Peacock et al., 1968; Aitken et al., 1979). However, a significant proportion of the clasts in these diamictons are 'Moine' (Grampian Group) psammitic granulites and quartzites together with minor amounts of local Permo–Triassic rocks, granite from Nairnshire, quartz porphyry and Jurassic limestone derived from the bed of the Moray Firth.

East Grampian Drift Group

Tills of this group are generally less than 5 m thick. They become patchy and are rarely thicker than 2 m across

central Buchan. They are generally very sandy because they incorporate much decomposed, grussified rock (Chapter 4). Many are deformation tills derived largely from local decomposed rock, but with well-dispersed pebbles derived from farther afield. Rubbly, clast-supported diamictons up to about 1.5 m thick are common where the ice has overridden relatively fresh, quartzitic rocks. The colour, matrix and clast composition of the tills of this group, more than any other, reflect the nature of the underlying bedrock, or rocks cropping out within a kilometre or so 'up-glacier' (generally west or south-west). They are generally yellowish brown in colour, but reddish brown, orange-brown and olive-grey tills occur locally depending on the colour of the local bedrock. The tills are pale grey to white, kaolinitic and contain many well-rounded pebbles of quartzite, vein-quartz and flint where ice has crossed outcrops of the Buchan Gravels Formation (Chapter 4). The tills are very gritty inland from Aberdeen, where they include much decomposed

Plate 8   Till.

a   Subglacial till of the Banchory Till Formation containing angular cobbles of granite and Dalradian metamorphic rocks, exposed in the bank of the Cowie Water [NO 8421 8845] (Z00410). Scale: compass shown in top right

b   Mill of Forest Till containing rounded cobbles of Devonian rocks at its type locality [NO 8620 8540] (Z00411). Scale: lens cap in centre

c   Till of the Banffshire Coast Drift Group at Oldmill [NK 0245 4407], showing downward transition from weathered to unweathered diamicton. (P104106)

d   Till of the Banffshire Coast Drift Group overlying cryoturbated gravel with vertical clasts [NK 0130 5491], near Kirkhill. (P104107). Section shown is about 1.5 m high granitic and psammitic bedrock (Figure 11). Where medium- to coarse-grained igneous rocks have been over-ridden, there are numerous boulders strewn across the surface that were probably formed as 'core-stones' within bedrock weathering profiles before being glacially trans-ported.

Banffshire Coast Drift Group

The tills of this group are typically clayey, bluish grey to brown in colour and contain erratics, fossils, microfossils and shell fragments derived from the Moray Firth, as well as more local rocks. They crop out extensively at the surface only between Fraserburgh and Peterhead, but they occur patchily at depth beneath tills of the Central and East Grampian Drift groups between Elgin and Aberdeen (Figure 4). Thicknesses of 10 m or more are common. In the Elgin area, the tills are typically dark olive-grey to greyish brown and in addition to psammitic granulites and quartzites, include clasts of calcareous sandstone, glau-conitic sandstone, black shale and shelly limestone from offshore (Peacock et al., 1968; Aitken et al., 1979). Fossils of Rhaetic to Early Cretaceous age are common. These dark tills of the Elgin area are almost certainly equivalent to diamictons within the Whitehills Glacigenic Formation, which has been recognised from Cullen eastwards (Chapter 8). Clasts of Jurassic mudstones and shales are common in these latter deposits, whereas the matrix has been formed partly from comminuted Mesozoic shales and partly from reworked Quaternary marine deposits from which the comminuted shell fragments have been derived. Fragments of Lower Cretaceous rocks and fossils are common in tills that outcrop farther east, towards Fraserburgh, where they include sparse clasts of chalk.

The diamictons of the Whitehills Glacigenic Formation are mostly deformation tills and glacitectonites. They commonly include lenses (boudins) and partially disaggre-gated masses of a range of Permo–Triassic, Jurassic and Cretaceous lithologies and Quaternary sediments (Plate 17). They include dark grey to black mudstones, shales, clays and lignite, and yellowish brown fine-grained sands. Lenses of white, friable, kaolinitic sandstone are found sporadically in the west of the district, whereas vivid reddish brown and orange marls occur to the east of Troup Head. The lenses of all the aforementioned rock-types range in size from a few centimetres up to large glacial rafts several hundreds of metres across (Chapter 7). The latter have been confused with bedrock in some site investigations. Contact relationships between the rafts and enclosing diamictons are locally highly complex and in places confusing owing to considerable glacitectonic dis-turbance (Appendix 1 Boyne Limestone Quarry and Gardenstown).

Logie-Buchan Drift Group

This group underlies the coastal lowlands to the north of Aberdeen, east of Ellon and south of Peterhead (Figure 50). It comprises a complex sequence of relatively poorly consolidated, thinly interbedded, vivid reddish brown, clayey, calcareous diamictons, waxy clays, muds,

sands and gravels. A few metres of stiff, stony lodgement till commonly occur at the base of the sequence, locally resting on yellowish brown lodgement tills of the East Grampians Drift Group. Otherwise the sequence is composed largely of laterally impersistent, and mainly cohesive, debris flow deposits (flow tills) that collectively may reach over 25 m in thickness, especially over hollows in the bedrock surface (Merritt, 1981; Munro, 1986) and close to contemporary ice sheet margins. The beds and lenses of sand within the sequence are typically fine to medium grained, silty and micaceous. Shell fragments are common. In addition to locally occurring rock-types such as amphibolite, psammite, quartzite and metagreywacke, the diamictons include rocks derived from the sea bed to the east, including dolomite, limestone, calcareous silt-stone, both white and red friable sandstones and red-stained, rounded quartzite pebbles from Old Red Sand-stone conglomerates to the south. Palynological analysis of the diamicton matrix has revealed complex assemblages, including Permo–Triassic pollen and Early Pleistocene dinoflagellate cysts (information from R Harland, 1980).

Mearns Drift Group

Tills of this group are invariably reddish brown in colour and contain a significant proportion of clasts derived from andesitic volcanic rocks, red sandstones, siltstones and mudstones of the Old Red Sandstone Supergroup cropping out in Strathmore. Unlike the Logie-Buchan Drift Group described above, the matrices are only weakly calcareous. Characteristically, they also contain rounded quartzite boulders, derived from Old Red Sandstone con-glomerates. These clasts of sedimentary and volcanic rocks predominate in tills developed within the outcrop of the Old Red Sandstone, but they become less numerous where the tills extend onto the outcrop of adjacent Dalradian and Highland Border Complex rocks. These latter diamictons, which commonly abut and, in places, interdigitate with glacigenic sediments of the East Grampian Drift Group, contain clasts derived from local Caledonian igneous and Dalradian metamorphic rocks in addition to Old Red Sandstone material.

The tills are generally less than 5 m thick and are typi-cally clayey, silty, pebbly diamictons that are matrix sup-ported. Extremely overconsolidated clayey and silty lodgement tills are developed abundantly at the base of thick glacigenic sequences. More friable, sandy diamic-tons, generally less than 2 m thick, occur as flow tills capping deltaic spreads of glaciofluvial sand and gravel. Both of these types of till resemble diamictons within the Logie-Buchan Drift Group, but the presence of offshore-derived shell fragments and exotic clasts of dolomite, limestone and calcareous siltstone, which are only present in these deposits, serve to distinguish them from diamictons within the Mearns Drift Group.

HUMMOCKY GLACIAL DEPOSITS

Hummocky glacial deposits form a distinctive sediment-landform association that has been identified on some of

Plate 9 Hummocky glacial
deposits: an extremely poorly sorted,
heterogeneous morainic deposit at
Blairdaff [NJ 7001 1797], near
Kemnay. The deposit comprises
bouldery gravel and sandy diamicton
that was laid down as debris flows at
the margin of the receding East
Grampian ice sheet. (P104108).
Scale: spade is 0.9 m long.

the more recently published maps. The deposits are highly
variable lithologically and include complex interdigitations
of matrix- and clast-supported diamicton, stratified and
unstratified silty boulder gravel, and lenses of sand, silt and
clay (Plate 9). Most are formed primarily of very poorly
sorted materials that were deposited during deglaciation.
They spilled from the ice-fronts as mud-slides and debris
flows, either subaerially, or into standing water. Some of the
deposits are constructional moraines formed at active ice
margins, but most examples in the district apparently
formed when large masses of ice stagnated. This situation
commonly occurred towards the margin of the East
Grampian ice sheet as it retreated westwards across
north–south-trending ridges. For example, sandy morainic
deposits derived largely from disaggregated granite occur
quite extensively on Sheet 76E Inverurie in the lee of the
Hill of Fare, between Banchory and West Cullerley
[NJ 766 030] (Map 8). These mounds are commonly
strewn with large boulders and blocks, but locally contain
better-sorted materials at depth.

Similar deposits occur within a col at Tillyfourie and
between the Hill of Fare and Cairn William, where they
form irregular mounds and ridges with intervening peat-
filled hollows (Map 8). These morainic deposits contain
a high proportion of silty, matrix-supported bouldery
diamicton as well as stratified sandy boulder gravel.
Those at Tillyfourie are well seen from the A994
Alford–Aberdeen trunk road.

Hummocky glacial deposits are also common within the
East Grampian Drift Group on Sheet 66E Banchory, where
ice stagnated in the lee of high ground (Map 10). They
occur extensively on the southern side of the Water of Feugh
catchment and on the interfluve between the Dee, Carron
and Cowie waters. Notable examples are present in the
valley of Burn of Greendams, south-east of Strachan and in
the valleys of the burns of Knock and Curran, to the south of
Banchory. Most of the deposits form boulder-strewn lateral

moraines, up to 10 m high in places, resting on granite
bedrock. They are composed of boulder gravel with a coarse
sandy matrix of disaggregated granite, derived from the
underlying bedrock. Good examples of lateral moraine
ridges, occur between Powlair [NO 621 912] and Greendams [NO 649 900]. Similar landforms are present on the
south-eastern side of the col between Craigbeg
[NO 770 913] and Cairn-mon-earn [NO 783 920]. They
were visible for many years from the A596 Banchory–Stonehaven trunk road (Slug Road), where it crosses the
Dee–Strathmore watershed, but most of these ridges are
now obscured by coniferous forest.

Few cross-valley retreat moraines have been recognised,
but a good example is the Lady's Moss Moraine
[NO 782 901] in the valley of the Black Burn, south of
Cairn-mon-earn (Auton et al., 1988). Other examples
include north-east-trending morainic ridges in the col
between Shillofad [NO 724 888] and North Dennetys
[NO 710 877], and a north-west-trending ridge up to 18 m
high, at Rouchan [NO 640 897] (Auton et al., 1990). The
cross-valley moraine at Lady's Moss and the morainic
ridges between Shillofad and Kerloch lie close to the limit
of the Aberdeen–Lammermuir Re-advance of Sissons
(1967; Figure 39). The former moraines occur at elevations of 170 to 200 m OD, the latter at 275 to 320 m OD.
They appear to mark still-stands in the late-stage retreat of
small glaciers in the cols rather than the limit of a regionally significant ice advance. Although the concept of
Sissons' Aberdeen–Lammermuir Re-advance is now
largely discredited, it was the recognition of morainic
features such as these that first led to its proposition.

Constructional moraines were formed locally at the
margin of the East Grampian ice sheet, especially where it
'actively' retreated from high ground and bedrock ridges
trending at right angles to ice flow. In central Buchan,
this situation led to the cutting of groups of north–south-
trending meltwater channels (Chapter 7), but the ice was

locally sufficiently active to form low morainic ridges parallel to the channels. Good examples of such ridges occur on Sheet 76E Inverurie in the vicinity of Burnhelvie [NJ 724 198] and Moss-side [NJ 710 183] (Map 8).

Push moraines appear to be rare in the district, but to the south of Ellon, two concentric ridges in the vicinity of Cross Stone Wood [NJ 955 278] are examples (Figure 50; Map 6). The ridges stand up to about 8 m high and are apparently formed mainly of sand and gravel with a capping of red diamicton. The ridges are asymmetrical in cross-section, with steeper slopes facing south-westwards. They were probably formed at the margin of the coastal ice stream that was responsible for laying down the deposits of the Logie-Buchan Drift Group. Other moraines of this type lie at the southern end of the Den of Boddam [NK 102 408], where the coastal ice pushed northwards towards the Hill of Longhaven (Map 7). The form of all these features strongly suggests that the coastal ice was pushing towards ice-free ground.

GLACIOFLUVIAL AND GLACIOLACUSTRINE DEPOSITS

GLACIOFLUVIAL DEPOSITS

Glaciofluvial deposits are sediments laid down primarily by waters issuing from ice sheets and glaciers. The source of the water also includes rainfall and run-off from ice-free slopes as well as melting ice. Two main categories of deposit have been distinguished on the basis of their geomorphology and shown on all the 1:50 000 maps of the district, but the name applied to these categories varies from map to map (Table 3). On the more recently published maps, moundy deposits are classified as *ice-contact deposits* whereas terraced spreads are termed *sheet deposits*. The two categories are described in general terms below, before details are given. As glaciofluvial deposits are commonly laid down some distance away from former ice margins, the link between ice-source and deposit is less strong than in tills, and it is therefore more practical to describe their areal distribution rather than by lithostratigraphical group.

Glaciofluvial ice-contact deposits

Glaciofluvial ice-contact deposits consist mainly of sand and gravel, but include subsidiary beds of diamicton, silt and clay (Plate 10). The sediments were laid down in supraglacial, englacial, subglacial and ice-marginal drainage systems. Hummocky topography is characteristic, but flat-topped plateaux are also included. Steep-sided ridges of gravel (*eskers*) are shown individually on the 1:50 000 geological maps where space permits, whereas rounded hillocks of sand and gravel (*kames*) and flat-topped mounds (*kame-plateaux*) are not delineated from other, more irregularly shaped mounds and undulating spreads. Ice-contact deposits are typically associated with *kettleholes*, which are the result of buried masses of ice melting out slowly after deposition of the sediment surrounding and overlying them. Steep slopes that were formerly in contact with the ice are commonly still recognisable. The kames and kame-plateaux

are composed typically of laminated muds that coarsen upward into sands and then gravelly deposits that formed as *fan-deltas* in ephemeral *ice-marginal lakes*. Lenses of diamicton and large boulders are common, having been deposited as debris flows, slurry and falls directly from the ice surface onto the accumulating glaciofluvial sediments.

The moundiness of ice-contact deposits is mainly the result of postdepositional collapse of the sediment bodies once the ice supporting them had melted. The sediments forming such mounds are commonly offset by normal faults. Reverse faults commonly occur where sediments have sagged into basinal structures (McDonald and Shilts, 1975). As glacier-ice rarely becomes totally stagnant as it melts out, many ice-contact deposits have been 'nudged', and in places more severely deformed, by minor glacial advances. Deposits formed at such active ice fronts are commonly disrupted by thrust faults and have been glacially overridden locally, leading to compaction and the deposition of till. Sequences such as these can be very complicated and therefore difficult to appraise and exploit commercially as aggregates.

The district does not include many large eskers and those that do occur are generally isolated, low-lying ribbon eskers that were laid down by subglacial meltwaters at retreating ice margins (Chapter 7). Eskers are formed mainly of gravel, but it is commonly very coarse and bouldery.

Glaciofluvial sheet deposits

Glaciofluvial sheet deposits were laid down mainly by braided streams in a proglacial environment leading to the accumulation of fan-shaped, elongate bodies and spreads of 'outwash' sand and gravel (*sandur*). Distinct terraces formed when parts of the outwash plain, or fan, were abandoned as streams graded to lower levels. Although sandar deposits generally become less coarse and more sandy downstream from the ice margin, many sheet deposits in the district coarsen upwards, which indicates that they accumulated during still-stands or minor glacial re-advances. Many sheet deposits pass upstream into glacial drainage channels.

Kettleholes do occur within sheet deposits, but they generally are not as large or as common as in the ice-contact deposits. The sheet deposits are also relatively more homogeneous and dense, less disrupted by faults and glacitectonic dislocations, contain fewer lenses of till, fewer large boulders and are consequently relatively easy to appraise and exploit commercially as aggregates.

DESCRIPTIONS OF GLACIOFLUVIAL DEPOSITS BY GEOLOGICAL SHEET

The resource potential of the glaciofluvial deposits is described more fully in Appendix 2.

SHEET 95 ELGIN

A large part of this sheet is underlain by glaciofluvial deposits that form significant resources of sand and gravel. The deposits are varied, both in composition and distribution, and the observations given below are based on the more detailed descriptions of Peacock et al. (1968) and Aitken et al. (1979).

a

b

c

d

Plate 10   Glaciofluvial sand and gravel.

a   Deltaic foresets dipping eastwards, capped by poorly stratified topset gravel at Lochinvar pit [NJ 185 613], near Elgin (P104109). Scale: spade is 0.9 m long

b   Climbing ripple-drift cross-lamination in deltaic sands at Kirkmyres pit [NJ 926 609], near Fraserburgh. (P104110). Face shown is about 2 m high

c   Channel-fill cross-bedded gravels forming glaciofluvial sheet deposits in Danshillock pit, south of Macduff (D6152). Hammer is 40 cms long

d   Topset gravels overlying sandy deltaic foresets in the Lochton Sand and Gravel Formation, Cammie Wood sand and gravel pit, Feughside (D4987)

Most of the deposits west of the Spey valley are moundy and were laid down during deglaciation at the margin of ice occupying the Moray Firth as it retreated west-north-westwards. Meltwaters draining that ice, together with streams draining the partly deglaciated areas to the south, were constrained to flow around the retreating margin, creating deep drainage channels such as the Black Burn valley, to the west of the River Lossie, and the Blackhills channel, south of Lhanbryde (Chapter 7; Map 1). The Spey valley would also have functioned as an overflow channel. Much of the sand and gravel lying between Elgin and the Spey was laid down as fans at the mouths of these, and other minor channels, or as deltas in temporary lakes held back by the ice cap (Plate 10a). Good examples of deltaic sand and gravel laid down by southeastward flowing meltwaters occur at Lochinvar Pit, near Elgin and at Rothes Glen pit [NJ 254 527], south of Elgin (Map 1).

Many of the deposits around Elgin were laid down on remnants of stagnant glacier ice that later melted to form kettleholes and brought about the generally moundy topography. The deposits of glaciofluvial sand and gravel tend to coarsen upwards, but to fine north-eastwards away from each source. Kameplateaux amongst the mounds are commonly formed of finegrained glaciolacustrine deposits capped by sand and gravel. Sand predominates to the north of the A96 trunk road, around Elgin and Lhanbryde, and it forms particularly large mounds towards the coast, for example at Binn Hill [NJ 305 656].

The glaciofluvial deposits to the north and north-east of Elgin become less moundy towards the coast where those occurring below about 35 m above OD appear to have been wave-washed locally. These undulating spreads are formed mainly of sand, silt and clay with subordinate gravel, and they merge with Late-glacial raised beach deposits, at about 30 m above OD, around Kinloss and Loch Spynie, indicating contemporaneous glaciofluvial and marine sedimentation at the former ice front (Chapter 7). To the east of the Spey, two gently undulating to moundy spreads of sand and pebbly sand lie between Fochabers and Buckie.

A distinctive suite of deposits lie between Kinloss and Lossiemouth where several east-north-east-orientated ridges, with flat tops descending towards the east-north-east, are separated by depressions. The sand and gravel forming the ridges tends to coarsen upwards into boulder gravel and the sequences are locally capped by till. The depressions are underlain by silts and clays, and parts of the ridges are formed of these lithologies too, commonly displaying complex softsediment gravitational deformation structures. The sands and gravels forming the ridges were probably laid down within icewalled chasms at the ice margin, which possibly connected with the sea. The silts and clays were deposited following retreat of the ice margin westwards. A similar style of sedimentation, possibly ice-proximal glaciomarine, has been proposed for the Ardersier Silts and Alturlie Gravels formations of the Banffshire Coast Drift Group occurring farther to the west (Merritt et al., 1995; Fletcher et al., 1996).

A prominent feature of the lower Spey valley is a series of gravelly terraces that formed during the whole period between the retreat of the last glaciers and the present day. The river floodplain and low-lying alluvial terraces are deeply incised into a set of older late-glacial ones. The most extensive 'Mosstodloch' terrace, and higher terraces to the south of Mosstodloch, are regarded as glaciofluvial because they contain kettleholes or merge into moundy glaciofluvial deposits (Map 1). Lower terraces are probably mainly of Loch Lomond Stadial age, or younger. The Mosstodloch terrace merges downstream into moundy ice-contact deposits around Garmouth, indicating that ice still lay to the north-west when the terrace formed. Ice probably completely blocked off the lower reaches of the Spey

valley downstream of Fochabers at an earlier stage, causing a large lake to be ponded up — the 'Fochabers Glacial Lake' (Peacock et al., 1968). Thinly laminated glaciolacustrine clays underlie parts of the Trochelhill and Balnacoul glaciofluvial terraces in the vicinity of Fochabers.

Sheet 96W Portsoy and Sheet 96E Banff

The glaciofluvial sands and gravels occurring on these sheets have been divided into two assemblages. The predominantly moundy deposits inland of Cullen, Portsoy, Whitehills and Banff have been placed in the Blackhills Sand and Gravel Formation of the Banffshire Coast Drift Group (Chapter 8). Deposits in the valley of the River Deveron, upstream of Bridge of Alvah, are not divided, and are placed in the East Grampian Drift Group, together with terraced deposits in the valleys of the burns of Boyndie, Durn, and Deskford (Map 3).

The Blackhills Sand and Gravel Formation includes shell fragments and sparse clasts of sedimentary rocks and fossils derived from the floor of the Moray Firth in addition to rocks occurring locally, or farther to the west and south-west. Many of the well-rounded clasts in these sediments have been derived from Old Red Sandstone conglomerates. Some deposits accumulated as outwash fans at the margin of ice retreating westwards. Other deposits accumulated as deltas in lakes held up by glacier ice that remained offshore (Figure 42), abutting the coast. A reference section occurs in Brandon Howe pit [NJ 667 637] within the Hills of Boyndie, south-west of Banff, where over 12 m of sand with minor lenses of gravel and widespread seams of silt was formed in a distal outwash fan that prograded south-eastwards. Sands and silts displaying soft-sediment deformation due to loading and dewatering are exposed in a small pit at Tipperty [NJ 671 608] (Map 3; Plate 11).

Plate 11 Convolute lamination in tectonised glaciolacustrine silts from the Blackhills Sand and Gravel Formation, Tipperty sand pit, south-west of Banff. (D6140). Hammer is 32 cm long.

Degraded glaciofluvial terraces lie within several of the valleys, particularly those of the Deveron and the Burn of Boyndie. Unlike the Blackhills Sand and Gravel Formation deposits, which were laid down mainly by meltwaters flowing south or south-eastwards from ice situated to the north, these deposits were laid down by meltwaters flowing towards the Moray Firth. The deposits therefore contain clasts derived from the south, including basic igneous rocks that are commonly weathered and friable. Such deposits are exposed at Danshillock pit (Map 3; Plate 10c).

SHEET 97 FRASERBURGH

About one third of this sheet is underlain by glaciofluvial and glaciodeltaic deposits, which occur mainly in a broad swathe between New Aberdour and Loch of Strathbeg (Map 4). Apart from fragmentary terraces in the valley of the North Ugie Water, most of the deposits have been assigned to the Blackhills Sand and Gravel Formation (Chapter 8). These deposits contain clasts and shell fragments (including *Nuculana pernula*) derived from the floor of the Moray Firth, in addition to locally occurring clasts. Fragments of porous limestone and spiculitic sandstone of Cretaceous age are common. Both moundy and terraced spreads are equally represented on the sheet.

The moundy deposits were laid down as fans and deltas within lakes dammed up against ice of the Moray Firth ice stream to the north (Figure 42b). Typical deltaic coarsening-upward sequences of sand and gravel occur at two pits in the vicinity of Blackhills [NJ 924 612] (Plate 10b), the type locality of the formation, and at Pitnacalder [NJ 872 628], near New Aberdour. Deposition was mainly from flow directed towards the south-east. Several of the moundy sediments were glacitectonically disturbed during, and shortly after deposition, by the ice to the north, which remained active and locally re-advanced to lay down a local capping of till (Merritt et al., 2000, fig. 26). One of the larger moundy deposits form the Sinclair Hills, south of Fraserburgh (Map 4).

There are two major spreads of terraced glaciofluvial sand and gravel on the sheet. Both were formed by braided meltwater streams flowing east or south, and then towards the sea, which, at the time, stood no higher than 5 m above OD (Peacock, 1997). Meltwaters also flowed south-eastwards into the valley of the North Ugie Water to lay down terraces there, and others formed the upper part of an outwash fan at Todholes [NJ 843 570], where an important series of events has been established at Howe of Byth pit (Appendix 1).

SHEET 86E TURRIFF

The glaciofluvial deposits on this sheet are mainly restricted to the degraded terraces within the valley of the River Deveron north of Turriff, the 'misfit' valleys of the Idoch Water and Burn of King Edward, and the upper reaches of the Ythan upstream of Fyvie (Map 5). It is likely that the present northward drainage of the Deveron was re-established only at a late stage in the deglaciation of the district. Before then, while ice still occupied the Moray Firth and the coast, meltwaters discharged southwards into the Ythan system via a major spillway at Towie [NJ 746 440]. Substantial terraced deposits of sand and gravel were laid down around Turriff, and underlie the site of the town's new sewage works. An excavation into a high terrace there revealed imbricate gravels deposited by southerly flowing water. Meltwaters also flowed southwards through the Afforsk spillway, south-east of Gardenstown (Map 4), where they fed into the catchment of the Burn of King Edward. Extensive terraces occur at the lower end of that valley, sloping towards the valley of the River Deveron. An isolated, but prominant terrace lies to the north of New Byth, at Tippercowan

[NJ 815 551], whereas several north-north-east or south-south-west orientated kames and eskers lie between the Deveron, north of Turriff, and Aberchirder.

SHEET 87W ELLON

Glaciofluvial deposits are not common on this sheet, especially the moundy variety. This is a paradox, because there is extensive evidence of the activity of glacial meltwaters in the form of drainage channels (Map 6). Furthermore, many of the deposits that do exist were laid down by meltwaters entering the district from the north, via the valleys of the River Ythan and North Ugie Water, at a relatively late stage in the deglaciation of the region. Part of the explanation for the apparent scarcity of glaciofluvial deposits may be that the meltwaters laid down sands and gravels in the deeper valleys where they are now concealed beneath alluvial, solifluction and gelifluction deposits.

Glaciofluvial terraces lie within the valley of the Ythan, downstream of Methlick, and in the valley of the North Ugie Water, downstream of Strichen. Some of the deposits of the Ythan are locally very coarse indeed, especially at Bellmuir [NJ 875 365] and Ardlethen [NJ 920 320], and might have been deposited during catastrophic flood events (Maizels and Aitken, 1991). The terraces in both valleys were partly built up by meltwaters entering the main valleys from tributaries, to the north and west, during westward retreat of the East Grampian ice sheet. The gravels in both valleys tend to become finer downstream, where the lower lying terraces rest on extensive fine-grained glaciolacustrine deposits and descend to the level of the main postglacial beach. Downstream of Ellon, a former valley of the River Ythan lies to the south of the present one and is largely filled with glaciofluvial deposits (Chapter 7). Glaciofluvial terraces also occur in the catchment of the South Ugie Water, in the vicinity of Stuartfield and Old Deer.

The deposits described above have all been assigned to the East Grampian Drift Group, together with a solitary east-north-east-trending esker near Auchorthie [NJ 923 523]. Other moundy glaciofluvial deposits are situated close to the boundary of the Logie-Buchan Drift Group deposits in the south-east part of the sheet. This boundary is most clearly defined to the south of Ellon in the vicinity of Cross Stone, where a pit [NJ 9523 2822] reveals about 5 m of poorly bedded gravel becoming more sandy upwards. The clasts include gneisses, quartzites, psammites, micaschist and granites, which are all typical of deposits derived from the East Grampian ice sheet, and the bedding appears to indicate an easterly palaeocurrent. There are glacitectonic structures present that have been caused by ice pushing from the east, and the deposit is locally capped by red till. It appears that the deposit formed at the margin of the 'Logie-Buchan' ice, which pushed inland to form the two arcuate, asymmetric moraine ridges that lie to either side of the deposit (Map 6).

Several moundy deposits occur at the boundary of the two groups in the vicinity of the Hill of Auchleuchries [NK 006 365]. However, the sandy deposits forming the hill (Auchleuchries Sand and Gravel Formation; Table 7) are capped by yellowish brown till of the East Grampian Drift Group, not red till, and at Tillybrex, 1.5 km to the south-south-west, at least 11 m of weathered gravels (Tillybrex Sand and Gravel Formation; Table 7) with no morphological expression, are capped by red till. The Tillybrex gravel was deposited by westward flowing meltwater. There are clearly complicated sequences preserved along the boundary of the two groups. For example, the gravels at Tillybrex almost certainly predate the Late Devensian, and other older glaciofluvial deposits crop out from beneath till in the vicinity of Leys pit [NK 004 524] and at Oldmill [NK 023 440] (Appendix 1 Kirkhill and Oldmill).

SHEET 87E PETERHEAD

The glaciofluvial deposits occurring on this sheet have been assigned to three groups (Map 7). Most of the deposits lie close to the boundaries of the former East Grampian ice sheet with the coastal ice streams that laid down the Banffshire Coast and Logie-Buchan drift groups (shown on the published map as inland, blue-grey and red series respectively). Extensive spreads of terraced sand and gravel lie at the confluence of the North Ugie and South Ugie waters, where they overlie up to 25 m of laminated silts and clays. The sediments were deposited by meltwaters flowing from the west into lakes held up against ice of the deflected Moray Firth ice stream to the east (Figure 42). This took place after the East Grampian ice sheet had begun to retreat westwards.

In the south part of the sheet, glaciofluvial deposits are commonly interbedded with red diamictons and silts of the Logie-Buchan Drift Group, but individual beds are generally too thin to map out, or they are concealed. Several, more extensive deposits occur around Hatton. A pit in the village [NK 054 371] revealed over 10 m of deltaic calcareous sand with seams of red diamicton beneath a drape of red clay. The sequence included a bed of brown diamicton possibly derived from East Grampian ice. A ridge of pebbly sand lying to the west of the village, at the boundary of the Logie-Buchan and East Grampian drift groups, is also capped by red till although the pebbles were probably derived from East Grampian ice. This body of sand coarsens upwards and is possibly similar to the deposit forming the Hill of Auchleuchries, on Sheet 87W Ellon, and both deposits rest on dark grey-coloured tills.

One of the best examples of a 'ribbon' esker in the entire district lies to the east of Meikle Loch (Plate 15). The Kippet Hills esker (Appendix 1) stands up to 15 m high and is partially buried beneath red diamicton. It appears to connect with three mounds of sand and gravel lying to the north. The easternmost mound at Knapsleask is flat-topped and formed as an outwash fan. The fan and esker were laid down by meltwaters flowing northwards through a lobe of ice that pushed onshore between Aberdeen and Peterhead (Figure 50). The deposits contain a significant proportion of dolomite, limestone and calcareous siltstone from the sea bed, together with shell fragments derived from the offshore Aberdeen Ground Formation.

Several isolated mounds of glaciofluvial and glaciodeltaic sand and gravel lie within the area formerly covered and weakly glaciated by the East Grampian ice sheet. Some of the deposits have poor surface expression and probably predate the last glaciation, like the one at Oldmill (Appendix 1). Others contain materials derived locally; for example, the deposit at Redleas [NK 093 430] is composed mostly of fragments of weathered granite, and gravelly deposits around the Buchan Ridge contain many rounded clasts derived from the Buchan Gravels Formation.

SHEET 76E INVERURIE

Glaciofluvial deposits are widespread on this sheet (Map 8). They were laid down mostly at the margin of the East Grampian ice sheet as it retreated westwards across the area. The pattern of retreat was greatly affected by the topography, especially by the high ground formed by Bennachie, Cairn William, Green Hill and the Hill of Fare in the west of the sheet. As the ice margin reached this crescentic ridge that reaches 300 to 500 m above OD, meltwaters carved out several major channels across cols and deposited sands and gravels on, and amongst, stagnant ice left stranded in the lee of the hills. Some of this stagnant ice blocked the engorged stretch of the Don valley upstream of Port Elphinstone, causing meltwaters to be diverted eastwards via another set of channels across the lower ridge lying between Kemnay and Inverurie. Water from

these channels debouched into the valley of the River Don in the vicinity of Kintore, giving rise to fans of sand and gravel. A similar ice blockage also occurred farther upstream, east of Castle Forbes [NO 622 191]. It caused meltwaters to be diverted northwards, where they cut a precipitous gorge into fresh granite bedrock called 'My Lord's Throat' [NJ 635 197].

A good example of one of the routes taken by meltwaters across the sheet can be demonstrated around Kemnay. Meltwaters passed through the col crossed by the A944 road at Tillyfourie, where they laid down moundy glaciofluvial gravels and cut channels through morainic deposits dumped there. The water passed through stagnant ice to the east of the col to lay down the gravels that now form eskers in the valley of the Ton Burn and beyond towards Kemnay. The main esker terminates to the north of Kemnay amongst moundy deposits of sand and gravel. From here, meltwaters passed through the deep channel at Tom's Forest and laid down the large fan of sand and gravel at Tavelty, just to the north of Kintore. At a slightly earlier stage in the deglaciation, meltwaters flowed directly eastwards via channels at Leschangie to lay down a sheet of sand and gravel to the south of Gauch Hill [NJ 787 151].

The gorge of the River Don upstream of Monymusk also served as a major conduit for meltwaters across the high ground. Initially, meltwaters flowed eastwards across the col (about 265 m OD) between Bennachie and Millstone Hill cutting channels through the morainic mounds in the valley bottom. At Pitfichie, they took a more direct route eastwards than the present river and laid down moundy deposits of sand and gravel around Blairdaff [NJ 703 180] (Appendix 1 Rothens). Other routes taken by meltwaters across the high ground led to the cutting of deep channels across several cols between the Hill of Fare and Green Hill [NO 636 098].

A similar pattern of westward ice retreat occurred in the south-eastern part of the sheet, where ice stagnated in the lee of the Hill of Fare. One route taken by subglacial meltwaters led to deposition of moundy deposits, including an esker, between East Finnercy [NJ 767 042] and Roadside of Garlogie. Another route is traced by the esker forming Horsewell Hillocks [NJ 779 019], west of Hardgate. The valley of the Corskie/Kinnernie Burn carried meltwaters that laid down moundy deposits of gravel, including eskers, upstream of Dunecht [NJ 754 091], and more sandy, terraced deposits downstream towards the Loch of Skene. Sandy glaciofluvial deposits also lie to the east of the loch, beneath alluvium, and they form an extensive spread in the valley of the Leuchar Burn downstream of Garlogie [NJ 782 057].

Glaciofluvial sands and gravels form mounds 5 to 15 m high and low-lying terraces around Torphins and Gallow Cairn [NO 639 010]. These deposits are associated with a major former route of glacial drainage along the valley of the Burn of Beltie. This drainage route continues south-eastwards, along the valley of the Burn of Canny, to meet the valley of the River Dee at Invercanny Waterworks, 2 km upstream of Banchory on Sheet 66E. Smaller mounds of gravel and sand south of Milltown of Campfield [NJ 648 004] almost block modern drainage along another former route of meltwater flow that also extends on to the adjoining Banchory sheet (see below).

The north of the sheet contains ground that is similar in many respects to that covered by the Ellon sheet, on which moundy glaciofluvial deposits are relatively sparse and terraced deposits are restricted to the main valleys. The valley of the River Urie was ponded up for a while during deglaciation by ice that blocked the valley of the River Don between Inverurie and Kintore. A large lake formed upstream of Inverurie in which deltas of sandy gravel were laid down. Many of these deltaic deposits now form terraces on either side of the Urie.

SHEET 77 ABERDEEN

Two contrasting suites of glaciofluvial deposits occur on this sheet (Map 9). Most have been assigned to the East Grampian Drift Group, but moundy red, silty sands and gravels of the Logie-Buchan Drift Group lie along the coast, extending inland for up to about 3 km to the north of Aberdeen. The latter deposits were laid down at the boundary between the East Grampian ice sheet and ice that pushed onshore from the North Sea. During deglaciation, the parting ('unzipping') of the two bodies of ice appears to have started in the north, leading to ponding of meltwaters between them in the ice-free enclave thus formed (Figure 42).

The deposits of the Logie-Buchan Drift Group in the north of the sheet include a ridge that extends from Drums [NJ 990 227] towards Newburgh, where it has been truncated by the River Ythan (Figure 50). A short fragment of the ridge lies on the northern bank of the Ythan at Waterside, and the meltwaters that formed it probably continued northwards to form the Kippet Hills esker at Slains, on Sheet 87E Peterhead. To the north of Drums the ridge is locally boulder strewn and probably formed at the landward margin of 'Logie-Buchan' ice. To the south, the ridge merges into a belt of moundy deposits formed of interbedded reddish brown sands, gravels, silts and diamictons. The belt arcs around low-lying ground at Balmedie [NJ 964 177], which appears to have been occupied by ice while the sequence was laid down.

Most of the glaciofluvial deposits on the sheet were laid down by precursors of the rivers Don and Dee. Those of the latter river occur mainly as terraces within the present valley, although deposits lying to the south of Torry indicate that the river formerly took a more direct route to reach the sea at Nigg Bay (Chapter 7; Figure 46). Meltwaters also flowed into the Dee valley from the north, but they did not lay down deposits quite as extensively as shown on the 1:50 000 map published in 1980. Likewise the belt of deposits lying between Cadgerford [NJ 835 056] and Blackburn, and between Craibstone [NJ 867 111] and Bankhead are also not as extensive as shown on that map (Appendix 2).

The River Don formerly took a direct route to the coast instead of turning southwards at Dyce. This was probably partly due to ice blocking the valley downstream of Dyce and partly because the river was flowing towards the sea at a higher level than today. Large resources of sand and gravel in the form of mounds and ridges formerly occurred between Corby Loch [NJ 925 145], a large kettlehole, and the coast at Blackdog [NJ 963 141], but many of the deposits have been worked out (Appendix 2). Gravels laid down from eastward flowing braided rivers predominate around Bishops' Loch and Corby Loch (Aitken, 1995, 1998), but many of the deposits to the east formed as deltas and consequently contain sequences that coarsen upwards from silty sands into coarse gravels. Particularly good sections in deltaic deposits were formerly exposed at Strabathie [NJ 955 136] (Appendix 1), in a pit that is now used as a landfill site.

A particularly large deposit of sand and gravel lies in the middle of the valley of the River Don in the vicinity of Liddell's Monument [NJ 869 153], where it has been exploited at Mill of Dyce pit (Appendix 1). This deposit formed as a fan delta at the margin of the East Grampian ice sheet when it was situated immediately to the west. The ice then retreated upstream and a lake was held back behind the deposits, which blocked the valley. Meltwaters entered the main valley from side valleys to form kame terraces, especially around Hatton of Fintray. The ice front probably stood in the vicinity of Straloch [NJ 860 211] at about the same time as the Mill of Dyce deposit formed. However, although terraced deposits of gravel were laid down between there and Newmachar, they are not as extensive as shown on the 1:50 000 map published in 1980 (Appendix 2).

SHEET 66E BANCHORY

The glaciofluvial deposits on the northern two thirds of the sheet have been assigned to the Lochton Sand and Gravel Formation of the East Grampian Drift Group, whereas those in the south belong to the Drumlithie Sand and Gravel Formation of the Mearns Drift Group (Chapter 8). The deposits of the former group were laid down at the margin of the East Grampian ice sheet as it retreated slowly westward (Map 10). Clasts of coarse- and fine-grained granite predominate, but gritty and micaceous psammite and schistose semipelite become more numerous in gravels in the valley of the River Dee, downstream of Banchory.

Much of the meltwater flowed towards the valley of the River Dee where substantial amounts of sand and gravel accumulated as kettled terraces. The thickest known deposits in the valley form a moundy spread in the vicinity of Park Pit, Drumallan [NO 797 978], where a minor still-stand occurred during ice retreat (Brown, 1993). Significant meltwater flow (directed towards the east and north-east) also took place to the north of the Dee valley, along channels linking linear ice-scoured basins. Moundy and terraced spreads of sand and gravel flank the basin north of Upper Lochton and others occur along the Bo Burn, south of the Raemoir Hotel; similar sands and gravels are also concealed beneath alluvial deposits.

Large amounts of meltwater entered the Dee valley from the south via the valleys of tributaries such as the Feugh and the Burn of Sheeoch. Once the ice front had retreated, meltwater drainage along the Dee valley was generally unrestricted. However, the valleys of the southern tributaries lay more or less parallel to the retreating ice margin and as a result drainage was impeded by masses of decaying ice. Meltwaters were ponded up and sequences of laminated silts, coarsening upwards into sand and cobble gravel, were laid down as fan-deltas. These deltaic sediments now commonly occur as kettled terraces and as flat-topped and irregular mounds. A most impressive kettled terrace (the Pitdelphin Wood Terrace), which stands up to 30 m above the level of the floodplain of the Water of Dye (Auton et al., 1990), occurs between Pitdelphin Farm [NO 654 912] and Bogarn [NO 657 903]. The upward coarsening glaciofluvial sand and gravel forming the terrace averages about 13 m in thickness and overlies glaciolacustrine silt and clay resting on till.

Large masses of stagnant ice were stranded in the lee of hills as the margin of East Grampian ice retreated westwards, giving rise to hummocky glaciofluvial and morainic deposits and many eskers. Eskers are most common along the southern side of the valley of the River Feugh, in the vicinity of Strachan [NO 674 923], and in the valley of the Burn of Sheeoch farther east; they are formed mainly of coarse, clast-supported gravel. A group of eskers that have an unusual south-east trend occurs on the northern side of the valley of the River Feugh (Map 10). They contain clasts of fine-grained basic igneous rocks, which are generally absent in the adjacent glaciofluvial gravels such as exposed in Cammie Wood pit (Plate 10d; Map 10).

The glaciofluvial deposits in the south of the area were formed mainly when the East Grampian ice sheet began to retreat north-westwards and the Strathmore ice stream retreated south-westwards. Meltwaters formed lakes between the two ice masses, which acted as sediment traps in which coarsening-upward deltaic sequences were laid down. At first the ponding took place high in the valleys to the south-east of the main watershed, for example, in the upper reaches of the Carron Water around Snob Cott [NO 801 885], where a considerable thickness of coarse-grained sand and gravel was laid down in contact with decaying ice. These deposits, assigned to the Lochton Sand and Gravel Formation, are characterised by clasts of granitic and metamorphic rocks. Ponding subsequently occurred at lower levels in minor valleys, such as those around Pitdrichie [NO 795 825], and the valley of the Burn of Caldcotts, upstream of Fettercairn, before the valleys of the Bervie Water and Luther Water were inundated. Deposition of sand and gravel derived from south-eastward flowing meltwaters from the retreating East Grampian ice sheet was concentrated around Auchenblae, where a large fan was laid down flanking the Luther Water.

Gravels of the Drumlithie Sand and Gravel Formation include moundy ice-contact sands and gravels interbedded with reddish brown flow-tills, which were laid down close to the glaciofluvial fan at Auchenblae. However, these moundy deposits were banked against the Strathmore ice as it retreated south-westwards. All of the sand and gravel deposited by meltwaters from the Strathmore ice stream in the Pitrichie, Glenbervie and Auchenblae areas, contains clasts derived from the adjacent Devonian bedrock and psammitic and granitic clasts, most of which were derived from the upland to the north-west. Most of the durable clasts within the gravels around Auchenblae consist of quartzite and psammite pebbles, the bulk of which were derived from nearby outcrops of Old Red Sandstone conglomerate. A significant proportion of the psammite and many of the weathered granitic clasts can be matched locally with adjacent outcrops of Dalradian and Caledonian granitic bedrock.

Much of the coarse gravel which forms mounds on the flanks of the western headwater tributary of the Burn of Caldcotts (the Burn of Balnakettle) is assigned to the Lochton Sand and Gravel Formation as it contains no Devonian clasts. A typical gravel deposit, 3.7 m thick, containing small intraformational ice-wedge casts, was exposed in a small working [NO 620 752] north-east of Mains of Balnakettle. Farther up the valley, a coarsening-upward sequence of reddish brown sands and gravels of the Drumlithie Sand and Gravel Formation, with abundant clasts of Devonian rocks, is exposed in the back-scar of a landslip [NO 618 757] on the south-western side of the burn (Appendix 1 Balnakettle). The sand and gravel has been tectonised and overridden by East Grampian ice which laid down a thin, moderate brown, silty till, containing a preponderance of local Dalradian schistose psammite and semipelite clasts. These topographical and stratigraphical relationships suggest that a minor re-advance of East Grampian ice took place. This occurred after the initial retreat of Strathmore ice and deposition of its outwash on the flanks of the upland.

The deposits of the Drumlithie Sand and Gravel Formation within Strathmore itself are typically vivid reddish brown in colour and relatively silty. They contain abundant clasts of friable sandstone and mudstone and decomposed andestic lava. Many deposits are concealed beneath red, clayey flow tills, as for example, to the east of Auchenblae and to the south-east of Drumlithie. The final south-westward retreat of the Strathmore ice stream was marked by the deposition of eskers in the vicinity of Bomershanoe Wood [NO 734 754], Meikle Fiddes [NO 799 808] and Little Wairds [NO 798 785].

SHEET 67 STONEHAVEN

As on the adjoining Sheet 66E Banchory to the west (see above), the glaciofluvial deposits on the northern half of the Stonehaven sheet have been assigned to the Lochton Sand and Gravel Formation, whereas those in the south belong to the Drumlithie Sand and Gravel Formation. Again, ponding occurred between the East Grampian ice sheet and the Strathmore ice stream, with the latter retreating south-westwards as deglaciation proceeded (Map 11). The ponding occurred mainly in the valleys of the Cowie and Carron waters, but several minor valleys descending toward the North Sea were also affected. This led to the deposition of coarsening-upward, sandy deltaic sequences in the vicinity of Cantlayhills [NO 881 905]. Meltwater drainage carved a deep drainage channel between Ury Home Farm [NO 859 882] and Logie

[NO 886 888] and laid down a fan of silty cobble gravel at its eastern end. The channel was formed near the front of the Strathmore ice stream, during a still-stand in its retreat, and shortly after its parting from the East Grampian ice sheet.

Thick, coarsening-upward, deltaic-style deposits form mounds on the sides of both major valleys. However, whereas in the Cowie catchment the deposits contain mainly durable clasts derived from metamorphic rocks and granites, those of the Carron catchment are dominantly more sandy and include friable sandstone and mudstone clasts and are commonly interbedded with reddish brown silt and clay. The terraced spreads of coarse gravel in the floor of the valley of the Cowie Water, which have been extensively worked inland from Stonehaven, were laid down by meltwater from the East Grampian ice sheet draining south-east towards the coast. These granite and psammite-dominated gravels were deposited on top of reddish brown glaciolacustrine silts and clays (Ury Silts Formation) of the Mearns Drift Group. The gravels contain isolated 'rip-up' clasts of reddish brown clay.

Meltwaters flowing north-eastwards from the retreating Strathmore ice stream deposited eskers that connect with moundy deltaic deposits in the vicinity of Brucklaywaird [NO 826 840], Muirtown of Barras [NO 837 813] and Fawsyde [NO 846 772]. Kame terraces and moundy deltaic deposits were also formed at several localities where the north-western boundary of the retreating Strathmore ice stream withdrew south-eastwards from high ground. For example, deposits occur in the vicinity of Foggie Brae [NO 827 845], Dunnottar Square [NO 864 848], to the west of Roadside of Catterline [NO 860 792] and at the mouth of the Catterline Burn [NO 866 777]. Moundy kame deposits occur at Uras Knaps [NO 876 810] and Greenden [NO 811 776], and terraced deposits lie on both banks of the Bervie Water at Inverbervie.

GLACIOLACUSTRINE DEPOSITS

Fine-grained sand, silt and clay laid down in standing water form part of many deposits categorised as 'glaciofluvial', especially those forming moundy topography. They are especially common at the base of coarsening-upward deltaic sequences, and interbedded with sands and gravels of distal outwash fans. However, only those occurring at the surface and that are extensive enough to justify mapping out separately have been identified as glaciolacustrine deposits. They are typically thinly laminated, micaceous, and commonly contain dropstones. These sediments were deposited mainly in ice-marginal, proglacial lakes that formed in the upper reaches of valleys when their lower reaches were blocked by ice. Such lakes formed extensively during deglaciation as the East Grampian ice sheet withdrew from the coasts while the coastal ice streams remained largely intact (Figure 42). Several of the more extensive glaciolacustrine deposits in the district have been worked in the past for clay for making bricks, tiles and drainpipes (Chapter 2).

DESCRIPTION OF GLACIOLACUSTRINE DEPOSITS BY GEOLOGICAL SHEET

SHEET 95 ELGIN

Fine-grained, glaciofluvial and glaciolacustrine micaceous sands, silts and clays underlie much of this sheet area (Map 1). Although glaciolacustrine deposits have been mapped only to the west of

Elgin, similar sediments underlie much of the ground farther east, mapped as moundy glaciofluvial deposits and as raised marine and estuarine alluvium. It is commonly difficult to distinguish between the various categories of fine-grained deposit as they tend to merge one with another. The pattern of deglaciation and sedimentation was notably complicated. The deposits give rise to a particular soil association (Grant, 1960).

Most of the mapped glaciolacustrine deposits occupy kettle-holes within moundy glaciofluvial deposits. Those occurring above about 30 m OD formed while ice remained in the Moray Firth, impeding drainage. Those at lower levels may have been laid down in brackish water at the time when the sea first transgressed into the area during deglaciation. The silts and clays vary in colour, from red to brown, yellowish brown and grey. They are commonly thinly interbedded with sands, gravels and pebbly, clayey diamicton.

Evidence of an earlier period of ponding can be found in the valley of the River Spey immediately upstream of Fochabers, where varved glaciolacustrine clays underlie gravelly river terrace deposits (Peacock et al., 1968). Ponding occurred extensively farther up the Spey valley to higher levels.

SHEETS 96W PORTSOY, 96E BANFF AND 97 FRASERBURGH

Glaciolacustrine deposits belonging to the Kirk Burn Silt Formation of the Banffshire Coast Drift Group occur along the coast between Portknockie and New Aberdour (Maps 2; 3). They generally underlie flat or gently undulating ground within 5 km of the coast and consist of ochreous to dark brown, thinly laminated silts and clays up to about 10 m in thickness. Bands of ferruginous nodules are common locally. The deposits were laid down after the East Grampian ice sheet had retreated from the coast, but when the Moray Firth ice stream remained offshore, forming a barrier against which meltwaters ponded to the south (Figure 42). Meltwaters entering the lakes from the land commonly formed deltas, as for example, in the valley of Burn of Boyne, near Drakemires [NJ 603 613], where sand and gravel overlies dark olive-grey silts and clays.

Most of the ponding took place during the final stages of deglaciation, but there is evidence on Sheet 96W that it also occurred during an earlier episode, before the last glacial advance across the hinterland. For example, 3.5 m of dark yellowish brown, thinly laminated silts and clays with dropstones crop out from beneath 2 m of stiff lodgement till in the valley of the Burn of Deskford near Inaltry [NJ 517 630] (Map 2). Additionally, red clay crops out beneath gravels and till in the valley of the Burn of Fordyce near Ardiecow [NJ 532 616], and laminated sands are overlain by red till in the valley of the Burn of Fishrie at the Mill of Minnonie [NJ 776 604].

Few glaciolacustrine deposits have been mapped on Sheet 97, but many of the moundy deposits there include silts and clays at the base of coarsening-upwards deltaic sequences. Terraced sands and gravels overlie extensive deposits of dark grey sandy silt and clay in the vicinity of Savoch [NK 047 588], and they probably pass beneath the alluvium surrounding the Loch of Strathbeg. The dark coloured deposits were possibly formed in a glacio-estuarine setting and are equivalent in age to the raised glaciomarine St Fergus Silts Formation occurring a few kilometres to the south-east.

SHEETS 86E TURRIFF, 87W ELLON AND 87E PETERHEAD

No silts and clays of glaciolacustrine origin have been mapped on Sheet 86E, but they do occur locally beneath some of the terraced glaciofluvial deposits, as for example, in the vicinity of Turriff. Glaciolacustrine deposits are most widespread on Sheet 87E and on the eastern margin of Sheet 87W, where

meltwaters were ponded up against the coastal ice that occupied the North Sea and laid down the Logie-Buchan Drift Group, after the East Grampian ice sheet had receded westwards (Figure 42). Most of the ponding occurred in the valley of the River Ythan downstream of Ythanbank, where red and brown laminated silts and clays up to about 10 m in thickness normally underlie glaciofluvial terraces and the floodplain alluvium (Map 6). Outcrops are rare, however, except on the southern side of the river immediately upstream and downstream of Ellon. The glaciolacustrine deposits extend eastwards beneath glaciofluvial sand and gravel infilling the former valley of the river, which lies to the south of Kirkton of Logie-Buchan (see cross-sections on Sheet 87W Solid-and-Drift). A small patch of red clay lying within a topographical depression in the vicinity of Littlemill of Esslemont [NJ 928 285] was formerly worked for the production of tiles and pipes (Chapter 2).

Ponding also took place in the valleys of the North Ugie and South Ugie waters, where up to about 25 m of interbedded silts, clays and very fine-grained sands are concealed beneath the glaciofluvial terraces, as at Denhead [NJ 998 521], and river floodplains. The deposits are commonly colour-banded, including reds, browns, yellowish browns, greenish greys, dark greys and blacks. Few deposits are exposed, but in stream sections near Ballus Bridge [NK 001 473], south of Mintlaw, 3 m of glaciofluvial gravel with an eastward palaeocurrent overlies over 2 m of yellowish brown sand, gravel and silt thinly interbedded with reddish brown and dark grey plastic clay. The multicolouring indicates that the two coastal ice streams and the East Grampian ice sheet were close by in the area. Furthermore, in a temporary section near the bridge [NK 003 471], the laminated sequence was brown, grey and black towards the base, becoming yellowish brown and vivid reddish brown towards the top, indicating that meltwater derived from 'Logie-Buchan' ice entered the lake relatively late (Appendix 1 Ugie valley). All the fine-grained sediments contained good assemblages of reworked Mesozoic palynomorphs, whatever their colour.

No glaciolacustrine deposits have been mapped on Sheet 87E, but they occur extensively beneath the ground mapped as 'undivided' red and blue-grey tills of the Logie-Buchan and Banffshire Coast drift groups, respectively. This ground is underlain by up to 25 m or more of thinly interstratified clayey diamictons, clays, silts and fine-grained sands that were deposited in lakes (possibly brackish) at the landward margins of the coastal ice streams as they began to break up during deglaciation (Merritt, 1981). Thick deposits of red, waxy clay lie inland of the Bay of Cruden where they were worked until recently at Errollston [NK 088 370] for making bricks (Appendix 1 Site 17).

SHEETS 76E INVERURIE AND 77 ABERDEEN

During deglaciation, as the East Grampian ice sheet retreated westwards across the area covered by these two sheets, several large proglacial lakes formed on the lower lying ground. Some of the present drainage routes were blocked by sediments and residual masses of ice, causing ponding locally, but it is also probable that the regional water table was elevated as a result of the relatively high sea level during deglaciation. Sediments in the vicinity of the Mill of Dyce (Appendix 1) caused ponding in the lower stretches of the Don valley where over 15 m of dark yellowish brown, olive-brown and grey sandy silts and clays underlie the floodplain. They form low-lying terraces near Wester Fintray [NJ 811 164], where a prominent mound of glaciofluvial sand and gravel stands in the centre of the valley. The mound formed as a delta and the sandy deposits there fine downwards into over 12 m of laminated micaceous silts.

Ponding also occurred upstream of Inverurie in the valley of the River Urie and its northern tributaries. Deposits of interlaminated fine-grained sand, silt and clay were laid down in the resulting lake, which stood at about 63 m above OD. Low-lying terraces underlain by fine-grained sands, silts and clays are also present on both sides of the River Urie farther upstream, in the vicinity of Old Rayne, where they lie at elevations of 95 m OD and above. They also form extensive deposits in the valley of the Lochter Burn. The silts and clays are commonly dark yellowish brown to grey, thinly laminated and micaceous. Moundy deposits within the main valley in the vicinity of Portstown [NJ 774 231] comprise of over 13 m of interbedded silts, clayey diamictons, sands and gravels and were probably laid down at the ice margin.

Brown, laminated silts and sands of glaciolacustrine origin are common around Kemnay, where they form low-lying terraces beside the River Don and lie within several topographical depressions. Ponding took place during deglaciation when the gorge of the Don remained blocked between Burnhervie and Port Elphinstone.

Few glaciolacustrine deposits are shown on the drift edition of Sheet 77 published in 1980, but they are commonly interdigitated with the glaciofluvial sand and gravel deposits stretching to the north-east of Belhelvie, which are now placed in the Logie-Buchan Drift Group. They also occur towards the base of many coarsening-upwards deltaic sequences within the moundy glaciofluvial deposits lying between Dyce and the coast. Isolated deposits of reddish brown, stiff, waxy clay and clayey silt occur at Tipperty [NJ 970 268] and Balmedie [NJ 965 166]. These sediments typically give rise to poorly drained, gleyed, brown, forest soils of the Tipperty/Carden soil association (Walker et al., 1982). Both deposits were possibly formed in lakes at the landward margin of ice retreating into the North Sea, but a glacio-estuarine origin is perhaps more likely (Simpson, 1955, p.195) (see below). The deposit at Tipperty was worked for making bricks and tiles (Chapter 2).

SHEETS 66E BANCHORY AND 67 STONEHAVEN

Glaciolacustrine deposits are quite widely distributed on these two sheets (Maps 10; 11). Those that were laid down in proglacial lakes at the margin of the East Grampian ice sheet, as it retreated north-westwards across the area, have been assigned to the Glen Dye Silts Formation of the East Grampian Drift Group (Chapter 8). Those formed at the margin of the Strathmore ice stream are assigned to the Ury Silts Formation of the Mearns Drift Group. A considerable thickness of waterlogged, thinly laminated sandy silt and clay is commonly present beneath the alluvium and peat in the elongate ice-scoured basins north of the River Dee (Chapter 7). These deposits, which are commonly grey or greyish brown in colour, have been proved to exceed 7.1 m in thickness in BGS Borehole NO79NW16, sited on the lacustrine alluvium within the Loch of Park basin. They have been interpreted as ranging between 7.9 and 9.2 m in thickness at nearby resistivity sites.

Many of the moundy and flat-topped spreads of glaciofluvial sand in the Water of Feugh catchment fine downwards into laminated sandy silt and clay. Notable spreads crop out on the western side of the Water of Dye near Bogarn and on the flanks of a small tributary stream at Miller's Bog [NO 637 861]. Up to 3.0 m of orange-brown and grey, laminated, stiff waxy clay was formerly exposed in the latter area. Similar deposits are present beneath flat-lying ground in the vicinity of Blairydryne [NO 749 926], in the valley of the Burn of Sheeoch, where they underlie deltaic glaciofluvial sands and gravels (Appendix 2). The deltaic sediments, which were formerly well exposed in Lochton Pit (Auton et al., 1988; Brown, 1994) prograded northwards into a glacial lake at an elevation of about 120 m OD (Brown, 1994). Glacial lakes may also have been present farther up the valley, at elevations of about 133 and 126 m OD, though

fine-grained sediment has been recognised in association with them only in BGS boreholes and trial pits (NO79SW6, 8 and 11).

The only notable outcrop of the Glen Dye Silts Formation on Sheet 67 occurs within a shallow ice-scoured hollow north of Rickarton [NO 816 891]. However, silty lacustrine deposits are probably concealed beneath peat and alluvium in several other ice-scoured basins in the northern part of the sheet. Fine sand with partings of laminated clay underlies glaciofluvial gravel at Bossholes [NO 812 884] indicating the presence of a small ephemeral proglacial lake.

Thinly laminated reddish brown fine sand, silt and clay of the Ury Silts Formation crop out extensively on the low-lying ground of Strathmore, where they also constitute a major part of the concealed succession recorded in boreholes and trial pits. Notable flat-lying spreads crop out east of Fettercairn, west of Fordoun and on the western side of Stonehaven (where they were formerly worked for making bricks and tiles). Smaller outcrops occur on the flanks of sand and gravel mounds north of Ury Home Farm and on the sides of the valley of the Carron Water. Red-brown silts and clays also occur between the ridges of sand and gravel that formed in contact with the south-westward retreating Strathmore ice stream, east of Auchenblae.

Within the valley of the Cowie Water concealed patches of red brown clay and silt extend up to at least 4.5 km inland from the coast. Moderate red, waxy clay, with partings of yellowish brown sand and silt occurs between 6.7 and 12.8 m depth in BGS Borehole NO88NW12 at Nether Findlayston (Auton et al., 1988). The clay occurs between two units of sand and gravel. The underlying silty gravel, which contains pebbles of quartz and fine-grained, dark igneous (possibly andesitic) rock as well as granitic, psammitic and semipelitic clasts, rests on semipelitic bedrock. The composition of the gravel indicates that it was deposited by meltwater sourced from the Strathmore ice stream. The moderate red colour and waxy nature of the clay are typical of lacustrine sediments within the Mearns Drift Group. It is suggested that a proglacial lake formed, fed principally by meltwaters draining north-westwards (up valley) from Strathmore ice that was retreating towards the coast. The 6.7 m of sand and gravel overlying the red clay fines downwards and becomes more silty and cohesive with depth suggesting an origin as a delta or fan-delta. Pebbles within the sand and gravel are predominantly granitic (with minor amounts of semipelite) indicating that the deltaic sediments were laid down by meltwaters from the East Grampian ice sheet draining south-eastwards down the valley into the lake.

Interbeds of red-brown silt and silty sand, 1 to 2 m thick, with sparse dropstone pebbles, are present within deposits of the Drumlithie Sand and Gravel Formation between Auchenblae and Glenbervie. In these sequences, however, all of the sediments were laid down by meltwater issuing from the Strathmore ice stream as it decayed.

PERIGLACIAL DEPOSITS

Head

Head deposits are poorly sorted and poorly stratified sediments that have formed mainly as a result of the slow, viscous, downslope flow of waterlogged soils (solifluction and gelifluction), soil creep and hillwash. Solifluction was most active while periglacial conditions existed in the district during the latter part of the Main Late Devensian glaciation and the Loch Lomond Stadial (Galloway, 1958). It occurred during the summer months when the uppermost 0.5 to 1 m of the soil, the so-called 'active layer', thawed, while the ground below remained permanently frozen. The thickness and potential mobility of active layers depend very much upon the cohesiveness of the sediments affected, hence the thickest head deposits tend to occur where thoroughly decomposed rocks, clayey tills and fine-grained glaciolacustrine deposits have been remobilised.

Head deposits are ubiquitous across the district (Plate 12), but generally it has not been practical to map them out. They are especially well developed across the Buchan plateau where decomposed rock materials have been affected by periglacial processes (Galloway, 1958; FitzPatrick, 1956, 1963, 1969, 1987). They commonly occur in the bottoms of glacial meltwater channels and valleys, locally concealing beds of gravel that were laid down during the formation of those features (Chapter 7). Head deposits redeposited from weathered granite bedrock typically consist of clayey coarse-grained sand, as, for example, around the Hill of Fare (Sheet 76E). Quartzites tend to produce rubbly deposits, as for example on Mormond Hill (Sheet 97), where angular fragments of quartzite are locally associated with sparse boulders of granite carried from the Strichen pluton to the west.

The outcrop of basic igneous rocks of the Insch intrusion on sheets 76E and 86E (Figure 2) is largely blanketed by head deposits composed of angular gravel derived from the underlying bedrock. Head has been mapped around the village of Old Rayne where it reaches several metres in thickness. A good exposure occurs on the southern side of a glacial drainage channel drained by The Shevock [NJ 664 286] (Map 8), where over 2 m of silty clast-supported angular gravel composed entirely of medium- and fine-grained basic rocks blankets fresh bedrock. Sandier gravel is seen to depths of over 1.5 m in several exposures in the village itself, and similar material mantles the norite bedrock near Oyne. Flatter spreads of clayey head, formed of remobilised till, have accumulated around the floodplains of The Shevock, near Insch, and the Gadie Burn around Buchanstone [NJ 656 261]; comparable head deposits flank lacustrine alluvium west of Myrebird [NO 742 990] on Sheet 66E (Map 10).

Head has been mapped relatively extensively on Sheet 96E Banff, where it applies to thick, weakly coherent masses of weathered bedrock or drift deposit prone to movement on waterlogged slopes, most commonly along steep cliff lines and along glacial meltwater channels.

Old, pre-Late Devensian head deposits have been located at several sites in the district where they provide an important part of the Pleistocene record (Chapter 8; Table 7). At least five periglacial episodes are represented at Kirkhill and Leys; other old gelifluctates and cryoturbates occur at Teindland, Crossbrae and the Moss of Cruden (Appendix 1).

Remobilised deposits of till

The remobilisation and solifluction of glacial diamictons following their deposition has been widespread in the lowlands of north-east Scotland (Galloway, 1958). However, it is only apparent at a few sites where solifluction deposits

rest on organic sediments dated to the Windermere Interstadial. These sites include the Rothes road cutting (Chapter 2, Landslips) and Garral Hill, near Keith (Godwin and Willis, 1959), both to the south of the district. Others occur at Woodhead on Sheet 86E (Connell and Hall, 1987), Moss-side Farm, near Tarves on Sheet 87W (Clapperton and Sugden, 1977), sites near New Byth and at Glenbervie (Appendix 1). These sites demonstrate the former instability of low-angle till slopes during the Loch Lomond Stadial and the widespread occurrence of slope-foot accumulations of periglacial diamicton. The frequency of former detachment slides involving former 'active layers' in the region is unclear, but their identification is important because slopes such as these are liable to be rendered unstable by engineering works.

Frost-shattered rock

Frost-shattered rock is widespread across the district and is particularly well developed on quartzitic rocks such as around Durn Hill (Sheet 96W) and Mormond Hill (Sheet 97). It reaches some 8 m in thickness on Sillyearn Hill [NJ 507 514], near Keith, to the south of the district (Galloway, 1958). Phyllites cropping out around the headwaters of the River Ythan (Sheet 86E) and the Glens of Foudland [NJ 605 348], farther west, also are shattered to depths of up to 5 m locally (Galloway, 1958). Some frost shattering occurred prior to the Main Late Devensian glaciation as shattered rock underlies till locally, such as at Newbigging [NJ 527 591], on Sheet 96W (Galloway, 1958).

Flint-quartzite head

Flint-quartzite head deposits have been mapped out only on Sheet 87W Ellon, whereas they are portrayed as 'float' on Sheet 87E Peterhead. They are probably also developed on the other outcrops of the Buchan Gravels Formation on Sheet 86E Turriff (Chapter 4).

Some outstanding examples of head deposits have formed on the flint and quartzite gravels of the Buchan Ridge and Windyhills (Appendix 1 Sites 13, 14). The gravels are invariably concealed beneath up to 1.5 m of 'cryoturbate', formed by stirring, churning and other processes of cryoturbation during repeated freezing and thawing cycles in former periglacial conditions. Vertically orientated elongate clasts are common at the top of these deposits, which are typically pale yellow/orange-brown to white in colour and consist of gravelly, sandy kaolinitic clays. The deposits are particularly susceptible to frost heaving and a road across the Moss of Cruden was rendered impassable as a result of this process one particularly cold winter in the late 1970s. Extensive iron pan is commonly developed at the base of the cryoturbated layer, causing poor drainage and the development of a peat cover at the surface.

A unit of moderate brown diamicton up to 1 m thick caps the cryoturbated Windy Hills Gravel Member at its type section near Fyvie (Map 5). The diamicton includes sparse striated clasts of fresh Dalradian metamorphic rocks as well as quartz and quartzite pebbles from the gravel below. The unit is intensely cryoturbated with excellent examples of erect clasts (FitzPatrick, 1987, fig. 13.5; Plate 4b). The unit is now interpreted to be a till that has subsequently experienced periglacial churning. Some of the flinty head deposits capping the Buchan Ridge probably have a similar origin.

ALLUVIAL AND AEOLIAN DEPOSITS

The alluvial deposits in the district are of five types: fluvial deposits underlying the floodplains and low-lying terraces of rivers, sediments within enclosed basins, lacustrine alluvium, alluvial fans, and river terrace deposits.

Alluvium of floodplains and low-lying river terraces

Ribbons of alluvium flank the courses of the major rivers in the district, forming low-lying ground potentially liable to flooding. River bank sections commonly reveal clast-supported gravel (shingle) capped by 'overbank deposits' consisting of one or two metres of laminated, humic, micaceous, silty sand (loam), locally intercalated with peat. In general, the clasts forming the gravel are subangular to well rounded, reasonably well sorted and exhibit a pronounced imbrication dipping upstream. Only the faster flowing rivers in the district, such as the Spey, Dee, Don and Deveron, have extensive gravelly beds. The River Spey is particularly fast flowing and is braided downstream of Fochabers, where large linear bars of cobble shingle form where river

Plate 12   Head deposits.

a   Rubbly clast-supported diamicton formed mainly of angular fragments of local frost-shattered metagreywacke and more far-travelled rounded clasts of metaquartzite [NJ 8214 5122]. Such deposits are widespread in central Buchan, typically about 0.5 m thick and are partially matrix-supported (P104111)
b   A gravelly head deposit with periglacial festoon structures overlying, and locally penetrating into, cryoturbated grussified granite in a pit near Kemnay [NJ 7502 1377]. The gravel locally truncates ice-wedge casts in the decomposed granite that are filled with weathered sandy diamicton of a pre-Late Devensian glaciation (P104112)

a

b

channels bifurcate. Abandoned braid-bars forming the floodplain are generally afforested or covered by rough, boulder-strewn scrubland (Lewin and Weir, 1977). Most other rivers are single thread and have cultivated flood-plains. Meander belts are common only along stretches of the Lossie, Ugie, Urie, Don and Feugh (Appendix 1 Nether Daugh).

Thicknesses of alluvium are difficult to generalise as they are very variable, both within, and between catchments. Downstream of Fochabers, the coarse, gravelly alluvium of the Spey is on average about 6 m thick, locally ranging up to 12 m or more. Near Banff, the floodplain of the Deveron is underlain by 2 to 3 m of silt overlying 7 to 15 m of alluvial deposits including sand and gravel. The flood-plain of the Ythan is underlain by 4 to 6 m of gravel down-stream of Methlick, whereas thicknesses of between 7 and 10 m of alluvial gravel are common in the Dee valley. Alluvial gravels of the Don typically range between 3 and 5 m in thickness downstream of Inverurie, whereas those in the catchment of the Ugie typically range between 1 and 5 m. In the lower reaches of the Ugie, Ythan, Urie and Don, the alluvial gravels commonly overlie thick sequences of fine-grained glaciolacustrine deposits. They commonly infill buried valleys (Chapter 7).

In most valleys, alluvial gravels would have accumulated either in cold climates following deglaciation and during the Loch Lomond Stadial, or, at the beginning of the Windermere Interstadial and Holocene, when little vege-tation was present to stabilise soils in the river catch-ments. Apart from the Spey and Dee, there can be few stretches of river in the district where gravels are current-ly accreting as opposed to being reworked from older deposits a short distance upstream. The floodplains of larger rivers are generally bounded by bluffs, which in the lower Spey valley are typically steep banks up to 10 m high. In minor valleys, however, and in valleys occupied by 'misfit' streams like that of the Idoch Water, south-east of Turriff (Map 5), and the Gadie Burn near Insch (Map 8), the former floodplains are commonly poorly defined because the bluffs have been degraded by periglacial processes since their formation. In these cir-cumstances, clayey solifluction deposits (head) commonly locally overlie alluvial sediments along the edge of the floodplains. In many small valleys, especially in central Buchan, there are no floodplains as such and little, if any, alluvium has been mapped. The floors of these valleys are underlain by up to 3 m or so of head deposits consisting of interbedded clayey sand and gravel and gravelly diamicton, locally overlying older river gravels.

Alluvium of basins

Alluvium has been mapped in numerous small, poorly drained, enclosed, or semi-enclosed basins across the district. Some of the basins are kettleholes, but the majority have been gouged out of rock or have been sub-glacially sculpted in till (Chapter 7). Basins are particu-larly common on Sheets 76E Inverurie and Sheet 77 Aberdeen, where they are typically lined by up to 3 m or so of thinly bedded, silty, medium- to coarse-grained, micaceous, quartzofeldspathic sand. Most depressions,

however, are underlain by a few metres of head deposits like those described in the minor valleys of the district above. Peat formerly occupied most of these depressions before being cut by man and drained artificially.

Lacustrine alluvium

Flat-lying spreads of interbedded humic sand, silt and clay lie within some poorly drained, enclosed basins that con-tained lochans before being drained artificially. In many instances it is difficult, however, to distinguish between lacustrine alluvium and the alluvium of enclosed basins described above, and the choice of category used varies somewhat from sheet to sheet. This is partly because many of the drainage schemes in the district were carried out following the Crimean War, in the 1860s, and it is no longer clear where the lochans once stood. Other spreads of lacustrine alluvium border existing bodies of fresh water, as, for example, around the Loch of Skene (Map 8). In general mapped spreads of lacustrine alluvium are finer grained than the alluvial deposits filling small basins.

Much of the lacustrine alluvium began to accumulate during Late-glacial times, as for example, that in the Loch of Park, on Sheet 66E Banchory. Organic sediments at this site (Appendix 1) preserve records of vegetational change that has occurred in the district since deglaciation.

Alluvial fan deposits

Fans composed of sand, gravel and gravelly diamicton have accumulated since deglaciation, where tributary streams with relatively steep gradients debouch into more major river valleys. Small, but notable examples occur on Sheet 67 Stonehaven north of Mill of Barras [NO 850 794] and near Greenden [NO 811 776].

River terrace deposits

Dissected remnants of former floodplains flank the alluvium of many of the larger rivers in the district. These form terraces that slope gently down-valley and are typically composed of several metres of stratified clast-supported gravel or sandy gravel. Terrace aggradation probably occurred mainly during Late-glacial and early Flandrian times when the gravel would have been deposited mainly as bars and channel infills by braided streams. The gravel is commonly overlain by spreads of sand and silt up to about 2 m thick, laid down as overbank deposits during the waning stages of periodic flood events. These fine-grained deposits are widespread on the broad terraces of the major rivers where they produce well drained, light sandy soils.

Most river terraces in the district are judged to have formed during ice-sheet deglaciation and consequently have been mapped as glaciofluvial sheet deposits. These terraces are kettled locally and commonly merge into spreads of moundy ice-contact glaciofluvial deposits. Others clearly formed as kame terraces banked up against ice stranded within valleys.

Particularly good examples of low-lying river terraces flank the floodplain of the Water of Feugh, upstream of

Strachan (Map 10). They generally rise no more than 3 m above the level of the floodplain and are composed entirely of interbedded sand and pebble gravel, capped by thin spreads of silt and clay. As many as three distinct terraces can be recognised, none standing greater than 1.5 m above the level of its neighbour and they preserve evidence of shallow palaeo-channels on their surfaces. These true river terraces lie adjacent to kettled glaciofluvial terraces, kames and esker ridges that may, in places, stand up to 25 m above the level of the floodplain.

Blown sand

Deposits of wind-blown sand occur in many coastal localities. They are most commonly found next to sandy beaches, from where most of the sand has blown, but sandy glaciofluvial deposits have been a source locally. Blown sand is generally well sorted and fine to medium grained. Its composition depends somewhat on the provenance of the sediment source occurring locally, but is generally quartzose with varying proportions of finely divided shell. Areas mapped as blown sand generally include stabilised dunes, links and cultivated aprons as well as active dunes. Long stabilised deposits have been mapped locally as 'older blown sand'. A detailed, systematic account of the coastal dune systems of north-east Scotland has been provided by Ritchie et al. (1978).

On Sheet 95 Elgin (Map 1), the largest spreads of blown sand face Burghead Bay where dune ridges tend to be orientated south-west–north-east, parallel with the prevailing wind, and reach up to about 15 m high. Within historical times, the migration of sand along the north and west sides of the Loch of Spynie materially affected local drainage conditions and accounted for the disappearance of several lochs in the area (Peacock et al., 1968). Most of the blown sand on the sheet is probably redistributed beach sand, carried westwards by longshore drift and swept north-eastwards by the prevailing winds, but glacial sands must also have provided a significant source.

Few deposits of blown sand have been mapped on Sheet 96W Portsoy and Sheet 96E Banff, although some of the sandy glaciofluvial deposits occurring along the coast probably have been considerably redistributed by wind in the past when arctic, arid conditions prevailed. Blown sand caps the cliffs at Whyntie Head [NJ 628 661] (Map 2), where the action of blowing sand has beautifully etched rocks in the vicinity (Read, 1923). Modern dunes fringe the bays of Cullen, Sandend and Boyndie, where they rest on deposits forming Flandrian raised beaches.

A continuous, generally 250 to 600 m-wide belt of blown sand backs the coastline for a distance of about 25 km between Fraserburgh and Peterhead on Sheet 97 Fraserburgh (Map 4) and Sheet 87E Peterhead (Map 7), where dunes reach 30 m in height locally (Peacock, 1983). The sand is generally quite shelly and was formerly used as a source of lime at St Fergus. The dunes are particularly active between Inzie Head and Rattray Head, where the combination of blowing sand and southward longshore drift finally blocked off a tidal inlet to create the Loch of Strathbeg in 1720 (Wilson, 1886). Less extensive deposits face the bays at Rosehearty and Phingask (east of Sand-

haven) on the northern coast (Wilson, 1882), and the bays of Peterhead and Sandford on the east coast. The blown sand fringing the Bay of Cruden is relatively more siliceous and locally contains grains of magnetite.

An extensive area of sand dunes lies between Collieston and the Ythan estuary on the boundary of Sheet 87E Peterhead and Sheet 77 Aberdeen. It is mostly designated as the Sands of Forvie Nature Reserve and in contrast to the other deposits described above, the sands there have spread over 2 km inland and reach up to 57 m above OD. Dunes are commonly 12 m or more in height and are particularly active towards the coast and across the peninsula at the southern end of the reserve. The dunes continue to the south of the Ythan estuary, where they form a belt up to 500 m wide that extends 14 km to the estuary of the River Don (Map 9). The dunes reach 15 m in height and back a sandy beach. They are relatively active with blow-outs mostly trending from north-west to south-east. Northward expanding deflation plains are common, where a veneer of sand conceals raised Flandrian beach deposits. Sand is banked up against, and largely obscures, the Main Postglacial Cliffline backing those deposits. It also extends inland to about 1 km from the coast, where it mantles till and glaciofluvial deposits; 'fossil' sand dunes occur in the vicinity of the coastal look-out at Menie Links [NJ 986 210].

The rugged coastline to the south of Aberdeen is largely devoid of mappable deposits of blown sand. This is mainly due to the high cliff line, but also because the beaches are mainly formed of shingle.

ORGANIC DEPOSITS

Peat

Deposits of basin peat in the district occur mostly within the sites of former lochans. Commonly, the peat contains tree boles and other woody fragments as well as the partially decomposed, acidic remains of sedges, reeds, rushes, *Sphagnum* and heather. Most raised peat mosses are a fraction of the original size because of their exploitation for fuel in historical times. Today, many deposits barely average even one metre in thickness, but they remain waterlogged and would partially regenerate in time. Many peaty hollows have been partly, if not completely infilled with boulders carted off the fields. Others have provided sites for dumping waste materials. Peat resources are discussed in Chapter 2 and the locality and extent of the main deposits are listed in Table 5.

Many basins are ice-scoured hollows in bedrock, such as Harestone Moss [NJ 932 195], which is underlain by ultrabasic igneous rocks north-east of Belhelvie (Map 9), and those formed on granite around New Pitsligo and Strichen (Map 6), which contain significant resources of peat. Many small basins are kettleholes in glaciofluvial sands and gravels or morainic deposits, such as large parts of Red Moss [NJ 747 014] and Leuchar Moss [NJ 788 047], on Sheet 76E Inverurie. The majority of basins, however, are ice-moulded hollows in till; abundant examples occur on Sheet 76E, for example Skene Moss [NJ 757 107] and Braigies Moss [NJ 758 047]. They are also common on

Sheet 87W Ellon (for example the Moss of Belnagoak [NJ 880 428] and Elrick Moss [NJ 954 418]), on Sheet 86E Turriff (for example between Cuminestown and the Ythan gorge) and on Sheet 77 Aberdeen (for example Burreldale Moss [NJ 829 239] and several mosses in the vicinity of Westfield [NJ 946 207]). Deposits of peat rest on raised marine and glaciomarine clays within hollows between Kinloss and Lossiemouth on Sheet 95 Elgin.

Some of the most extensive spreads of low-lying blanket peat in the district rest on particularly clayey and impermeable deposits of till, such as those of the Banffshire Coast Drift Group beneath Rora Moss [NK 045 515] and St Fergus Moss [NK 055 536] (Map 7), and several mosses around New Pitsligo (Map 6). Peat has developed on stiff red till of the Logie-Buchan Drift Group at Lochlundie Moss, 5 km south-west of Cruden Bay on Sheet 87E Peterhead.

Hill peat is most extensive on the granite outcrops in the south of the district, where tree roots and stumps, particularly of Scots pine, are very common. Blanket peat covers parts of Bennachie and the Hill of Fare on Sheet 76E Inverurie and on hilly areas underlain by the Mount Battock granite on Sheet 66E Banchory, where it is relatively thick and widespread. For example, up to 4 m of peat overlying decomposed granite is exposed beside a forestry track on the south-west side of Kerloch [NO 697 879]. Other spreads include those on the flaggy quartzites forming the Hill of Stonyslacks, 8 km south of Buckie on Sheet 95 Elgin, and the Moss of Fishrie, on Old Red Sandstone sandstones and shales between Gardenstown and New Byth, on Sheet 96E Banff. Small spreads occur on Bracklamore and Windyheads hills to the south of Pennan, on Sheet 97 Fraserburgh, which are formed of Old Red Sandstone breccias and conglomerates.

Blanket hill peat is also widespread on the relatively high ground of central Buchan formed by the 'Buchan Ridge' and the Hill of Longhaven, which lie to the south-west of Peterhead (Map 7). This ground is either underlain by the clayey, kaolinitic deposits of the Buchan Gravels Formation, or on bedrock that has commonly decomposed to sandy kaolinitic clay (Chapter 3).

Woody peat deposits probably developed extensively around the coasts of north-east Scotland in the early to mid-Holocene, prior to the 'Main Postglacial' marine transgression, when most were destroyed by coastal erosion. However, some deposits have survived in sheltered situations, such as beneath the alluvium of the Water of Philorth, 3 km south-east of Fraserburgh (Map 4; Appendix 1). Remnants of this so-called Boreal forest are exposed on the foreshore in Burghead Bay at low tide (Map 1). Remnants of older peat deposits formed during the Windermere Interstadial are more sparse (Tables 7; 8; Appendix 1 Rothens, Mill of Dyce, Loch of Park, Glenbervie, Howe of Byth). Deposits of early Devensian age or older are even rarer (Appendix 1 Crossbrae, Moss of Cruden, Burn of Benholm).

Shell marl

Thin, discontinuous spreads of freshwater shell marl interbedded with basin peat and fluviatile silt overlie raised marine deposits around Loch Spynie, 4 km north-north-east of Elgin (Map 1). A 35 cm-thick bed of shell marl resting on Late-glacial marine clay was formerly visible in a brick pit situated in another alluvial depression at Gilston [NJ 206 662], nearby.

RAISED MARINE DEPOSITS

In addition to present-day coastal sediments, two distinct sets of raised marine deposits are present in the district. The sets were formed during periods of relatively high sea level, in Late-glacial times and in the mid-Holocene (Chapter 7; Figure 48). Sea level was appreciably lower than it is today between these periods. Each set of deposits is capable of being divided lithologically into 'shoreface and beach deposits' (mainly shingle and sand) and quiet-water sedimentary facies formed in tidal-flat, brackish lagoon and estuarine environments (mainly fine-grained sand and silt). Only the raised glaciomarine deposits are described here; the raised beaches and associated deposits, rock platforms and other features of coastal erosion are described in Chapter 7.

Late Devensian raised marine and glaciomarine deposits

Raised marine beds, mainly silty clay, are known between Elgin and Lossiemouth, from north of Peterhead, and possibly between Ellon and Stonehaven. In the Elgin area, the *Spynie Clay Formation* underlies much of the low-lying ground around Loch Spynie [NJ 237 667] (Chapter 8; Map 1). It occurs also below Lossiemouth Airfield where it is overlain by sand and gravel forming Late-glacial and Flandrian raised beaches. The maximum recorded thickness is 12.5 m (Peacock et al., 1968). North of the Spynie basin the formation attains a minimum level of 10 m above OD, and possibly reaches over 20 m OD, but on the south side of the basin the mapped level is only a little above present sea level, probably because the sea was partly excluded by stagnant ice from this area when sedimentation had begun elsewhere.

North of Peterhead, on Sheets 87E and 97, the dark grey silts, clays and fine-grained sands of the *St Fergus Silt Formation* extend up to about 16 m above OD (Chapter 8). Most of the formation is concealed by lacustrine alluvium (peat and silt) or blown sand. It borders red diamicton, clay and sand of the Logie-Buchan Drift Group to the west, whereas seawards, it appears to be banked against, or to have been deformed into, an end moraine (Hall and Jarvis, 1989; Anderson in Scott, 1890; Glentworth and Muir, 1963; Map 4). Their lithology suggests that the St Fergus Silts are glaciomarine in origin. Most of the marine bivalves, foraminiferids and ostracods in it belong to boreal or boreal-arctic taxa, and two adjusted radiocarbon dates of about 14.9 and about 14.3 ka BP indicate that it was laid down prior to the Windermere Interstadial (Appendix 1 St Fergus).

Some of the isolated deposits of red clay that have been mapped as glaciolacustrine deposits along the coast between Peterhead and Stonehaven may also be partly glaciomarine. These deposits, the *Tullos Clay Member* of the Logie-Buchan Drift Group (Chapter 8), contain a macrofauna and microfauna that is predominantly derived from reworked older deposits (Jamieson, 1882b; Bremner, 1916; Simpson, 1948; Munro, 1986). The clays lie between present sea level and about 30 m OD and may correlate with the Errol Clay Formation of eastern Scotland (Peacock, 1999). Early workers recorded the skull of a seal at Westfield [NJ 993 308], near Kirkton of Logie-Buchan on Sheet 87W (Nicol, 1860; Jamieson, 1882b), a fish at Tipperty [NJ 971 268] (Map 9), and remains of *Ophiura* from the clay at Clayhills [NJ 941 055], in Aberdeen. The red clays on the south side of the city of Aberdeen reach over 18 m above OD at Tullos (Simpson, 1948), where they were worked for making bricks [NJ 949 052] (Chapter 2).

PRESENT-DAY MARINE DEPOSITS

Shoreface and beach deposits

The physical characteristics of beaches vary considerably over short distances and only broad generalisations can be offered here. The following notes are based on a detailed, systematic account of the beaches of north-east Scotland by Ritchie et al. (1978). The composition of beach deposits is intimately associated with the configuration of the coastline and with available sources of material. Broadly speaking, there are three types of coastline in the district.

i *Long sandy bays of gentle arc backed by extensive sand dunes* Typically, these occur between Aberdeen and Fraserburgh, where the beaches have generally accreted outwards by processes of shoreline 'regularisation' and northerly longshore drift. The beaches are relatively unstable in that there are marked seasonal changes in profile and in the position of the high tide mark. This instability is also manifested in the presence of dynamic bars and troughs in the intertidal beach zone that cause waves to break farther from the coast than is usual and tend to generate strong rip currents.

ii *Long bays of gentle arc like those above, but formed mostly of shingle* Beaches of this type back Spey Bay and Burghead Bay on Sheet 95 Elgin. Fluvial inputs have been relatively important in the evolution of this type of beach, together with the dominant westerly longshore drift. The beaches are relatively unstable and are associated with rapidly developing spits and offshore bars.

iii *Isolated cliff-foot and bayhead beaches along rugged, indented rocky coastlines* These predominantly shingly beaches occur along the northern coast between Buckie and Rosehearty and less commonly along the eastern coast between Boddam and the mouth of the Ythan (with the exception of the Bay of Cruden) and along the coast south of Aberdeen. The beaches are relatively stable because they occur in sheltered coves and inlets, and because they normally lie on rock platforms. They are rarely backed by sand dunes of any great extent. The size and orientation of the beaches along the northern coast reflect the strike of the rocks and the dominant input of wave energy from the north-east. The largest beaches are superimposed on remnants of raised Flandrian beach deposits.

Saltmarsh deposits

Several small saltmarshes occur within the estuary of the River Ythan downstream of Kirkton of Logie-Buchan on Sheet 87W Ellon. The features are underlain by tenacious silty clay penetrated by decaying rootlets. The only other saltmarsh that has been mapped in the district lies to the west of Kinloss on Sheet 95 Elgin.

Sea-bed sediments

Sea-bed sediments are depicted on three BGS 1:250 000 scale maps of the continental shelf listed in *Information sources*. The sediments lying off the northern coast of the district are described in the BGS offshore regional report covering the Moray Firth (Andrews et al., 1990), whereas those occurring off the eastern coast are given in another report in the series covering the central North Sea (Gatliff et al., 1994).

SEVEN

Geomorphological features

A set of generalised maps showing the distribution of drift groups and a selection of deposits, landforms and localities mentioned in the text is provided (Maps 1–11).

FEATURES OF GLACIAL EROSION

Glacial striae, *roches moutonnées* and glaciated rock knolls

Evidence of glacial erosion, in the form of scratched and abraded bedrock, indicates the former direction of ice movement. Striated surfaces are not common in north-east Scotland because there are few outcrops of fresh bedrock. Those that do occur are generally restricted to the coasts where glacial erosion has been concentrated (Chapter 3). Most striae are assumed to have been produced by the last major glaciation, namely the Main Late Devensian, but this is not always the case. Some have survived from earlier glacial phases, protected beneath a mantle of till, as, for example, beneath the Craig of Boyne Till Formation at the Boyne Limestone Quarry site (Appendix 1). Furthermore, striae probably only represent particular phases of glaciations when the ice was wet-based and able to glide over rock outcrops. Most of the striae depicted on the geological maps of the district were recorded during the primary survey and few are now visible, but those that are, together with several discovered more recently, generally confirm the earlier observations.

On Sheet 95 Elgin (Map 1), there are two distinct sets of striae trending east-south-east and south-south-east respectively, with a few directed in between these points. The trend of the former set accords with the general direction of ice movement deduced from the distribution of glacial erratics in the area (Figure 45). The direction of travel indicated by the latter set is less clear, but they were possibly formed by ice moving inland from the Moray Firth during the final phase of glacial activity in the area, the 'Elgin Oscillation' (Peacock et al., 1968). Glaciated pavements are well preserved on siliceous sandstones between Carden Hill [NJ 142 622] and Quarry Wood [NJ 190 640], where plucked surfaces generally confirm the east-south-east direction indicated by the striae recorded thereabouts.

South-east- to south-south-east-trending striae are dominant on Sheets 96W Portsoy and Sheet 96E Banff, but some striae are directed towards the north-east. There is also a record of east-south-east-orientated striae crossed by northward-directed striae in the south-west corner of Sheet 96E (Read, 1923; Map 3). Read (1923) linked the south-east- to south-south-east-directed striae with the glaciation that brought shelly tills and Jurassic erratics from the Moray Firth, and the north to north-east set with a major ice movement towards the coast (see

also Bremner, 1934). The former agrees with the views expressed here (Chapter 5), but the latter may be related to more limited movements of 'inland' ice at a late stage of the Main Late Devensian glaciation, following retreat of the Moray Firth ice stream towards the coast (Peacock and Merritt, 1997, 2000).

Striae are relatively common along the foreshore between Fraserburgh and Rattray Head (Map 4). *Roches moutonnées*, grooves and striations are also found immediately inland where there is a gently undulating, glaciated surface developed on gneiss, granite and metabasic igneous rock. Most evidence indicates ice movement towards the east-south-east (Wilson, 1886; Milne, 1892b), but a later set, found on shore sections to the north of a line from the mouth of the Burn of Philorth [NK 028 650] to Inzie Head, suggest a later south-south-east movement (Peacock, 1997, 2000, fig. 27). Inland from these glaciated surfaces there is a transition, over a distance of a few kilometres, from rock that is little weathered at the surface to a wide area where the rocks are generally deeply decomposed with only sparse outcrops of fresh rock. Mormond Hill is one such outcrop, where striae are preserved on ice-smoothed, brecciated quartzite at the entrance to a quarry [NJ 950 568] at about 137 m above OD. Striae have also been observed on a smooth quartzite surface on the north side of the hill [NJ 982 581] at about 155 m OD. When ground as high as this was affected by east-south-eastward-moving ice, it

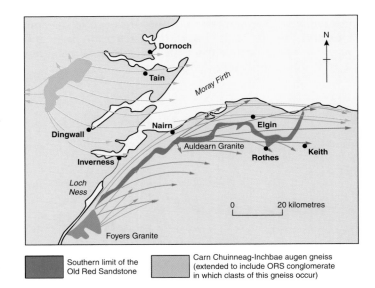

Southern limit of the Old Red Sandstone

Carn Chuinneag-Inchbae augen gneiss (extended to include ORS conglomerate in which clasts of this gneiss occur)

Figure 45   Map showing the transport paths of some indicator erratics across the southern part of the Moray Firth (after Mackie, 1905; Sissons, 1967).

strongly suggests that most, if not all of Buchan is likely to have been glaciated in the Late Devensian. However, large erratic boulders of Strichen granite are buried beneath thick gelifluctate on the northern and western slopes of the hill [e.g. NJ 9585 5690], suggesting that a prolonged period of periglacial conditions followed glaciation.

Striae are rare on Sheet 86E Turriff and Sheet 87W Ellon. Those that do occur are mainly orientated towards the south-east or south-south-east, and were probably associated with the deposition of overlying blue-grey tills of the Banffshire Coast Drift Group relatively early in the Main Late Devensian glaciation, if not earlier (Figure 4; Wilson, 1886; Read, 1923). However, some north- to north-north-east-trending striae located to the north-east of Huntly, on Sheets 86W and 96E, are associated with a later ice movement, as at two localities, they cross, and are therefore younger than, striae directed toward the east (Read, 1923, fig. 11). Several sets of striae and *roches moutonnées* in the south of Sheet 87W relate to a later eastward movement of East Grampian ice.

On Sheet 87E Peterhead, striae are mainly restricted to the coast where several north-north-east-orientated markings were reported by Wilson (1882) in former granite quarries around Stirling Hill and near Yoags' Haven [NK 116 394], some 2 km to the south. These striae were possibly created by East Grampian ice that flowed north-east towards the North Sea prior to the incursion of the coastal ice responsible for laying down the Logie-Buchan Drift Group.

Striae are sparse in the glaciated upland on the western side of Sheet 76E Inverurie. They have been recorded to the north of Torphins and on the northern side of Hill of Fare, where they indicate eastward flow of the East Grampian ice sheet. Glaciated streamlined rock knolls are more numerous, particularly on the lower ground and suggest slightly divergent ice-flow directions (generally towards the north-east and south-east). This divergent flow probably occurred at a relatively late stage when ice movement was influenced by the topography. Such flow directions accord well with retreat positions of the ice front indicated by moraines in cols north-east of the Hill of Fare (see below).

Striae and *roches moutonnées* are relatively abundant on Sheet 77 Aberdeen reflecting the cumulative amount of glacial erosion throughout the Quaternary (Chapter 3). Inland, striae are mostly orientated towards the east, but towards the coast they swing towards the north-east, especially to the south of the Don (Bremner, 1938; Munro, 1986, fig. 31). Jamieson (1882b) noted that striae directed towards the north-north-east were superimposed on ones orientated towards the east-north-east in a former quarry south-east of Cove Bay and striae of roughly similar orientations are reported to the north of Banchory-Devenick [NJ 915 005] (Munro, 1986). Several swarms of ice-smoothed, elongated, streamlined knolls (rock drumlins) occur on the sheet, notably around Newmacher, Peterculter, to the north-west of Belhelvie, and in the vicinity of Ythan Lodge [NJ 995 270] (Map 9). Most of the features on high ground to the north of the Don are orientated towards the east, but those on the flanks of the Potterton

Burn drainage channel are orientated south-east. These orientations accord with that of nearby rock knolls suggesting, as on Sheet 76E, that early ice-flow was towards the east, but as the ice thinned and the higher ground became ice-free, flow directions were influenced by the underlying topography. To the south of Nigg Bay, at Doonies Hill [NJ 960 030], there are several northward-directed *roches moutonnées*, reinforcing the evidence from striae that ice moved in that direction along the coast.

Few striae have been recorded on Sheet 66E Banchory. To the north of the Dee and the Feugh they show a consistent north-east alignment, but north-north-east-orientated striae have been recorded south of Kirkton of Durris (Map 10). Both sets indicate that the flow of the East Grampian ice sheet was directed parallel to the alignment of the Dee valley. On the watershed between the catchment of the Dee and the Bervie Water, striae trending east-south-east are present on a glaciated surface of granite exposed in a newly constructed forestry track, north-east of the Wild Mare's Loup drainage channel. In Strathmore, a single *roche moutonnée*, on the northern side of the Glen of Drumtochty (Map 10), indicates south-east directed ice movement. Crossing south-east- and east-south-east-orientated striae have been recorded on an exposure of andesite north-east of Knockbank Farm [NO 746 798]. These observations suggest that the East Grampian ice sheet impinged onto the north-western flank of Strathmore, before the north-east flowing Strathmore ice stream became established (Figure 4).

Striae are much more numerous on Sheet 67 Stonehaven. Those occurring to the south of the Carron Water are consistent with ice moving north-eastwards along Strathmore. It is apparent that the northern margin of the Strathmore ice stream passed offshore just to the north of Muchalls, from where it 'hugged' the coastline closely towards Aberdeen. It encroached onshore near Portlethen, where distinctive red-brown deposits of the Mearns Drift Group were laid down.

Most of the striae and glaciated rock knolls developed on the Dalradian outcrop in the northern part of the sheet imply ice movement towards the south-east and accord with the alignment of the majority of those recorded by Bremner (1920a). However, his three records to the north of Muchalls of north-north-east-orientated striae crossed by a later south-east-orientated set, have not been confirmed during this survey. Although the trend of the north-north-east set of striae is coincident with the dominant foliation in the Dalradian rocks, the precision of his quoted measurements leave little reason to doubt Bremner's original observations. Based on his records of crossing striae, distribution of erratics and the outcrop pattern of glacial deposits, Bremner (1920a) suggested that at some point Strathmore ice moved several kilometres inland onto the upland areas to the north and west of Stonehaven, probably reaching elevations of at least 330 m above OD (Figure 4).

Drumlins, drumlinoid ridges and large-scale glacial gouges

Drumlins are one of the most distinctive features of glacial moulding, yet they are relatively rare in the district.

A drumlin is typically a smooth, oval-shaped hillock of glacial drift with a steeper, blunter end pointing up-glacier and a gentler sloping, pointed end in the former down-glacier direction. Most glaciologists agree that drumlins are bedforms streamlined in the direction of ice movement and produced mainly as a result of subglacial deformation of soft sediments (Benn and Evans, 1998). Apart from a few elongate till ridges of uncertain origin occurring in the Elgin district (Peacock et al., 1968; Map 1), drumlins appear to be restricted in distribution to the east coast. Even there, however, the features are probably best described as 'drumlinoid ridges' because few are perfectly formed.

The general absence of drumlins probably results from the relatively sluggish flow of ice across most of the district, rather than to any lack of deformable substrate, although drift deposits are relatively thin. The East Grampian ice sheet may also have been 'cold-based' during most of its existence, in which case little subglacial deformation is likely to have occurred (Benn and Evans, 1998). However, there are local exceptions, as for example to the south-east of Lurg Hill [NJ 506 575] on Sheet 86W Huntly, where there is a swarm of drumlinoid ridges and peat-filled gouges glacial sculpting of Old Red Sandstone bedrock is apparent o the south-western flank of the Hill of Finden [NJ 803 637] (Plate 13), south of Gardenstown (Map 3).

A swarm of drumlinoid ridges and intervening, anastomosing gouges extends from a kilometre or so west of a line from Strichen to Maud, on Sheet 87W Ellon, eastwards towards the coast (Maps 6; 7). The ridges typically have a relief of 10 to 25 m, and are formed mostly of deeply decomposed bedrock and till. The trend of the moulding suggests a direction of ice-movement between east and east-north-east, which is slightly at variance with east-south-east- directed striae recorded nearby on Mormond Hill (see above) and two sets of north-east-orientated striae on Sheet 87E Peterhead. The moulding affects, and therefore postdates, most of the deposits at the important glacial/interglacial succession at Kirkhill (Appendix 1), including the blue-grey Corse Diamicton Formation. The patchy uppermost till at Kirkhill, the Hythie Till Formation, has a fabric indicative of ice flow towards the east-north-east, suggesting that it was laid down during the streamlining event. The blue-grey till and rafts of the Banffshire Coastal Drift Group occurring at Oldmill (Appendix 1) has been streamlined similarly. There is little or no evidence of glacial streamlining on the 'Buchan Ridge' and across central Buchan, but there are distinct east-south-east to south-east-directed gouges in the vicinity of Ellon, depicted on Sheet 87W (Map 6). These features might have been produced by the ice that pushed inland from the North Sea basin, laying down the deposits of the Logie-Buchan Drift Group, but it seems unlikely that the margins of this ice mass would have had the power to create such features.

Glacial streamlining is evident across large parts of Sheet 76E Inverurie and Sheet 77 Aberdeen, and it generally becomes more pronounced eastwards, especially to the east of a line between Oldmeldrum and Inverurie. Many of the ridges have rock cores, particularly at their stoss (up-glacier) ends and only a thin covering of till. Around Newmachar (Map 9), for example, there is a gradation between small south-east-trending rock-cored drumlinoid till ridges and ice-moulded rock knolls, which is difficult to delineate by surface mapping alone. Limited drilling and trial pitting in the area (Auton and Crofts, 1986) has shown that till on the crests of ridges can vary in thickness between 0.6 and 2.6 m.

Plate 13
Glacial sculpturing of Old Red Sandstone bedrock, Hill of Findon, south of Gardenstown (D6142).

The orientation of the drumlinoid features, particularly those on higher ground, generally indicates an eastward flow of ice towards the North Sea. West–east-orientated drumlins are well developed between Belhelvie and Udny Station where several examples of crescent-shaped depressions filled with peat occur at the stoss end of the features (Map 9). In the southern part of Sheet 77, the orientation of the features swings around to the east-north-east, but again indicates seawards flow of the East Grampian ice sheet. The south-east trend of ridges on the lower ground around Hatton of Fintray, Newmachar and south-west of the Potterton Burn may be preserving evidence of an early south-east directed ice movement (Bremner, 1928). Alternatively, the features may have formed late in the glaciation when flow directions of the thinning East Grampian ice sheet became more influenced by the local topography.

On Sheet 66E and Sheet 67E, drumlinoid ridges are well developed only within the outcrop of the East Grampian Drift Group. Their orientation is compatible with the ice flow directions that can be inferred from nearby striae and glaciated rock knolls (see above).

Ice-scoured depressions

These typically oval-shaped hollows scoured out of bedrock have a broadly similar distribution to the drumlinoid ridges described above and tend to occur in the areas where there has been most cumulative glacial erosion (Chapter 3). They are commonly peat filled because of poor drainage, and are especially common in the area underlain by the Strichen Granite on the margin of Sheet 87W Ellon and Sheet 97 Fraserburgh. Smaller, more irregularly shaped features are common to the west of Maud (Map 6) where highly weathered gabbro and norite crop out. Harestone Moss occupies an ice-scoured hollow on fresher ultrabasic rocks of the Belhelvie Pluton on Sheet 77 Aberdeen. Numerous alluvium-filled depressions lying to the east of Bennachie and the Hill of Fare on Sheet 76E were scoured out by eastward flowing ice, the largest one being occupied by the Loch of Skene (Map 8).

Large elongate elliptical hollows have also been eroded by eastward flowing ice north of the River Don on Sheet 76E. They are linked by west–east-trending drainage channels, and are filled with fine-grained alluvial sediments and peat. Marshes occupy the largest of the hollows, where former lochs, such as Loch of Park and Loch of Leys [NO 702 978] have been drained. South-east-flowing ice has also scoured large irregular hollows in resistant Dalradian bedrock on Sheet 67, north of Stonehaven. Many hollows are occupied by peat mosses, the largest of which is Red Moss [NO 860 940].

FEATURES OF GLACIOFLUVIAL EROSION

Glacial meltwater channels

Channels cut by glacial meltwaters are very common throughout the district (Bremner, 1928, 1934; Clapperton and Sugden, 1977). Known locally as 'dens', many of them are not associated with any significant modern drainage and may be referred to as 'misfit' valleys (Plate 14). Three broad generic types of channel have been recognised on some maps: *subglacial channels, ice-marginal channels and proglacial spillways*. Although the best examples of each type are distinctive, most channel systems formed time-transgressively and distinctions between them are commonly blurred. Furthermore, many have a long and complicated history spanning more than one glaciation. For example, the Tore of Troup on Sheet 97 Fraserburgh (Map 4) and Tom's Forest channel on Sheet 76E Inverurie (Map 8) are both deeply incised valleys that evidently predate the last glaciation because they are partly filled with till.

It should be born in mind that many of the largest and most impressive drainage channels in the district are not identified as such on the geological maps because they carry modern drainage and contain deposits mapped out within them.

Subglacial, ice-directed channels

Channels of this type, also known as 'Nye' channels (Benn and Evans, 1998), formed while most of the district remained buried beneath ice. Subglacial meltwaters were constrained by the regional hydraulic gradient to flow parallel to the direction of ice movement, irrespective of the local subglacial topography. The hydraulic gradient caused meltwaters to flow uphill within some subglacial channels, giving them their characteristic 'up and down' long profiles. They are most commonly preserved in cols cutting across topographical barriers that were orientated at an oblique angle to the former direction of ice movement. The channels commonly begin at the crests of such barriers and spurs where the hydraulic gradient and discharge was greatest (Sugden and John, 1976). Where cut into bedrock they are typically steep-sided, winding features that branch and reunite repeatedly. Subglacial channels cut into drift are generally broader, more open features, with a gently undulating long profile.

Ice-directed channels are particularly prominent to the south of Troup Head (Maps 3; 4) where they have been incised into the Middle Devonian conglomerates around the Tore of Troup (Merritt and Peacock, 2000, fig. 22). Most channels are straight, like the Den of Muck [NJ 817 607], but some are winding, like one to the south of Pennan Head [NJ 864 643].

Several subglacial channels aligned parallel to the direction of ice movement cut across the high ground linking Bennachie and the Hill of Fare (Map 8) and across relatively high ground to the west of Stuartfield [NJ 973 458] (Map 6). Particularly good examples include 'My Lord's Throat', cut into fresh Bennachie Granite to depths greater than 20 m and the Bandodle channel, of similar depth, cut into the Hill of Fare Granite (both on Map 8). The latter channel extends east-north-eastwards from a col near West Bandodle for a distance of 3 km. Inland from Aberdeen, several channels cut across Tyrebagger, Elrick and Brimmond

Plate 14   Glacial drainage channels.

a   Blackhills subglacial drainage channel cut into Dalradian psammitic rocks of the Glen Lethnot Grit Formation, north of Stonehaven (P104113)
b   Ice-marginal channels formed at the retreating margin of the Moray Firth ice stream of the Main Late Devensian ice sheet in the valley of the Burn of Fishrie [NJ 775 579], looking southwards. (P104114)

a

b

hills (Munro, 1986). Many of the major east-draining valleys such as the Don, Dee and Ythan probably also acted as major conduits for ice-directed drainage. In places they have irregular longitudinal profiles due to both ice and meltwater scour.

Ribbon channels, linking ice-scoured bedrock basins, occur north of the River Dee on Sheet 66E Banchory and numerous ice-directed channels are cut into Old Red Sandstone bedrock in Strathmore. A small, but impressive example of the latter is the Paldy Fair Den, near Glenfarquhar Lodge [NO 722 813] (Map 10); large examples include the group of north-east-trending channels around Woodburnden [NO 766 731]. Similar

large features, such as the Glasslaw and Barras channels occur on Sheet 67 Stonehaven. A particularly good example of a subglacially formed 'up and down' channel runs north-eastwards from Craig Den [NO 803 831], on the boundary between Sheet 66E Banchory and Sheet 67 Stonehaven, from where it crosses a col and links with an esker in the vicinity of Lindsayfield [NO 821 844] (Map 11). This channel, and others on the southern half of Sheet 97 are depicted on the sand and gravel resource map of Auton et al. (1990). Another impressive example is the Devil's Kettle, north-west of Stonehaven, which is cut into resistant, gritty psammitic bedrock to depths that exceed 30 m in places.

Ice-marginal and submarginal channel systems

Complex networks of shallow (1 to 5 m deep) channels are encountered across most of the district, even where clear evidence of glacial erosion is absent, such as in central Buchan. Two distinct sets of channels occur that tend to be orientated at right angles to each other, most commonly north–south and east–west respectively, particularly on Sheet 87W Ellon (Map 6). They were created mainly during deglaciation, as the East Grampian ice sheet receded westwards, and they are particularly common where bedrock is decomposed. Most systems exhibit increasing conformity with local topography at decreasing elevation, with ice-directed channels on the cols, passing at lower levels into ones increasingly orientated downslope in the lee of topographical barriers. The higher elements would have been created first, controlled by the overall configuration of the ice sheet, whereas the lower, slope-related channels would have been cut below thinning ice in the later stages of deglaciation (compare with Sissons, 1958, 1961a, 1961b). Good examples of these types of channels with their characteristically curved or crescentic lower courses include the Pitcowdens and Mount Shade channels on Sheet 66E (Map 10) and the Burn of Muchalls and Burn of Elsick channels on Sheet 67 (Map 11). Dendritic networks of lee-side channels occur locally, as for example, to the north-north-east of New Deer and to the west of Auchnagatt (Map 6).

The origin of the 'lee-side', slope-directed channel networks is not absolutely clear, but they were probably formed in the submarginal zone of the retreating ice sheet, particularly where stagnant ice was stranded on the lee side of topographical barriers as they became exposed during ice-retreat. The channels were probably cut initially into the surface of the ice (compare with Price, 1983; Clapperton and Sugden, 1977), then into the frozen ground beneath. This assumes that the ice sheet was cold-based, which is likely, especially as the lee-side channels are intimately related to suites of generally north–south-orientated, 'ice marginal', stoss-side channels that commonly occur on the west-facing slopes of the topographical barriers. Ice marginal channels such as these are generally thought to be associated with cold-based, polar ice sheets (Benn and Evans, 1998).

North–south-orientated ice marginal channels are particularly well developed along the eastern side of the valley of the Little Water, which joins the Ythan valley at Chapelhaugh [NJ 843 393], on Sheet 87W Ellon (Map 6). At this locality, the marginal channels generally have arched longitudinal profiles, suggesting that the meltwaters that formed them eventually drained westwards beneath the ice margin into a major submarginal channel that is now occupied by the Little Water. Higher ice-marginal channels are commonly truncated by, or feed into, lower ones, indicating that they formed progressively as the ice margin retreated. Other good examples of flights of channels such as these occur in the vicinity of Berryhillocks [NJ 683 597], south of Banff, and around Cook [NJ 800 574], south of Gardenstown (Map 3).

The orientation of the two sets of channels across Buchan indicates general east-north-east to east meltwater flow within the East Grampian ice sheet, at least in its retreat phase, a direction similar to that indicated by the orientation of drumlinoid features (see above), which almost reach the east coast. The pattern thus lends support to the hypothesis that most, if not all, of northeast Scotland was glaciated during the Main Late Devensian Glaciation (Clapperton and Sugden, 1977). However, individual ice-marginal and lee-side channel systems are coupled locally, and both sets of channels formed successively as the ice front retreated westwards, the drainage being continued both through topographically controlled marginal or submarginal channels and along precursors of the present stream system (compare with Bremner, 1928, 1934). The evidence does not support the existence of an integrated, contemporaneous network across the whole region as proposed by Clapperton and Sugden (1977).

Extensive ice-marginal and interlobate channel systems

Large-scale ice-marginal channels, commonly tens of metres deep, occur along the Banffshire coast on Sheet 96W Portsoy and Sheet 96E Banff, where they arc north-east towards the Moray Firth (Maps 2; 3). They formed at the margin of the combined central Highland and Moray Firth ice stream as it retreated westwards (Bremner, 1934). Similarly orientated channels occur along the southern margin of Sheet 95 Elgin and on Sheet 85E Knockando to the south, the largest being the Blackhills channel (Peacock et al., 1968) (Map 1). The relatively large scale of these features compared with those associated with the retreating East Grampian ice sheet across Buchan, may result from later formation when the climate was comparatively warmer and there was greater meltwater production. Unlike the subglacial channels described above, they generally have regular long profiles, sloping gently towards the coast.

South-eastwardly orientated channels north of Aberdeen, such as those drained by the Potterton and Blackdog burns, were probably initiated as ice-directed subglacial features (Map 9). However, most of the meltwater drainage along them took place when they emerged from the margin of retreating East Grampian ice sheet. The Culter Burn, Silver Burn and Den of Murtle channels near Peterculter carried drainage around the margin of the East Grampian ice sheet into the Dee valley as the ice-front retreated inland.

Several channels were carved out at the former margins of abutting ice masses. For example, on Sheet 87E Peterhead, a prominent channel system links the upper reaches of the Water of Cruden, west of Hatton, northwards to the valley of the River Ugie, to the east of Longside. The route follows the 'misfit' valleys of the Laeca Burn, Den of Aldie, Dens and West Den (Map 7). The system was probably formed close to the landward margin of ice that pushed in from the coast on more than one occasion, when meltwaters were diverted across the high ground linking the 'Buchan Ridge' and the Hill of Longhaven. The Den of Boddam [NK 114 415] is partly infilled with gravels of the Buchan Gravels Formation and certainly has a long history. Meltwaters also flowed eastwards at the margin of the Moray Firth ice stream. For example, several major channels 'loop' inland and back again along the coast between Buckie and New Aberdour (Maps 2 to 4). To the east of New Aberdour, on Sheet 97, meltwaters flowed farther inland via several channels, towards Rathven and the Loch of Strathbeg.

The Strathfinella–Glen of Drumtochty channel on Sheet 66E Banchory (Map 10) and the south-west–north-east-trending Ury Home Farm–Limpet Burn channel, on Sheet 67 Stonehaven (Map 11), were eroded between the abutting margins of the East Grampian ice sheet and the Strathmore ice stream (Simpson, 1955). A group of secondary interlobate channels is present on the northern flank of the former channel. These channels drained eastwards into the main channel in the vicinity of Drumtochty Castle [NO 699 801]. The White Hill, Blackhills and Den of Cowie channels on Sheet 67 were formed by drainage at the margin of the Strathmore ice as it retreated (Plate 14a). This meltwater probably included drainage from ice-dammed lakes that developed between the two adjacent ice masses.

Proglacial spillways

In contrast to the subglacial and ice-marginal types of drainage channels described above, proglacial spillways generally have regular longitudinal profiles. When cut into bedrock, they also, typically, have a pronounced V-shaped cross-section, and interlocking spurs. Many formed as 'overflow' channels where ice-dammed lakes drained across cols (compare with Charlesworth 1926, 1956; Synge, 1956). Others formed when major drainage routes became blocked by glacier ice. Some would have been cut catastrophically as a result of glacier outburst floods or 'jökulhlaups' (Maizels and Aitken, 1991; Brown, 1994).

A good example of a spillway, the 'Towie Spillway', occurs on Sheet 86E Turriff between Darra [NJ 744 474] and Towie Castle [NJ 744 439] (Map 5). It formed when the valley of the River Deveron was blocked by ice downstream of Turriff, causing all drainage in the upper catchment of that valley to be diverted southwards into the Ythan valley (Figure 42). The Blackhills channel on Sheet 95 Elgin probably also functioned as an overflow channel, as did the valley of the Black Burn, which joins the River Lossie at Elgin (Map 1). Two prominent spillways draining former water-filled basins have been identified on Sheet 77 Aberdeen, near Blackburn and Kingswells (Munro, 1986).

The Westhills channel, west of Belhelvie, may have acted as a spillway for meltwater draining from the East Grampian ice sheet, before ice-marginal drainage became established in the nearby Potterton Burn channel (Map 9). Although no discrete channel-feature is preserved, it is clear that the extensive spreads of sand and gravel around Corby Loch were laid down by eastward flowing meltwater draining along the Don valley and across the interfluve towards the sea. This drainage route existed before flow was re-established in the lower reaches of the valley of the present river (Aitken, 1998).

A major spillway, Slack Den breaches the watershed between Glen Dye and Strathmore (the 'Slack of Birnie' of Synge, 1956) on Sheet 66E (Map 10). The spillway is cut, locally to a depth of more than 100 m, into resistant hornfelsed interbanded psammites and semipelites. The intake of the spillway occurs at about 350 m above OD, about 1.6 km north of the present watershed. The spillway was cut by drainage from a glacial lake that was ponded behind East Grampian ice blocking the lower reaches of Glen Dye. The south-west-directed drainage also dissected ice-contact sands and gravels laid down at the margin of the Strathmore ice stream during its retreat, depositing a fan of sand and gravel (containing clasts predominantly of psammite and semipelite) in the vicinity of Clatterin Brig [NO 664 782]. The surface of the fan stands some 30 m above the level of the present valley floor and has been dissected by the Slack Burn. The south-western end of the spillway also bisects the group of secondary interlobate drainage channels on the northern flank of the Strathfinella–Glen of Drumtochty channel (see above), indicating that meltwater drainage along the spillway commenced after the initial parting of the East Grampian and Strathmore ice streams.

Buried valleys

There are several examples of deep, buried valleys around the coasts of north-east Scotland. These channels were probably formed by subaerial drainage when sea level was relatively low during glacial stages, but when the land was not greatly depressed isostatically. Most are unlikely to date from the early Late-glacial period because sea levels were relatively high during deglaciation. They are commonly capped by glaciofluvial deposits, thus ruling out formation during the low sea-level stand of the early Holocene (Chapter 5). Moreover, many are partly filled with till and clearly predate the Main Late Devensian glaciation. Some may be very old features indeed, like one underlying the lowest reaches of the valley of the River Spey (Map 1), which is partly filled with unusual pale greenish grey sand (Aitken et al., 1979).

Deep channels have been located at the mouths of both the Don and Dee at Aberdeen (Figure 46). Both are largely filled with till and predate the Main Late Devensian glaciation. The River Don flows through a gorge cut in rock immediately upstream of the Bridge of Don [NJ 940 097], yet boreholes in the vicinity of Seaton Park, some 600 m to the south of the gorge, show that rockhead is well below sea level (Lumsden, 1958; Munro, 1986). A buried channel bottoming at about 15 m below OD lies beneath the floodplain of the River Dee between the Bridge of Dee [NJ 928 036] and the Wellington Suspension Bridge [NJ 943 050] (Peacock et al., 1977; Munro, 1986). Downstream of the latter bridge, a drift-filled channel diverges from the present valley and extends eastwards beneath the valley between Torry and Tullos Hill (Simpson, 1948). It reaches the coast at Nigg Bay (Map 9), where it bottoms at a depth of at least 40 m below OD, and extends for more than a kilometre beneath the bay (Munro, 1986). Geophysical data suggests that the feature is cut in bedrock and is narrow and gorge-like (Law, 1962). Another buried channel has been located beneath the centre of Aberdeen (Figure 46).

A further buried valley has been located beneath the valley of the River Ythan at Ellon (Merritt, 1981) (Map 6). It diverges from the present valley downstream of the town, passing south of the Hill of Logie [NJ 978 297] before reuniting with the present course at the estuary (see cross-sections on the Solid and Drift edition of Sheet 87W). Like those at Aberdeen, the buried channel of the Ythan is gorge-like and contains till. Seismic evidence suggests that it descends to some 40 m below OD at the Snub [NK 002 282] (Quaternary Research Association, 1975).

It is likely that many buried channels remain to be discovered, particularly along the coast. For example, the River Ugie flows through a gorge cut in bedrock near Inverugie [NK 102 482] (Map 7), yet boreholes drilled upstream failed to reach bedrock at 2 m above OD (McMillan and Aitken, 1981). This suggests that a buried channel occurs somewhere between Inverugie and St Fergus village [NK 098 520]. Inland, segments of old channels cut in rock and locally up to 10 m deep are known, for example, at Kirkhill Quarry, where the protection from glacial erosion provided by the features has helped preserve the complex Middle to Late Devensian succession there (Connell et al., 1982; Connell, 1984a; Appendix 1   Site 7).

FEATURES OF GLACIAL OR GLACIOFLUVIAL DEPOSITION

Eskers, kames and moraines

Descriptions of individual eskers and kames are given in Chapter 6 in the notes on glaciofluvial ice-contact deposits, and their resource potential is discussed in Appendix 2. The most prominent eskers in the district are located on Maps 1 to 11. Perhaps the most studied and photogenic esker is that forming the Kippet Hills beside Meikle Loch at Slains (Plate 15; Appendix 1).

Moraines are not common within the district (Charlesworth, 1956; Synge, 1956). Those that do occur have been described in Chapter 6 in the description of 'hummocky glacial deposits'. Most moraines are of a rather chaotic nature and formed when large masses of ice stagnated in the lee of major topographical barriers. There are some examples of recessional moraines formed during 'active' glacial retreat, particularly on Sheet 66E Banchory and Sheet 67E Stonehaven, and a significant terminal moraine close to the limit of the Logie-Buchan Drift on Sheet 87W Ellon.

Erratics

It has long been recognised that the occurrence of boulders in areas far removed from their source is good evidence of the former direction of glacial transport (Plate 16). At times too much emphasis has been placed on the distribution of distinctive 'indicator erratics' in

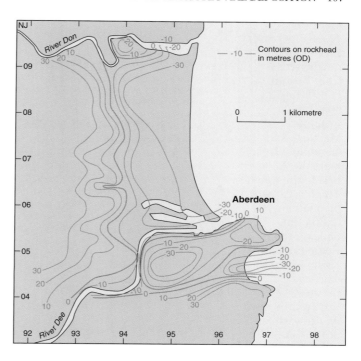

Figure 46   Contoured rockhead surface beneath the City of Aberdeen (after Munro, 1986).

early reconstructions of former ice sheets and transport pathways (e.g. Jamieson, 1906; Read, 1923; Bremner, 1934a, 1938; Synge, 1956). In addition, some of the observations made in the older literature are questionable. However, with caution, the known distribution of erratics still provides valuable evidence for determining the extent and interrelationships of former ice sheets in the district (Sissons, 1967). They also form an important element in the lithostratigraphical division of the Quaternary sequence at group level (Chapter 8). In more modern work, the complete description of clast types is made as 'stone counts' rather than relying on single far travelled clasts that may have complicated transport histories.

The paths determined for the transport of erratics in the Elgin area are reasonably consistent, revealing a predominantly easterly flow of ice, although there is also evidence of an older movement towards the south-east (Figure 45). The distribution of erratics of the granitic 'Inchbae' augen gneiss from central Ross-shire is particularly important, indicating the dominant ice flow into the Moray Firth from the north-west Highlands. The picture that emerges in the rest of north-east Scotland is more difficult to understand (Figure 47). The south-easterly movement of erratics both from the Boyndie metagabbro and the 'lower zone' cumulate basic and ultrabasic rocks of the West Huntly and Barra Hill intrusions, is consistent with the distribution of other erratics and rafts of sediment and fossils derived from the bed of the Moray Firth (see below and Whitehills Glacigenic Formation, Chapter 8). The radial movement of erratics away from the Netherly Diorite and West Huntly intrusion is more difficult to explain. One explanation could be that the

Plate 15
The Kippet Hills Esker looking northwards from Broom Hill [NK 0332 3058], near Collieston. Meikle Loch occupies a large kettle hole beside the esker, which is capped by red clayey diamicton typical of the Logie-Buchan Drift Group. The esker merges northwards into an outwash fan at Knapsleask (D2791).

outcrops of indicator rocks may be more extensive than thought hitherto because of poor exposure. For example, concealed masses of gabbro have been identified by geomagnetic surveys between Huntly and the coast (Munro, 1970; Munro and Gallagher, 1984). Another explanation involves the probable cold-based nature of former ice sheets over the interior of north-east Scotland, and their inability to scour away the products of previous glaciations. The interrelationship of East Grampian ice with more powerful ice streams occupying the Moray Firth basin may also be responsible for such patterns (Peacock and Merritt, 2000).

The analysis of ice-movement direction is further complicated by the reported south-westward carry of erratics of the Peterhead granite (Bremner, 1928, 1943; Synge, 1956; Figure 47). However, the occurrence of little known sheets and small bodies of granite, including red granite, between Fraserburgh and Peterhead on Sheet 97 Fraserburgh and Sheet 87E Peterhead casts some doubt on the observation. The boulders do, however, lie along the margin of the ice that moved onshore from the North Sea basin giving rise to the Logie-Buchan Drift Group (Chapter 8). It is difficult to account for this direction of movement unless ice crossed the North Sea basin from Scandinavia. Some foundation to this argument has been derived from the discovery of erratics of rhomb-porphyries and larvikites also thought to have been carried from southern Norway (Read et al., 1923; Campbell, 1934; Bremner, 1939, 1943; Ehlers, 1988; Hall and Connell, 1991, fig. 74). These include a boulder on the beach at Portsoy, and several findings around Ellon and Aberdeen

(Figure 47; Sissons, 1967), including Nigg Bay (Appendix 1). However, the majority of confirmed Scandinavian erratics have been found inland of where deposits of the Logie-Buchan Drift Group occur at surface and are thus likely to have been transported onshore during a pre-Devensian glaciation.

Large rectangular blocks of hornfelsed sillimanite-bearing pelitic gneiss, many weighing several tons, form an erratic train eastward from its outcrop around Green Hill [NO 638 162], on the western margin of Sheet 76E Inverurie (Map 8). The erratic train extends as far as Monymusk, and isolated erratics many metres in diameter have been found farther east, in the vicinity of 'The Horner' [NO 748 156], east of Kemnay.

A south-eastward movement of erratics of andalusite schist and an eastward carry of mafic and ultramafic rocks from the Insch intrusion have been recorded in the northwest part of Sheet 77. The Insch erratics result from the final eastward flow of the East Grampian ice sheet, but the schists were probably carried by an earlier movement. At the coast, erratics of sandstone and conglomerate have been carried north-westwards by the onshore movement of ice that laid down deposits of the Logie-Buchan Drift Group. This movement possibly also carried erratics of the mafic and ultramafic Belhelvie intrusion, which extends offshore, rather than the north-eastward movement depicted on Figure 47. An earlier south-east movement is required to account for the Belhelvie erratics found in the lowest exposed gravels at Nigg Bay (Bremner, 1928).

Bremner (1920a) reported the presence of erratics of quartzite and sandstone that had been carried north-east

Plate 16 Glacial indicator erratics.

a Rhomb porphyry erratic from the Oslo Graben, Norway, identified in 1979 by Dr J A Dons, Mineralogisk-Geologisk Museum, Oslo. The erratic was found in the Sandford Bay Till Member of the Hythie Till Formation at Sandford Bay [NK 1245 4348], south of Peterhead (P015377)

b Tabular erratic of sillimanite-bearing pelitic hornfels from an erratic train to the west of Monymusk (P104115)

from Strathmore onto the watershed between the Dee and Cowie Water (Map 10). In particular, a quartzite boulder was found in the col between Craigbeg [NO 769 911] and Mongour [NO 757 900], over 330 m above OD. This north-eastward transport has been confirmed by the recent discovery of a rounded erratic of porphyritic andesite, many metres in diameter, 'wedged' in the bottom of a small glacial drainage channel occupied by the Burn of Anaquhat. The erratic was found [at NO 746 881], 260 m above OD and almost 5 km inland of the nearest outcrop of andesite in Strathmore. Bremner (1920a) postulated that the Old Red

Sandstone erratics had been transported by an advance of Strathmore ice, prior to the arrival of 'Dee Valley Ice' (East Grampian ice sheet), which 'completely cleared out or covered up all traces of Strathmore drift, in places where one is safe to infer that it was once present'. The age of the early advance of Strathmore ice is not known, but it possibly occurred during the maximum development of the Main Late Devensian glaciation of the district, prior to 22 ka BP (Appendix 1   Balnakettle).

A distinctive suite of erratics occurs in the deposits of the Logie-Buchan Drift Group, including Permian and Mesozoic limestones and dolomites in addition to Old

Figure 47
Transport paths of some indicator erratics in north-east Scotland (after Read, 1923; Bremner, 1928; Synge, 1956; Sissons, 1967; Ehlers, 1988).

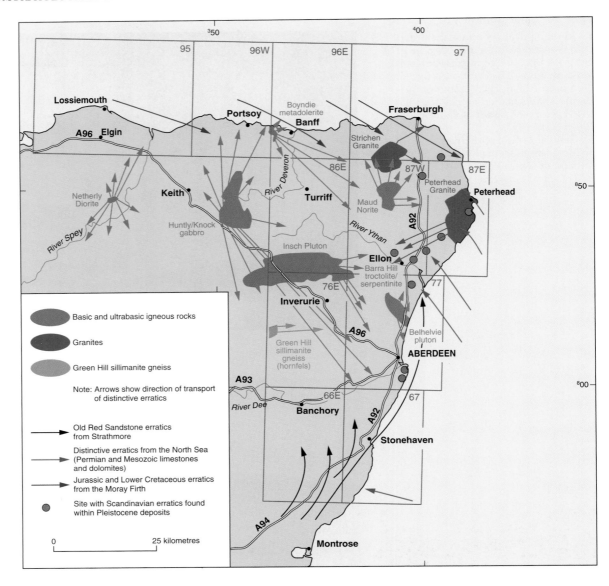

Red Sandstone and conglomerates, indicating a source to the south and offshore (Appendix 1 Kippet Hills and Sandford Bay)

Glacial rafts

Glacial rafts or 'megablocks' are dislocated slabs of rock and unconsolidated strata that have been transported from their original position by glacial action (Aber et al., 1989). Erratics and large glacial rafts of Mesozoic rocks derived from the floor of the Moray Firth and the western North Sea are widespread in coastal areas, where they occur in deposits of the Banffshire Coast Drift Group and Logie-Buchan Drift Group respectively (Figure 4). Rafts are particularly common in the White-hills Glacigenic Formation of the former group, where they include reddish brown Permo–Triassic marls, dark grey to black Jurassic mudstones, Cretaceous glauconitic sandstones and Quaternary marine clays, silts and sands

(Plate 17). The rafts range upwards in dimension from fist-size fragments to slabs that are probably several tens of metres thick and a kilometre or more in width. Several very large rafts of mudstone, silt and clay were worked for making bricks and tiles (Chapter 2).

The known localities of rafts are concentrated along the northern coast of the district where glacial striae also indicate that ice moved onshore towards the south-east at one stage. This suggests that the rafts were emplaced in zones of longitudinal compression, either proglacially or subglacially, as ice occupying the Moray Firth passed from deformable marine sediments on to bedrock, and was forced to override the high coastline and interior valleys (Peacock and Merritt, 1997, 2000). The rafts were detached in the Moray Firth basin, probably as high pore-water pressures built up in confined aquifers (sands interbedded with clays), a situation that has been invoked for the emplacement of similar rafts at Clava, near Inverness (Merritt, 1992b). Rafts were once

Plate 17   Glacial rafts.

a   Tectonic lamination in a raft of sand (possibly glaciomarine) disrupted by low-angle shears at Ardglassie, near Fraserburgh (P104123). Spade is 0.9 m long

b   Attenuated, fractured and sheared lenses of sand within the Whitehills Glacigenic Formation at the Boyne Limestone Quarry (P104124).

assumed to have been moved while 'frozen on' to the base of cold-based ice sheets, but their intimate association with deformation tills suggests that failure may also have occurred along *décollement* surfaces in a subglacial deforming layer (Benn and Evans, 1998).

Rafts have been described at several localities in the district (Appendix 1   Boyne Limestone Quarry, Gardenstown, Oldmill and Moss of Cruden). At the time of writing, several rafts could be seen in a quarry at Ardglassie [NK 012 617], 5 km south-south-east of Fraserburgh on Sheet 97 (Merritt and Connell, 2000). The rafts include sheared pale yellowish brown fine-grained sands with traces of ripple cross-lamination and dark grey clay. One raft of black lignitic clay contained a rich Late Jurassic palynoflora (information from Dr L A Riley, 1998). The rafts rest on relatively fresh bedrock and are capped locally by a thin unit of brown till. The quarry is situated on a south-west–north-east-orientated ridge, and most of the shear planes preserved in the rafts dip towards the north-north-west, suggesting ice push from that direction against the ridge. Many characteristic features of rafts are displayed, including shear zones, conjugate microfaulting, brecciation and folding.

Plate 18 Glaciofluvial terrace gravel infilling an ice-wedge cast at Inverurie [NJ 7678 2255]. The structure has developed in glaciolacustrine silty sands and sandy diamicton, both derived mostly from local decomposed mica schist (P104116).

Plate 19 Earth-pillar in the valley of the Allt Dearg, near Fochabers, formed in weathered Devonian conglomerate (Spey Conglomerate Formation) at the Craigs of Cuildell (D2144).

PERIGLACIAL PHENOMENA AND FEATURES OF MASS WASTAGE

Periglacial features

A wide range of phenomena that formed in former cold, periglacial environments has been recognised across the district (Galloway, 1958; FitzPatrick, 1956, 1963, 1969, 1987). Many elements of the present landscape are essentially relict periglacial landforms. The main effect of periglacial mass wasting has involved the downslope movement of till deposits and regolith and the subsequent accumulation of 'head' deposits as spreads on lower slopes and valley floors (Chapter 6). These processes have created the ubiquitous rounded hill crests of the region with valley sides that are commonly convexo-concave in profile (Ballantyne and Harris, 1994). Solifluction terraces are very common, as is frost shattered bedrock and structures indicative of the former presence of ground ice, notably cryoturbation structures, ice-wedge casts and tundra polygons (Plate 18). Although mostly occurring at the surface,

features such as these have been identified at several stratigraphical positions within the Pleistocene sequence (Chapter 8). Aeolian periglacial features such as ventifacts, 'cover sands' and loess are likely to occur, but apparently have not been reported.

Intense periglacial activity last occurred during the Loch Lomond Stadial. Marked slope instability, with destruction of soils and remobilisation of diamictons by gelifluction, led to slope-foot accumulations of significant thickness, locally burying organic sediments of Windermere Interstadial age (Chapter 8). It is very likely that permafrost had been established previously during ice-sheet deglaciation, because ice-wedge casts and polygon networks affect the sediments of low river terraces in the lower Ythan and Ugie valleys (Clapperton and Sugden, 1977; Gemmell and Ralston, 1984; Armstrong and Paterson, 1985). These sediments were formed after the retreat of the coastal ice that laid down the Logie-Buchan Drift Group, possibly during a subsequent intensely cold period late in the Main Late Devensian glaciation (see Appendix 1   Ugie Valley).

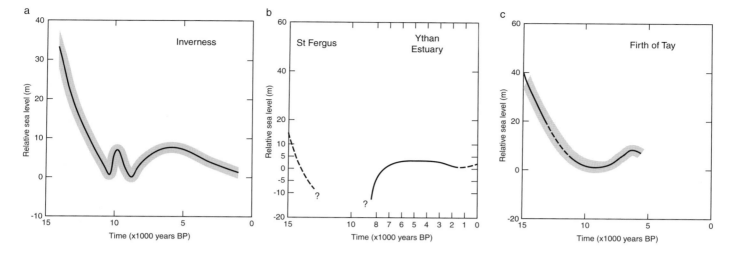

Figure 48   Inferred sea-level change over the past 15 000 years, based on height-age relationships of raised shorelines. Curves a and c from Lambeck, 1993a; b from Smith et al., 1999 with speculative Late-glacial part of curve extrapolated from St Fergus.

Earth-pillars

Earth-pillars have developed on the slopes of several valleys in the district from spurs and ribs of material containing durable clasts in a friable, easily eroded matrix. Large boulders cap these features, which have been eroded mainly by rain-wash on steep slopes. Good examples occur on the eastern side of the valley of the River Spey near Fochabers at Craigs of Cuildell [NJ 332 553] (Map 1), where they have been carved out of weathered conglomerate (Plate 19).

FEATURES OF MARINE EROSION

Two distinct sets of raised beaches, clifflines and rock platforms are common around the coasts of north-east Scotland. The sets were formed during periods of relatively high sea level, in Late-glacial times and in the mid-Holocene (Figure 48). Sea level was appreciably lower than it is today between these periods. Raised glaciomarine deposits have been described in Chapter 6.

During the past twenty thousand years or so, sea levels in north-east Scotland have been influenced by both the isostatic depression of the land under the ice load, and by global (eustatic) sea levels (Chapter 5). The largest amount of isostatic depression occurred in the western Highlands where the ice load had been greatest. That region has consequently been the centre of uplift since the ice melted. The rate of uplift was greatest during and immediately after deglaciation and has fallen exponentially since. The result of this pattern of isostatic recovery is twofold. Firstly, all raised marine deposits and associated erosional features in north-east Scotland are tilted eastwards and northwards away from a centre of uplift positioned in the vicinity of Rannoch Moor. Secondly, the younger, lower lying deposits and features are tilted less than the older ones. It follows that beaches of the same height in different places are not necessarily of the same age.

The older, Late Devensian set of raised features formed during Late-glacial times, that is, between the initial decay of the Main Late Devensian ice sheet in this region and the end of the Loch Lomond Stadial (Gray and Lowe, 1977). These remnants of shorelines lie up to about 30 m above OD, and are backed by discontinuous, degraded clifflines. The deposits locally merge into glaciofluvial sediments that were laid down contemporaneously. The Late-glacial set of shorelines and deposits are commonly truncated by an extensive cliffline, the *Main Postglacial Cliffline*. This feature formed during the mid-Holocene, at the climax of a major marine transgression, the *Main Postglacial Transgression*, which culminated between 6300 and 5700 BP in the Fraserburgh area (Appendix 1   Philorth valley) and at about 4750 BP in the lower Ythan valley (Smith et al., 1999). The Main Postglacial Cliffline commonly backs the younger set of Flandrian (Postglacial) raised marine deposits, which lie up to about 10 m above OD.

Late Devensian raised beaches

Late Devensian raised beach deposits are present on the Moray Firth coast between Forres and Banff, but there is little or no record of such features between the latter locality and Fraserburgh. They are best developed between the mouths of the River Findhorn and the River Spey on Sheet 95 Elgin, particularly on the seaward side of the Covesea–Roseisle ridge, which would at one time have been an island (Peacock et al., 1968, fig. 18; Map 1). They are less evident on the landward side of the ridge, where they are replaced in part by undulating, kettled glaciofluvial gravels, probably because the sea was excluded by stagnant glacier ice left behind during the westward retreat of the Moray Firth ice stream. Stagnant ice may also have been present on the southern side of the Spynie basin where Late-glacial beaches and raised marine sediments are again absent. The highest beaches occur as short stretches to the west of Burghhead (about 24 m above OD), Hopeman (about 24 and 19.8 m OD), and Easter Covesea (about 24 and 21 m OD). The hill on which Lossiemouth stands formed an island, and beach features at 22.5, 19.5 and 14.6 m OD were formerly seen in the western suburbs. These stretches of beach are formed of gravel, bouldery in places. At Greenbrae Quarry [NJ 137 692], the '19.8 m' beach overlies a striated glaciated pavement. Detailed levelling has shown that the back feature of another beach descends eastwards from about 14.3 m OD south of Grange Hall [NJ 064 606] to about 13.7 m OD by Milton Brodie House [NJ 092 629] over a distance of 3 km.

There is no certain evidence for Late-glacial beaches between Lossiemouth and the mouth of the Spey, but farther east a raised beach at 14 m OD merges southwards into the Mosstodloch Terraces of the River Spey (Map 1). Eastwards, there is a gravel beach up to 2.4 m thick (top surface about 15 m OD) resting on till on the low quartzite cliffs at a locality [NJ 450 675] immediately south of Craig Head, west of Findochty (Map 2). A Late-glacial beach feature at about 22 m OD is seen south-east of Whitehills, and again in Banff, where it is followed by the A96 trunk road (Map 3). These beaches are generally formed of a metre or so of poorly sorted gravel and boulders. Spreads of glaciofluvial sand and gravel appear to merge with raised beaches (at about 25 m OD) at the mouth of the Boyndie Burn, forming a raised delta. Other degraded Late-glacial shoreline features were identified between Cullen and Tochieneal [NJ 522 653] (Map 2) by Read (1923), but they have not been confirmed by the recent revision. At Rosehearty (Map 4), an area of bare rock below 9 m OD, which extends for about a kilometre west of the village, may have been scoured by the Late-glacial sea, but no deposits are known.

At St Fergus, a raised beach has been identified at about 15 m OD (Appendix 1), but apart from this no Late-glacial raised beaches have been identified with certainty between Fraserburgh and Aberdeen. Farther south, Ritchie et al. (1978) reported a possible raised beach terrace at 10 to 12 m OD on the north-east side of the estuary of the River Ythan, above the level of Flandrian raised beach terraces at 3 to 6 m OD. However, terraces around 10 to 12 m OD are glaciofluvial (Merritt, 1981). Raised beaches may be expected adjacent to the low coastline north of Aberdeen, because raised marine clays are known at Aberdeen itself (see below) and the trend of isobases of Late-glacial beaches to the south suggests that the contemporary sea level would have been above 15 m OD (Sissons, 1983, fig. 4). The absence of Late-glacial beaches suggests that the Late-glacial sea here extended into an area of sediment-covered ice, the subsequent decay of which destroyed any evidence of beaches (compare with the Elgin area, above), the subsequent decay of which destroyed any evidence of beaches.

Bremner (1920a) was the first to question the extent of raised beach deposits shown as underlying much of the town of Stonehaven on early editions of Sheet 67. He recorded the presence of red till and 'brick clay' on the surface of the supposed '100 foot raised beach' and an

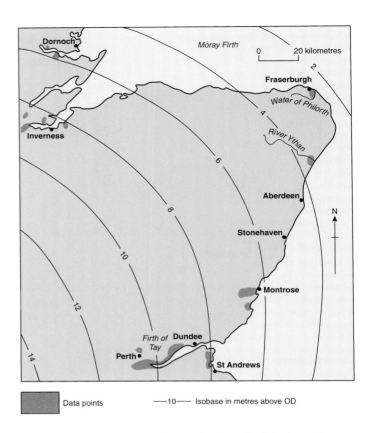

Figure 49 Isobases for the Main Postglacial Shoreline in eastern Scotland (after Cullingford et al., 1991).

'absence of all appearance of sifting and arrangement by sea waves' of the supposed beach gravels. These observations have been confirmed during the resurvey of Sheet 67. The gravels underlying the town that were formerly assigned to the 'Second Raised Beach' are shown as glaciofluvial sheet deposits on the current geological map.

Fragmentary Late Devensian raised shorelines have been recognised at six localities between Stonehaven and Inverbervie (Cullingford and Smith, 1980, table 3, fig. 2) and have been grouped with more extensive fragments that extend as far south as St Cyrus. The Stonehaven–Inverbervie shorelines occur at elevations ranging from 7.5 to 38.8 m OD between Garron Point and the mouth of the Cowie Water, at Downie Point, Bowdun Head, north of Crooked Haven, north of Upper Mill and at Kinghornie near Inverbervie (Map 11). Raised beach deposits have been mapped south-westward from Garron Point, at Downie Point and at Kinghornie. The deposits appear to be heterogeneous. A resistivity sounding at Garron Point indicated that the deposit is composed mainly of clayey silt, but in a degraded bluff 350 m south-west of Mains of Cowie [NO 877 868], 3 m of well-rounded cobbly beach gravel was formerly exposed, overlying red-brown till (Auton et al., 1988).

The shoreline fragments occur at distinctly higher elevations than the majority of those found farther south, between Dundee and Montrose (Cullingford and Smith,

1980, fig. 7) and have not been directly correlated with these well established sequences. Because of their high elevation and proximity to spreads of kettled glaciofluvial outwash, Cullingford and Smith (1980) postulated that the Stonehaven–Inverbervie shorelines developed early in the deglaciation of Strathmore, while ice was still present in the vicinity. These Late-glacial shorelines are therefore probably older than any of the features in the Montrose Basin and in the Forth and Tay valleys.

Flandrian raised beaches

The highest and best developed Flandrian raised beaches in the district were formed about six thousand years ago. They generally equate with the 'Main Post-glacial Shoreline' of eastern Scotland, which is tilted gently north-eastwards and drops to its lowest level between Fraserburgh and Peterhead (Figure 49). Along the coast of the Moray Firth it falls from about 9 m OD at Inverness, to 7.5 m OD near Elgin, and to near present high-tide level at Fraserburgh (Peacock et al., 1968; Smith et al., 1982, 1983; Firth and Haggart, 1989). Along the North Sea coast, it rises southwards to about 4 m OD in the Aberdeen area, and to between 5 and 5.5 m at Milton Ness [NO 772 647], 8 km north of Montrose, and 6.4 m at Montrose (Smith and Cullingford, 1985; Cullingford et al., 1991). At Fraserburgh, however, the youngest Flandrian shoreline is about one metre higher than the Main Postglacial Shoreline, indicating that sea level is continuing to rise hereabouts (Smith et al., 1982).

The most extensive Flandrian raised beach deposits in the district occur on Sheet 95 Elgin (Map 1), where they form storm beaches and shingle ridges that stretch from Kinloss to Burghead and from Lossiemouth to Portgordon (Ogilvie, 1923; Steers, 1937; Peacock et al., 1968). The western belt ranges from 180 m to 1.6 km in width and comprises shingle ridges that are more or less parallel to the beach and that are separated from one another by strips of peat. The ridges are locally buried beneath blown sand. The eastern belt ranges from 700 m to 2 km in width and contains similar features, except that the ridges bend southwards away from the shore at the mouth of the River Lossie (Plate 20). The ridges are formed of openwork, well-sorted, well-rounded, pebble to cobble grade gravel with sparse beds of sand. Although generally spherical, wave action has created local accumulations of tabular or bladed clasts. Shell fragments are common locally in the lower ridges. The gravel is generally very durable and has been exploited widely for aggregate (Appendix 2). The highest ridges occur at the back of the two belts, where they locally reach about 9 m OD. These features probably formed during the creation of the Main Postglacial Shoreline, whereas those at lower levels towards the sea formed subsequently, while sea level was falling.

There is a belt of peaty and sandy flats that stretches between Kinloss and Loch Spynie, behind the beach ridges (Map 1). The flats have been variously mapped as raised marine deposits, alluvium and peat, depending on what sediment occurs at the surface. They mark the sites of former brackish lagoons and tidal flats that existed when Roseisle and Branderburgh were islands in the mid-Holocene. The marine deposits generally consist of olive-grey or brown silty clay and yellowish brown mica-ceous sand. The River Lossie seems at various times to have found outlets northwards near its present mouth at Lossiemouth, and westwards south of Burghead. In historical times, there was a seaport beside the Palace of Spynie [NJ 231 658] until the late 15th century.

Flandrian raised beaches, formed mainly of shingle, occur discontinuously between the mouth of the River Spey and Troup Head, mainly at the heads of inlets and coves along this rugged, indented stretch of coast. The back feature of the raised beaches locally coincides with the base of the cliffline backing the pre-Late Devensian rock platform.

Raised Flandrian estuarine deposits are widespread in the Fraserburgh area, where they consist mainly of brown silty clay. The clay locally overlies peat, within which two laterally persistent layers of minerogenic material occur (Appendix 1 Philorth valley). The lower layer consists of grey sand and lies at about 1 m below OD. It has been dated to about 7 ka BP and is attributed to a regionally significant tsunami ('tidal wave'). The upper layer consists of micaceous sandy silt that was laid down between 6.3 and 5.7 ka BP during the Main Postglacial Transgression. The brown clay at the surface began to accumulate after about 4.8 ka BP during a subsequent transgression that probably continues today.

Shingle ridges and sand dunes, some of which were formed in historical times, extend along the coast between St Combs and Rattray Head, on Sheet 97 Fraserburgh. They separate the Loch of Strathbeg (formerly a tidal lagoon) and the historical port of Old Rattray from the sea (Peacock, 1983). Between Rattray Head and Peterhead (Map 7) the highest Flandrian beach is only a little above high tide level, and is backed by a low cliff cut in glacial deposits. A borehole drilled on the raised beach 350 m south-east of Annachie [NK 1085 5280] proved 2.6 m of sandy gravel resting on till at an elevation of 1.1 m below OD (McMillan and Aitken, 1981). The gravel is well rounded with a matrix of medium- to coarse-grained shelly sand. However, this raised beach, and others facing Sandford Bay and the Bay of Cruden, is largely concealed by blown sand. A borehole positioned amongst sand dunes behind the beach at the Bay of Cruden [NK 0858 3555] proved 8.1 m of grey sand and sandy gravel with shell fragments resting on till at 9.5 m below OD.

Several small deposits of grey silty clay flank the estuary of the River Ythan downstream of Kirkton of Logie-Buchan on Sheet 87W Ellon. These deposits resemble estuarine carse clays of the Forth estuary and are associated with the highest Flandrian shoreline at about 4.5 m OD (Smith et al., 1999). At Waterside [NK 003 273], a layer of grey sand attributed to a major tsunami (see above) has been found lying at about 0.5 m OD (Long et al., 1989). The main raised Flandrian shoreline is extensive between the Ythan estuary and Aberdeen, but it is generally buried beneath blown sand, like the low cliffline backing the feature. However, information from boreholes and trial pits in the vicinity of Menie Links indicates that the shoreline is mainly under-lain by over 2 m of dark grey, micaceous fine-grained sand and sandy silt (Auton and Crofts, 1986). The deposit is

Plate 20   Postglacial raised beach deposits bordering Spey Bay, looking westwards towards Binn Hill. The long, parallel ridges of beach shingle reach 7 m above OD (MNS 2113).

peaty at the top and becomes gravelly towards the base. The peaty top is probably equivalent to several beds of peat that have been recorded below present-day beach deposits in the vicinity (Jamieson, 1858; Bremner, 1943). Extensive building development has taken place on the raised beach deposits between the mouths of the Don and the Dee, which are mainly composed of sand and gravel (Munro, 1986).

A broad raised beach, at about 5 m OD occurs immediately inland of the present high-water mark around Stonehaven Bay where it forms the flat-lying ground beneath Stonehaven sea front and shopping centre. The raised beach is generally about 200 m wide and is backed by a degraded cliffline, trending north–south, cut in bedrock capped by thin till and head (Barclay Street follows the break of slope at the base of the cliff). The raised beach reaches some 350 m in width near the mouth of the Cowie Water, where the flat ground has been developed as holiday caravan parks. Records from site investigations near Stonehaven harbour indicate that the raised beach deposits comprise pebbly sand and gravel, containing small amounts of shell debris locally. The deposits generally range in thickness from 2 to 5 m and overlie either red-brown laminated clay or bedrock. The nature of the stratification within the raised beach is unknown, but surface topography suggests that Allardice Street follows a north–south-trending beach ridge (Carroll, 1995a).

Flat-lying raised beach deposits have been mapped, inland of the high water mark between Bervie Bay (Inverbervie) and the southern margin of Sheet 67. They extend to between 5 and 10 m OD. Some of the highest deposits may be of Late-glacial age, but most of the lower spreads are products of marine transgression during the Holocene. The deposits consist of coarse-grained pebbly sand and gravel, with minor amounts of shell debris (Carroll, 1995b). The raised beaches are backed by the steep Main Postglacial Cliffline, which is cut into till and bedrock. Widespread landslides are developed along the cliff face and in many places, the raised beach deposits are concealed beneath slipped material at the foot of the cliff.

Raised wave-cut platforms and associated features

Wave-cut rock platforms with back-features a few metres above present high water mark are common along the rocky northern coast of the district on sheets 95, 96W and 96E. Between Burghead and Lossiemouth, at Cullen and at Gardenstown, a prominent rock platform and cliff is cut into the relatively weak Permo–Triassic and Devonian sandstones. There are 'fossil' sea-stacks on the platform above high tide level to the east of Covesea (Map 1), and also at Cullen. These features are probably partly postglacial in age, but between Portsoy and Troup Head, the platforms probably occur at more than one level and some of them are partially overlain by glacial deposits of the Whitehills Glacigenic Formation (Chapter 8). The main platform reaches some 300 m in width between Whyntie Head and Whitehills (Plate 21), but where the cliffs are particularly high, such as at Troup Head, it is either missing or is represented only by rock reefs with concordant levels. There is little evidence of correlative platforms along the eastern coast of the district. One does occur at the 'Bridge of One Hair', south of the Bay of Nigg, on Sheet 77, where a sea-stack is joined to the mainland by a plug of red till some 30 m thick (Synge, 1956).

A raised rock platform standing at an elevation of about 23 m OD extends between Ruthery Head [NO 888 872] and the Glen Ury distillery [NO 871 869], north of Stonehaven. Bremner (1920) suggested that the platform is a 'preglacial' feature because it is capped by till. It is more likely to have been formed following a major pre-Late Devensian glaciation(s) when sea level was

Plate 21   Raised wave-cut rock platform at Whyntie Head [NJ 626 658], near Portsoy, looking westwards. The platform was cut prior to the deposition of shelly sands of the Whitehills Glacigenic Formation, which overlie it locally (P104117).

relatively high as a result of glacio-isostatic depression of the land. The till-covered feature is, after all, capped by Late Devensian raised beach deposits. Other fragmentary rock platforms occur to the south of Stonehaven at Downie Point, Bowdun Head, near Crooked Haven by Kinneff, and at Kinghornie, just north of Inverbervie.

Submerged wave-cut platforms

Along the eastern coast, the sea cliffs locally descend to a topographical break, possibly a wave-cut platform, at 55 to 60 m below OD (Crofts, 1975). A series of till-covered platforms have been detected from seismic records offshore between Stonehaven and Inverbervie (Stoker and Graham, 1985). These occur at about 30, 45 to 50 and 60 to 70 m below sea level. Wave-cut platforms such as these have been attributed to severe marine erosion during periods of lowered sea level coincident with periods of glaciation (Sutherland, 1984a). Submerged geos and

stacks partially enclosed by till occur to the north of Newburgh and south of Aberdeen (Synge, 1956; Walton, 1959).

'Pre-glacial' raised beaches

Deposits of angular and rounded boulders, gravel, and sand lie locally between bedrock and till along the Banffshire coast. They were ascribed by Read (1923) to a 'pre-Glacial raised beach'. However, re-examination of these sediments at Portessie [NJ 4510 6722], in a gully [NJ 4710 6815] between Findochty and Portknockie, and at Portnockie harbour [NJ 4882 6858] (Map 2), suggests that most, if not all, are of glacial or glaciofluvial origin (Peacock and Merritt, 2000). The deposit at the first locality is interpreted as 'immature' till whereas the gravel at the other two localities is almost certainly glaciofluvial, laid down or reworked immediately prior to the advance of the Moray Firth ice stream of the Main Late Devensian ice sheet across the area.

EIGHT

Quaternary lithostratigraphy and correlation

INTRODUCTION

Some of the more recently published drift editions of sheets covering the district accommodate lithostratigraphical units of glacigenic deposits in addition to the conventional morpho-lithogenetic categories (Figure 5). For example, several glacigenic formations have been mapped on Sheet 66E Banchory and Sheet 67 Stonehaven. On other sheets, such as 87W Ellon and 76E Inverurie, conventional morpho-lithogenetic categories of deposit have been assigned to one or more lithostratigraphical group on the basis of gross lithological characteristics. It is beyond the scope of this publication to define and describe every lithostratigraphical unit, but some discussion of stratigraphical correlation is included, especially of the tills, because it is important for deciphering the history of events that have taken place. A set of generalised maps showing the distribution of drift groups and a selection of deposits, landforms and localities mentioned in the text is provided at the back of this publication (Maps 1 to 11).

Definition of terms

Lithostratigraphy involves the description, definition and naming of rock units (Whittaker et al., 1991). It is fundamental to stratigraphy because accuracy in *biostratigraphy* (based on fossils), *geochronometry* (based on measurements of age in years) or *chronostratigraphy* (apportioning units of strata to intervals of time) relies on the correct recognition of the relative spatial relationships of rock units, as does subsequent palaeogeographical reconstruction (see Chapter 1). Individual units are normally described and defined using their gross lithological characteristics and by their interrelationships with adjacent units, but this is not always possible for Quaternary glacigenic sequences, especially when they have been glacitectonised and/or occur as large glacial rafts. Such units are defined primarily on the basis of their unconformable bounding surfaces, rather than on lithology alone (e.g. Whitehills Glacigenic Formation, see below). Strictly speaking, the result is a hybrid of lithostratigraphy and *allostratigraphy* (Hedberg, 1976; Owen, 1987), the latter being based on the recognition of units bounded by discontinuities and their correlative conformities, such as in offshore *seismic sequence stratigraphy*.

In the Quaternary succession of north-east Scotland, discrete lithological units are named, as far as possible, after localities lying close to characteristic sections. They are then ranked in a formal hierarchy of *Bed, Member, Formation* and *Group* (Table 7). The *Formation* is commonly defined as the basic mappable unit in the hierarchy. *Type sections* (stratotypes) are chosen to illustrate typical lithologies, characteristics and relationships to adjacent units, but in complex and commonly poorly exposed areas a single section or borehole sequence rarely suffices and several *partial type sections* and *reference sections* may be required. Where formations and groups embrace numerous lesser ranked units, or there is poor exposure, a *type area* is given instead of a type section. Many type sections and type areas are defined in the descriptions of important localities in Appendix 1. All units established by BGS have been entered into the *BGS Lexicon of named rock units and the index of computer codes* (BGS, 2002).

Numerous units have been named in the literature; most are informal and as far as possible each one has been allocated here to one of the groups and assigned formation, member or bed status (Table 7). Names generally have not been changed unless ambiguity exists, where they have been used already for other lithostratigraphical units, or cardinal rules have been broken (e.g. the use of 'lower' and 'upper' qualifiers).

Since setting up the scheme outlined here, a revised correlation of Quaternary deposits in Scotland has been published by the Geological Society of London as part of a wider review of the British Isles (Sutherland, *in* Bowen, 1999). No groups are recognised in that publication, the 'member' becomes the basic mappable unit and formations are really 'type sequences'. For example, the 'Kirkhill Formation' embraces all units at Kirkhill irrespective of lithology, provenance, genesis and age. Sutherland's scheme is not adopted here for several reasons. The main objection is that most of his formations are not mappable units. Some include all drift deposits occurring in an area (e.g. Kirkhill) whereas others include only minor elements of the local drift sequence (e.g. Castleton and Camp Fauld). The omission of lithological qualifiers makes it difficult to comprehend what the units are and how they interrelate. There are also numerous spelling mistakes (e.g. Elton instead of Ellon Member) and inconsistencies (e.g. similar locality names appear at both formation and member level). Another problem is that several important papers have been published since the Geological Society Report was in preparation and recent mapping has helped clarify spatial relationships of many deposits away from type localities. For example, new information at Boyne Bay and Gardenstown indicates that the 'Castleton Formation' is almost certainly a collection of glacial rafts and not in situ, and the organic deposits at Camp Fauld and Crossbrae are now placed in a different stratigraphical context (Appendix 1). In general, Sutherland's nomenclature has not been adopted here unless new names are required for formerly inappropriately named, or unnamed units.

Table 7 Correlation of lithostratigraphical units in north-east Scotland.

Oxygen Isotope Stage	Teindland/Elgin	Boyne Limestone Quarry/Keith	Gardenstown/Banff	Byth/Crossbrae	Kirkhill/Leys	Peterhead/Cruden	Eilon/Fyvie	Aberdeen	Banchory	Stonehaven
Flandrian (Holocene) — 1										
Loch Lomond Stadial — 2a	Garral Hill Gelifluctate Bed			Todholes Gravel Bed			Woodhead Gelifluctate Bed			
Windermere Interstadial — 2b	*Garral Hill Peat Bed*			*Thinfolds Peat Bed*				*Mill of Dyce Peat Bed*	*Loch of Park Gyttya Bed*	*Glenbervie Peat Bed*
Dimlington Stadial — 2c	Spynie Clay Formation	Kirk Burn Silt Formation	Kirk Burn Silt Formation		*St Fergus Silt Formation*	St Fergus Silt Formation	Woodhead	Tullos Clay Member	Lochton Sand & Gravel Formation	Drumlithie Sand & Gravel Formation
	Waterworks Till Formation	Arnhash Till Member		Crossbrae Gelifluctate Bed	Manse Gelifluctate Bed	Ugie Clay Formation	Ugie Clay Formation	Drumlithie Sand & Gravel Formation	Glen Dye Silts Formation	Ury Silts Formation
		Blackhills Sand & Gravel Formation		Auchmedden Gravel Formation	Kirkhill Church Sand Formation	Essie Till Formation	Kippet Hills Sand & Gravel	Glen Dye Silts Formation	Mill of Forest Till Formation	Mill of Forest Till Formation
	Tofthead Till Formation	Old Hythe Till Formation	Crovie Till Formation	Byth Till Formation	East Leys Till Formation	Hatton Till Formation	Hatton Formation	Mill of Forest Formation	Banchory Till Formation	
					Hythe Till Formation	Sandford Bay Till Member	Bearnie Till Member	Nigg/Kingswells Till members		
							Auchleuchries Sand & Gravel Formation	Ness Sand & Gravel Member		
								Den Burn Till Member		
early Late Devensian glaciation	Altonside Till Formation	Whitehills Glacigenic Formation	Whitehills Glacigenic Formation		Corse Diamicton Formation	Rafts at Oldmill	Pitturg Till Formation	Anderson Drive Diamicton Formation		
3				Howe of Byth Gravel Formation	Corsend Gelifluctate Bed					
4	Woodside Diamicton Formation		Pishlinn Burn Gravel Bed			Aldie Till Formation				
						Hardslacks Gelifluctate Bed				
5a-c	*Badentinian Sand Bed*			*Crossbrae Farm Peat Bed*		*Berryley Peat Bed*				*Burn of Benholm Peat Bed*
Ipswichian Interstadial — 5e	*Teindland Palaeosol Bed* / *Truncated palaeosol*				*Fernieslack Palaeosol Bed*					
	Orbliston Sand Bed					Moreseat Farm Sand Bed				
6?	Deanshillock Gravel Formation	Crag of Boyne Till Formation		Crossbrae Till Formation	Rottenhill Till Formation	Camp Fauld Till Formation	Pitlurg Till in part?			Benholm Clay Formation
	Red Burn Till Formation				West Leys Sand & Gravel Formation		Tillybrex Sand & Gravel Formation			Birnie Gravel Formation
					Camphill Gelifluctate Bed		Bellscamphie Till Formation			
					Swineden Sand Bed					
7?					Kirkhill Palaeosol Bed					
					Pitscow Sand & Gravel Formation					
8?					Kirkton Gelifluctate Bed					
					Denend Gravel Formation					
					Leys Till Formation					
References	Hall et al. (1995)	Sheet 96W; Godwin and Willis (1959); Peacock and Merritt (2000a)	Sheet 96E; Peacock and Merritt (1997)	Hall et al. (1995); Whitington et al. (1998)	Connell and Hall (1987)	Sheet 87E; Connell and Hall (1987); Whitington et al. (1993)	Sheet 87W; Connell and Hall (1987); Hall and Jarvis (1995)	Bremner (1931, 1943); McLean (1977); Munro (1986); Murdoch (1977)	Sheet 66E; Vasari (1977)	Sheet 67; Auton et al. (2000)

Dated unit

NOTE: In general, minimal ages are shown. For example, Crossbrae Gelifluctate Bed may be OIS 2c to 4, Anderson Drive Diamicton may be OIS 6, Kirkhill Palaeosol Bed may be OIS 9 or 11.
All Peat and Palaeosol beds are assigned to the group of the underlying or enclosing deposit.
Italicised units are informal: they have not been entered into the BGS Lexicon.

Central Grampian Drift Group   East Grampian Drift Group   Banffshire Coast Drift Group   Logie-Buchan Drift Group   Mearns Drift Group

The groups

Five lithostratigraphical groups of predominantly glacigenic (glacial, glaciofluvial, glaciolacustrine) deposits have been established in north-east Scotland. They reflect lithology and provenance rather than age and may encompass units formed during several glaciations. Although some nonglacigenic materials such as palaeosols and periglacial deposits are included, these units are closely associated with the glacigenic sequence and represent important events in the geological record of Quaternary events in the district. A 'top down' approach has been adopted in order to allow sufficient room in the classification for complex successions and to accommodate more information in the future. It also allows traditional morpho-lithogenetic units to be embraced on sheets where formal lithostratigraphical division has not been attempted.

Traditionally, it has been the practice in north-east Scotland to relate the glacigenic deposits to one of three distinct bodies of ice that are thought to have existed in the region (Figure 4). Three 'series' have been informally recognised (Sutherland, 1984a; Hall and Connell, 1991). Ice moving north-eastwards along the east coast led to the deposition of the *Red Series* (Jamieson, 1906; Synge, 1956), which include a variety of materials of a typically vivid reddish brown colour. Ice moving eastwards along the Moray Firth impinged onto the coastal lowlands and was responsible for laying down a suite of dark grey deposits assigned to the *Blue-grey Series* (Synge, 1956). In order to complete the picture, the typically sandy, yellowish brown diamictons laid down in the interior by ice flowing from the Grampian Highlands were assigned to an *Inland Series* (Hall, 1984a).

The tripartite division is retained in the present scheme, but the series are renamed as groups ('series' is a chronostratigraphical term). The Blue-grey series becomes the *Banffshire Coast Drift Group* and the Inland series becomes the *East Grampian Drift Group*. The Red series warrants division in order to separate deposits that were laid down by ice that moved onshore from the North Sea basin (*Logie-Buchan Drift Group*) from those that were derived entirely from ice flowing within Strathmore (*Mearns Drift Group*). Additionally, the deposits laid down by ice emanating from the Central Highlands and entering the area via the Spey valley are assigned to the *Central Grampian Drift Group*. Clast composition is the most important attribute in deciding from which ice stream a deposit was derived. Matrix colour remains a useful parameter for general classification. However, both methods require care in their application because distinctive clasts may have been reworked from older deposits and incorporation of local rock or sediment can locally change the typical colours of matrices.

The distribution of the five drift groups is shown approximately in Figure 4 and on Maps 1 to 11. The relative positions of the ice streams changed with time during the last glaciation, and the ice streams of earlier glaciations probably did not cover exactly the same

Table 8   Radiocarbon dates from Late-glacial sites in the district.

Site	Grid reference	Laboratory number	Age (years BP)	Dated material and setting	Reference
Rothes cutting	NJ 277 498	Beta-86532	11 110 ± 70	peat under remobilised till	Appendix 1
Garral Hill, Keith	NJ 444 551	Q-104	10 808 ± 230	peat under remobilised till	Godwin and Willis (1959)
Garral Hill, Keith	NJ 444 551	Q-103	11 098 ± 235	peat under remobilised till	Godwin and Willis (1959)
Garral Hill, Keith	NJ 444 551	Q-102	11 308 ± 245	peat under remobilised till	Godwin and Willis (1959)
Garral Hill, Keith	NJ 444 551	Q-101	11 888 ± 225	peat under remobilised till	Godwin and Willis (1959)
Garral Hill, Keith	NJ 444 551	Q-100	11 358 ± 300	peat under remobilised till	Godwin and Willis (1959)
Woodhead, Fyvie	NJ 788 384	SRR-1723	10 780 ± 50	peat under remobilised till	Connell and Hall (1987)
Howe of Byth	NJ 822 571	SRR-4830	11320	peat beneath gravel	Hall et al. (1995)
Moss-side, Tarves	NJ 833 318	I-6969	12 200 ± 170	peat under remobilised till	Clapperton and Sugden (1977)
Loch of Park	NO 772 988	HEL-416	10 280 ± 220	kettlehole infill	Vasari and Vasari (1968)
Loch of Park		HEL-417	11 900 ± 260	kettlehole infill	Vasari and Vasari (1968)
Mill of Dyce	NJ 8713 1496	SRR-762	11 550 ± 80	kettlehole infill	Harkness and Wilson (1979)
Mill of Dyce	NJ 8713 1496	SRR-763	11 640 ± 70	kettlehole infill	Harkness and Wilson (1979)
Glenbervie	NO 767 801	GX-14723	12 460 ± 130	peat under remobilised till	Appendix 1
Glenbervie	NO 767 801	SRR-3687a (humic)	12 305 ± 50	peat under remobilised till	Appendix 1
Glenbervie	NO 767 801	SRR-3687b (humin)	12 340 ± 50	peat under remobilised till	Appendix 1
Brinzieshill Farm	NO 7936 7918	SRR-387	12 390 ± 100	peat under remobilised till	Auton et al. (2000)
Rothens	NJ 688 171	SRR-3803	10 680 ± 100	kettlehole infill	Appendix 1
Rothens	NJ 688 171	SRR-3804	11 640 ± 160	kettlehole infill	Appendix 1
Rothens	NJ 688 171	SRR-3805	11 760 ± 140	kettlehole infill	Appendix 1

Plate 22   Southern face of Kirkhill Quarry [NK 011 528] in 1979 (P528224). The sands and gravels resting on bedrock belong to the Pitscow Sand and Gravel Formation. The conspicuous pale band at their surface is the podzolic Kirkhill Palaeosol Bed. Periglacial sediments of the Swineden Sand Bed and the Camphill Gelifluctate Bed succeed this. The base of the Rottenhill Till Formation is at shoulder level with the figure. It is weathered towards the top (Fernieslack Palaeosol Bed) and is overlain by the pale coloured Corsend Gelifuctate Bed. The overlying Hythie Till Formation has a sharp erosional base. Note, the deformation structure affecting sediments beneath the till (above and to the left of the figure). Approximately 8 m of sediments are exposed.

Plate 23   Partially glacitectonised, podzolic Teindland Palaeosol Bed below sandy diamicton of the Woodside Diamicton Formation at Teindland quarry [NJ 297 570]. Luminescence dates of 79 ± 6 ka and 67 ± 5 ka have recently been obtained from the sands overlying the palaeosol bed to the left of the scale card. (P104118). Scale bar showing 1 cm intervals.

ground as in the Late Devensian (Chapter 5). There is consequently local, but stratigraphically significant interdigitation of deposits belonging to different groups.

Organic deposits, buried soils and periglacial sediments and structures

The lower relief areas of north-east Scotland are notable for the occurrence of a significant number and variety of Middle and Late Pleistocene nonglacial, terrestrial sediments. These serve as important stratigraphical marker horizons and provide evidence for ice-free episodes within the Pleistocene succession. Additionally, they

provide important, and in some cases unique, evidence from which to reconstruct the climate of these periods.

Organic deposits

Peat, organic mud and organic-rich sand occur at numerous sites buried by later sediments. Most commonly found are Late-glacial organic deposits (Gunson, 1975), chiefly lake or pond muds and thin peat beds, organic muds and sands (Table 8 Garral Hill, Byth, Woodhead, Mill of Dyce, Loch of Park and Glenbervie). The pollen record from these sites reveals that in the Windermere Interstadial, opening at around

13 000 BP, an initial phase of open habitat vegetation was succeeded by closed heath and scrub vegetation, with local growth of tree birch and pine. After around 11 500 BP, the temperature began to decline and tundra vegetation had become established by the start of the Loch Lomond Stadial at 11 000 BP.

No deposits have yet been found in north-east Scotland representing the last nonglacial period preceding the Main Late Devensian Glaciation. This 'Sourlie Interstadial' (Jardine et al., 1988), dated to around 30 000 BP, has been recognised from sites in western and central Scotland, where good evidence of tundra flora and fauna is preserved. The buried soil at Teindland and the peat at Crossbrae (Table 7; Appendix 1) were formerly thought to date, at least in part, from this period, but the original relatively young radiocarbon dates from the sites probably result from contamination.

Evidence of older organic sediments has been found, but these deposits are beyond the range of radiocarbon dating. As only a few uranium series and luminescence dates have been obtained, the ages of these deposits are poorly known. However, detailed pollen investigations, and an increasing number of studies of Coleoptera allow tentative biostratigraphic correlation. Evidence of Early Devensian interstadial conditions is found at Camp Fauld, Crossbrae and Burn of Benholm (Table 7; Appendix 1). The organic materials at these sites appear to be equivalent in age to Oxygen Isotope Stages (OIS) 5c and 5a, but correlation with one or other interstadial remains tentative. Birch and pine woodland was probably extensive in the region during the warmest parts of these interstadials. Episodes with dominant dwarf birch and willow scrub mark cooler episodes and the terminations of the interstadials are presaged by a decline into tundra conditions.

Organic muds and sands associated with podzolic buried soils at Teindland and Kirkhill contain pollen of temperate woodland. At Teindland, the pollen record shows an early phase of pine, alder and hazel woodland with grassland clearings, followed by vegetation indicative of increasingly colder conditions. This appears to mark cooling at the close of the last interglacial, OIS 5e. At Kirkhill, an older interglacial (possibly OIS 7) is represented by thermophilous pollen of pine and alder and very rare grains of birch, lime and elm (Appendix 1). This pollen is contained within a unit of organic-rich sand (Swineden Sand Bed), which is believed to have been redeposited from soil horizons in the vicinity. The pollen spectra are dominated by open grassland communities, which were probably present during the redeposition (information from J J Lowe, Royal Holloway University, London, 1990). This unit of sand is conformably overlain by sandy and gravelly gelifluction deposits (Corse Gelifluctate Bed) that formed in a periglacial environment (Table 7; Appendix 1). As at Teindland, the organic sand contains evidence of climatic deterioration.

Interglacial and interstadial buried soils

Buried soils and associated weathering profiles occur at several sites and record significant intervals of temperate conditions (Plates 22; 23). Truncated podzolic soil profiles

of interglacial status, developed in sand and gravel parent materials, occur in association with organic muds and sands at both Teindland (OIS 5e) and Kirkhill (possibly OIS 7) (Table 7; Appendix 1). A second, younger, truncated soil profile at Kirkhill (Fernieslack Palaeosol Bed) is considered to represent a gleyed 'brown earth' of interglacial status (OIS 5e) developed in diamicton parent material (Connell and Romans, 1984). Weathering profiles, with disintegration of clasts and mobilisation and movement of iron and manganese, are developed on the surface of older tills, gravels and sands of the East Grampian Drift Group at Teindland, Boyne Limestone Quarry, Howe of Byth, Moreseat, Tillybrex (Ellon), and Kirkhill (Table 7; Appendix 1). Around this last site, weathered till and gravel can be traced over an extensive area and represents both OIS 5e and earlier warm intervals.

Plate 24   Attenuated rafts of sand and brecciated dark grey pebbly clay within the Whitehills Glacigenic Formation at Oldmill pit [NK 0243 4389]. Greyish brown diamicton of the Banffshire Coast Drift Group caps the section above a metre-thick unit of grey, sheared sandy diamicton (a penetrative glacitectonite). The ice that overrode the sequence flowed towards the left (east-south-east) (P104119).

Periglacial sediments and structures

A range of periglacial features have been recognised in units at several positions within the Pleistocene sequence. These include mass movement deposits, frost shattered bedrock, fluvial sands and gravels and structures indicative of the former presence of ground ice, notably cryoturbation structures and ice-wedge casts. It is clear that periglacial processes have had a significant impact on the landscape of north-east Scotland during most of the Quaternary, particularly across central Buchan (Chapter 7). Intense periglacial activity last occurred during the Loch Lomond Stadial when marked slope instability, with destruction of soils and remobilisation of diamictons by gelifluction, led to slope-foot accumulations of significant thickness, locally burying organic sediments of Windermere Interstadial age.

Evidence of older periods of periglacial activity includes a bed of sandy gelifluctate with ice-wedge casts that lies beneath rafts and diamicton of the Whitehills Glacigenic Formation, and on slightly weathered gravel at Oldmill (Appendix 1). These features probably represent one of the cold periods of the Middle or Late Devensian. Organic sediments of probable OIS 5e age at Teindland and either 5c or 5a age at Camp Fauld and Crossbrae (Appendix 1) record a decline to tundra conditions and the increasing slope instability is manifest as increased accumulation of sand, gravel and diamicton.

Undoubtedly the clearest record of recurring events is from Kirkhill, where quarrying activities over a period of 10 years allowed detailed resolution of multiple periglacial phases (Table 7; Appendix 1). Extensive mass movement occurred towards the end of the Main Late Devensian glaciation and possibly also in the Loch Lomond Stadial (Manse Gelifluctate Bed). Cryoturbation, solifluction and ice wedge growth is also recorded in the Corsend Gelifluctate Bed lower in the sequence. The latter bed contains elements that probably formed both at the close of the last interglacial (OIS 5e) and prior to the complex phase of glaciation that deposited the Corse Diamicton and overlying Hythie Till (Table 7). 'Silt-droplet fabrics' have been revealed by micromorphological examination and probably record the development of an immature soil, or soils, close to a periglacial landsurface (J C C Romans, personal commnication *in* Connell, 1984c). These events may also be represented in an equivalent bed at Oldmill (Appendix 1).

Another sandy gelifluctate bed in the Kirkhill/Leys sequence (Corse Gelifluctate Bed) is associated with an ice-wedge cast network. It has been dated by luminescence to 142 ± 19 ka yr BP (Duller et al., 1995) and if correct, marks an important periglacial period within OIS 6. The oldest periglacial sediment recognised at Kirkhill is the angular felsite rubble of the Kirkton Gelifluctate Bed, which must date to OIS 8 or an older cold period.

BANFFSHIRE COAST DRIFT GROUP

This group of deposits contains clasts derived from the Moray Firth basin in addition to more locally occurring rock types. They occur from Inverness eastwards towards Peterhead (Figure 4). The diamictons, silts and clays are typically dark grey, calcareous and contain ice-worn Quaternary shell fragments in addition to abundant reworked Early Jurassic to Early Cretaceous microfossils. Permo–Triassic and Lower Jurassic to Upper Cretaceous strata crop out within the Moray Firth basin (BGS, 1977) and both clasts and large glacial rafts of these rock types are common in the tills of this group (Chapter 7). Locally, both the colour and composition of the tills stongly reflect the lithology of the underlying bedrock. The sands and gravels generally contain similar suites of clasts within the tills, and shell fragments are common.

TILL

The most distinctive tills of the Banffshire Coast Drift Group are bluish grey in colour, clayey and contain erratics and rafts of Mesozoic rocks and Quaternary sea floor sediments derived from the Moray Firth basin, together with reworked fossils and shell fragments. They have a discontinuous distribution at the surface and some older units occur at depth interbedded with deposits of the other groups (Figure 4; Table 7). Though it has long been concluded that most of these bluish grey diamictons were deposited by ice moving landwards from the Moray Firth basin (Jamieson 1906; Read, 1923; Bremner, 1928, 1934), direct evidence for such ice movement, such as striations or clast fabric measurements, are rare (Chapter 7). Furthermore, there has been little consensus concerning when, and how many times, ice moved onshore to produce these distinctive deposits (Chapter 5).

Whitehills Glacigenic Formation

Blue-grey diamictons occur widely, but not extensively near, or at, the surface between Elgin, Peterhead and Ellon. Most of them probably belong to one laterally discontinuous, allostratigraphical unit, the Whitehills Glacigenic Formation (Peacock and Merritt, 1997, 2000). The formation is generally capped by brown, red or grey till containing clasts derived from the west or west-north-west (Jamieson, 1906; Read, 1923; Peacock et al., 1968; Aitken et al., 1979; Hall et al., 1995a) and is almost certainly of Late Devensian age (Peacock and Merritt, 1997, 2000). It is named after the village lying 4 km to the west of the type section in the Boyne Limestone Quarry [NJ 612 658] on Sheet 96W Portsoy (Map 2; Appendix 1 Boyne Bay, Gardenstown and King Edward).

The Whitehills Glacigenic Formation is notable for its inclusion of a relatively large proportion of glacial rafts of Mesozoic strata and Quaternary sediments derived from the Moray Firth basin (Plate 17b). The rafts include mudstones, friable sandstones, clays, silts, sands and gravels (Chapter 7). Jurassic material predominates in the west, whereas some Lower Cretaceous rocks and fossils are also present in the east. Some rafts crop out at the surface and are large enough to be mapped out individually on Sheet 96W and Sheet 96E, such as the one at

Whitehills itself (Map 3), which has been largely removed for making bricks and tiles (Chapter 2). A 'blue oolite clay' with striated stones, probably a till including rafts of clay, was formerly worked at Lumbs [NK 028 578] (Milne, 1892a), on Sheet 97 (Map 4), and further occurrences of locally shelly 'blue clay' have been reported to the south and west of Fraserburgh (Jamieson 1906). Large Jurassic erratics were formerly exposed at Atherb and Brucklay Castle, to the south-west of Strichen (Map 6; Jamieson, 1906) and around Plaidy and Turriff (Map 5; Read, 1923). Dark grey to black till, locally shelly, has been recorded at depth at several localities in the catchment of the River Ugie (McMillan and Aitken, 1981). There are few exposures, but at the time of writing several rafts could be seen in a quarry at Ardglassie [NK 012 617], 5 km south-south-east of Fraserburgh (Chapter 7), and others were exposed at the Oldmill site (Plate 24; Appendix 1).

Several other units of blue-grey till that have been recognised in the district may correlate with the Whitehills Glacigenic Formation (Table 7). For example, in the valley of the Red Burn, south of Fochabers, Hall et al. (1995a) reported a dark till (*Altonside Till*) with Mesozoic erratics and a fabric indicating derivation from the northwest (Map 1; Appendix 1 Teindland). They correlated the Altonside Till with other dark tills recorded by Aitken et al. (1979) in the lower Spey valley. At the King Edward site (Map 5; Appendix 1), south-west of Gardenstown, stones in an unnamed unit of dark till retain striations suggesting ice movement from the west or north-west (Jamieson, 1906). Although that unit of till is apparently absent at the Howe of Byth site to the east (Map 4; Appendix 1), the *Howe of Byth Gravel Formation* there contains some Mesozoic erratics and might have been laid down contemporaneously with the Whitehills Glacigenic Formation. The *East Leys Till* that occurs between Fraserburgh and Peterhead is another correlative, although it is not overlain by younger tills (Appendix 1 Kirkhill). It is typically decalcified to a depth of about 4 m and includes red granite, probably from the Peterhead area, and local Inzie Head Gneiss, in addition to chalk and flint. The dark grey silty clay matrix of the till contains dinoflagellate cysts of Late Jurassic to Early Cretaceous age (Hall and Jarvis, 1993a). The East Leys Till is generally thought to have been laid down during the Devensian because it lies stratigraphically above the Fernieslack Palaeosol Bed (formerly Kirkhill Upper Buried Soil, which is likely to be Ipswichian in age or older (Duller et al., 1995). The *Pitlurg Till*, which occurs between Ellon, Cruden Bay and the Buchan Ridge (Appendix 1 Ellon) also includes materials that were apparently carried westwards from the North Sea basin in addition to Jurassic erratics and fossils derived from the Moray Firth (see below).

Tills overlying the Whitehills Glaciogenic Formation

At the type section of the Whitehills Glacigenic Formation in the upper part of the Boyne Limestone Quarry (Appendix 1) the shelly till and rafts are generally overlain by a unit of brown stony till containing clasts derived mainly from the west. This unit, named as the *Old Hythe Till Formation* (Peacock and Merritt, 2000) is correlated with the *Crovie Till Formation* (Peacock and Merritt, 1997) occurring at Gardenstown to the east (Appendix 1). At both localities the contact of the brown till with deposits of the underlying Whitehills Glacigenic Formation is gradational. Furthermore, at both localities the brown tills are judged to have been laid down by ice flowing towards the east following a change of direction from the south to south-easterly movement that brought about the emplacement of the Whitehills Glacigenic Formation and its included rafts. It is apparent that the Crovie Till Formation and Old Hythe Till Formation are likely to have been laid down during the same event, but by East Grampian ice and ice from the central Grampians respectively (Figure 4; Table 7).

A similar sequence occurs at the Oldmill site (Appendix 1; Plate 24), on Sheet 87E Peterhead, where an un-named unit of brown till is tentatively correlated with the Hythie Till Formation at Kirkhill. The brown till has erratics derived from farther west and a weak 'west to east' fabric (Merritt and Connell, 2000). It rests on dark grey till containing rafts of Jurassic shale, Quaternary marine clay and sand (Table 7). At Kirkhill, dark clay with Mesozoic erratics (*Corse Diamicton*) also locally underlies brown till with clasts derived from the west and south-west, the *Hythie Till* (formerly Kirkhill Upper Till). The former quarry at Kirkhill is situated within a drumlinoid ridge trending west–east suggesting that ice last flowed from the west. Other ice-moulded features in the area are similarly orientated (Map 6). The history of events in Buchan thus appears to match that elucidated at the Boyne Limestone Quarry and Gardenstown, with ice first flowing across the area from the north-west, transporting rafts from the Moray Firth, before changing towards an easterly direction.

Essie Till Formation

Calcareous blue-grey tills occur at the surface between Rosehearty and Peterhead on Sheet 97 and Sheet 87E. Most glacial striae, ice moulded features and erratics (Maps 4; 7) indicate that the last movement of ice was towards the east-south-east or south-east across the area. There is also evidence of a minor re-advance of ice from the north and east that affects a narrow coastal zone between Cairnbulg Point and St Combs (Peacock, 1997). The tills are commonly mottled grey and brown at the top and contain a variety of materials eroded from the bed of the Moray Firth. They cannot easily be distinguished lithologically from those of the Whitehills Glacigenic Formation described above, but they do not appear to have been affected by any subsequent ice movement from the west as at Boyne Bay, Gardenstown, Kirkhill and Oldmill (Appendix 1). In the absence of overlying tills derived from inland, it is uncertain whether these coastal blue-grey tills, assigned here to the Essie Till Formation, are strictly equivalent to those of the Whitehills Glacigenic Formation (or the East Leys Till); they probably relate to a later event.

The type locality of the Essie Till Formation is around South Essie farm, near St Fergus (Map 7) where several pipeline trench exposures revealed more than 2.5 m of dark grey, calcareous silty clayey diamicton with shell fragments and erratics of pink granite, quartz and metasedimentary rock (M. Munro *in* British Geological Survey Records). The unit generally overlies a reddish brown, calcareous silty clayey diamicton that probably correlates with the Hatton Till of the Logie-Buchan Drift Group. The Essie Till Formation was also exposed in a BGS Registered trial pit NK05NW6 [NK 0280 5778] dug in the vicinity of the former Lumbs claypit (Map 4) where a 'blue oolite clay' with striated stones was formerly worked (Milne, 1892a). The pit exposed over 3.8 m of stiff, dark grey clay and pebbly clay diamicton with lenses of reddish brown diamicton, small masses of red clay and very sparse fragments of shell and chalk. The dark diamicton contained sparse pebbles and cobbles of marl and red sandstone whereas the red diamicton contained granules of red sandstone and mudstone. The red materials and black mudstones and chalk are derived from outcrops of Permo–Triassic and Cretaceous strata lying within the Moray Firth basin to the north (Figure 2). The trial pit was dug into an isolated, drumlinoid hill trending north-west to south-east. Both the glacial streamlining of the feature and the composition of the till indicate ice flow towards the south-east.

Several boreholes in an area of some 15 km² to the north of the River Ugie, on Sheets 97 and 87E, proved dark shelly diamictons that are probably equivalent to the Essie Till Formation (McMillan and Aitken, 1981). For example, Borehole NK05SE9 drilled at Kinloch Farm [NK 0989 5093], 4 km north-west of Peterhead, proved 5.4 m of stiff, dark bluish grey pebbly clay diamicton with clasts of pink granite, quartz and schistose metasedimentary rock. As in the pipeline trenches, the underlying unit (3.2 m thick) is similar in lithology, but is dark reddish brown in colour and contains sparse boulders of granite. It rests on 9.8 m of crudely laminated, red, reddish brown and yellowish brown, micaceous silt of probable glaciolacustrine origin and equivalent to the Ugie Clay Formation of the Logie-Buchan Drift Group. Both diamictons are probably deformation tills and result from an advance of coastal ice across glaciolacustrine deposits (Figure 42a). The latter were laid down in lakes ponded against coastal ice after ice from inland had retreated westwards.

The clasts of pink granite in the Essie Till Formation are probably derived from a small outcrop of coarse-grained pink granite to the north-west of Crimond (Map 4) described by Wilson (1882). Commercial records indicate that pink granite also underlies part of the gas terminal at St Fergus.

Arnhash Till Member

There are several localities along the northern coast of the district where glaciofluvial deposits have been disturbed by, or locally overridden by a late-stage re-advance of the Moray Firth ice stream (Peacock and Merritt, 2000). The readvancing ice locally laid down a thin deposit of reddish brown to bluish grey gravelly diamicton, for example on the sands and gravels at Arnhash (now Pitnacalder) gravel pit [NJ 872 628], south of New Aberdour (Map 4). This unit is named here as the Arnhash Till Member of the Essie Till Formation. Other sites include the former Gallows Hill pit [NJ 512 665] near Cullen (Bremner, 1928), on Sheet 96W, Troup Head on Sheet 96E (Peacock, 1971) and Broomhead pit [NJ 983 640] south-south-west of Fraserburgh on Sheet 97. At Pitnacalder and Broomhead, deltaic deposits of the Blackhills Sand and Gravel Formation have been pushed by ice nudging inland from the north-west and north-north-east respectively (Peacock and Merritt, 2000e, fig. 26).

Older blue-grey tills

At Camp Fauld, on the Moss of Cruden (Appendix 1), excavations revealed greenish grey diamicton (Aldie Till) stratigraphically overlying a dark grey diamicton (*Camp Fauld Till*). The Aldie Till contains clasts of flint, quartzite, fresh red granite, red sandstone, mica-schist and weathered basic rocks (Hall and Jarvis, 1994; BGS records) whereas the Camp Fauld Till contains quartzite, flint, red granite and Lower Cretaceous sandstone; the last-named rock was probably derived from the nearby Moreseat outcrop (compare with Hall and Connell, 1982; Hall and Jarvis, 1994), rather than from offshore. Elsewhere on the Moss, black clay with clasts of granite and quartzite was seen near Moreseat [NK 053 403], where it underlies rafts of Lower Cretaceous sandstone and clay (Jamieson et al., 1897). Dinoflagellates of possible Late Jurassic to Cretaceous age have been recovered from black clayey till, probably also Camp Fauld Till, at a locality about 500 m west of Moreseat (Harland *in* Hall and Connell, 1982).

Whittington et al. (1993) concluded that both the Aldie and Camp Fauld tills were deposited by ice moving from west to east or north-west to south-east. The sequence therefore seems at first sight to be similar to that occurring at Kirkhill, Oldmill, Gardenstown and the Boyne Limestone Quarry. However, an Early Devensian or latest Eemian age for units of peat overlying the Camp Fauld Till at Camp Fauld (Whittington et al., 1993; Whittington, 1994; Duller et al., 1995) indicates that the till was laid down in a pre-Devensian glaciation predating the deposition of the Whitehills Glaciogenic Formation (Table 7). The possibility of two 'dark coloured boulder clays' in the Ellon district was first suggested by Bremner (1928, p.163).

Further complications occur to the south of the Buchan Ridge where three and locally four till units are separated by deposits of sand and gravel (Merritt, 1981; Hall and Jarvis, 1995). From Ellon north-eastwards towards the Buchan Ridge, dark grey till (the *'indigo'* till of Jamieson, 1906 and/or the *Pitlurg Till* of Hall and Jarvis, 1995) underlies red diamictons of the Logie-Buchan Drift Group. At Ellon, the red deposits rest on grey till of western provenance whereas at Bearnie, they rest on a till of westerly derivation that incorporates material derived from the Pitlurg till (Appendix 1 Bellscamphie; Figure A1.20). The igneous and metamorphic clasts in both the Pitlurg Till and 'indigo' tills are likely to have been derived from the gneisses, gabbros and norites that crop out to the north-west of Ellon. The ammonite in dark indigo clay recorded by Jamieson (1906) at Ellon, and dinoflagellate cysts found in the Pitlurg Till by Hall and Jarvis (1995), indicate derivation ultimately from the Moray Firth, as rocks of this age are exposed only on the sea floor in an narrow outcrop extending westwards from Fraserburgh (Gatliff et al., 1994). A weak north–south fabric has been detected in the Pitlurg Till by Hall and Jarvis (1995), but direct transport from the north-west through to the north-east is difficult to accept because the dark grey tills contain few, if any, erratics of flint and white quartzite from the Buchan Ridge. An amino-acid ratio derived from a shell fragment in the Pitlurg Till (Appendix 1 Bellscamphie) suggests that it is the equivalent of the Whitehills Glacigenic Formation rather than the older Camp Fauld Till. However, most ratios obtained indicated older ages.

Similar diamictons to the 'indigo' till have been reported at South Anderson Drive in Aberdeen (Bremner, 1934, 1943; Hart, 1941) close to the Bridge of Dee [NJ 927 036]. Bremner reported that 'dark shelly boulder clay', named here as the *Anderson Drive Diamicton Formation*, was overlain by a 'grey boulder clay' with characteristic erratics derived from Belhelvie and Barra Hill to the north-west (Appendix 1 Nigg Bay). The latter till is probably equivalent to the Den Burn Member of the Banchory Till Formation of the East Grampian Drift Group (Table 7).

Several other units of grey diamicton containing Mesozoic erratics that have been recorded in the district are probably significantly older than the Whitehills Glacigenic Formation. For example, the lowest stratigraphical unit recognised by Hall et al. (1995a) in the vicinity of the Teindland site (Appendix 1) is a reddish brown till that includes a few erratics of Mesozoic siltstone and sandstone. The presence of these erratics derived from the Moray Firth basin suggests that this diamicton, named as the *Red Burn Till*, should be included in the Banffshire Coast Drift Group despite its colour. The age of the till is not known, but it is likely to predate the Teindland Buried Soil Bed, for which an Ipswichian age is likely (Hall et al., 1995; Table 7).

Origin of the blue-grey tills

The Essie Till Formation rests on red diamictons of the Logie-Buchan Drift Group in the vicinity of Peterhead. It occurs at the surface and is most probably Late Devensian in age. Most other blue-grey tills, with their erratics and rafts of Mesozoic rocks, are buried beneath thin brown tills of inland derivation and are conveniently regarded as part of a single lithostratigraphical unit, the Whitehills Glacigenic Formation. As the brown tills generally overlie the blue-grey ones with little evidence of a break, both were probably laid down during a single glacial episode, during which the direction of ice movement across Buchan changed from south-east to east. This episode either occurred in the Middle Devensian or, more likely, the early part of the Late Devensian. On entering the North Sea basin to the east of Fraserburgh, part of the Moray Firth ice stream may have been deflected westwards again onto the coast south of Peterhead, probably by Scandinavian ice lying offshore.

It appears also that ice moved onshore from the Moray Firth during at least one major glaciation before the Ipswichian to lay down the Camp Fauld Till near the crest of the Buchan Ridge and the Red Burn Till at Teindland (Table 7). The bluish grey Benholm Clay Formation at the Burn of Benholm site was possibly laid down by Moray Firth ice deflected onshore to the south of Inverbervie, but its distinctive palynoflora suggests derivation from the south (Appendix 1).

SANDS AND GRAVELS

Typically, these glaciofluvial deposits include shell fragments and clasts of sedimentary rocks derived from the

Plate 25 Brittle star (*Ophiocten sericium*) in glaciomarine clay of the Spynie Clay Formation. Collected from the former brick pit at Spynie [NJ 232 672], near Elgin (Royal Museum of Scotland specimen 1913.22.1 GN 035).

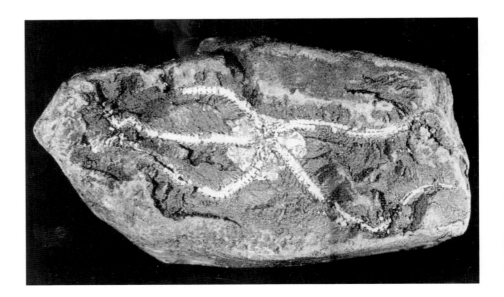

Moray Firth basin in addition to locally occurring rocks. They occur mainly at the surface and are considered to be of Late Devensian age, although older deposits are known at depth locally (Table 7). The surficial deposits lie between Elgin and Fraserburgh and extend inland for no more than about 10 km. They are assigned here to the *Blackhills Sand and Gravel Formation* (Peacock and Merritt, 1997). Two distinct suites of deposits can be distinguished on the basis of landform: moundy 'ice-contact deposits' and terraced 'sheet deposits'. The two may be regarded as informal members of the formation.

Other units include the *Auchmedden Gravel* of Hall et al. (1995b), which forms the upper part of the outwash fan being exploited at the Howe of Byth gravel pit (Appendix 1). It probably correlates with the Blackhills Sand and Gravel Formation and overlies the Byth Till, also probably of Late Devensian age. The latter rests on a lower unit of gravel, the *Byth Gravel*, which is of Mid-Devensian age or older (Hall et al., 1995b; Table 7). A unit of the Kirkhill sequence (Appendix 1) is assigned tentatively to the group, the Kirkhill Church Sand Formation (formerly Kirkhill Upper Sands). This unit is probably Late Devensian in age.

GLACIOLACUSTRINE DEPOSITS

Spreads of clays, silts, sands and gravels extend along the coastal hinterland between Portknockie and New Aberdour on Sheets 96W, 96E and 97. They were first described in detail by Read (1923), who named them the 'Coastal Deposits'. He also included dark grey shelly tills and clays in the unit, but these lithologies are now known to belong to a separate, underlying formation composed mainly of till (Peacock, 1971), since named as the Whitehills Glacigenic Formation (Peacock and Merritt, 1997). The main bodies of sand and gravel within the 'Coastal Deposits' are now included in the Blackhills Sand and Gravel Formation. The remaining glaciolacustrine silts and clays are defined as the *Kirk Burn Silt Formation*, the type section of which occurs at Castle Hill, Gardenstown (Peacock and Merritt, 1997; Map 3; Appendix 1).

The main outcrops of the Kirk Burn Silt Formation lie between Portknockie and Portsoy, at Macduff, between Gardenstown and Pennan, and to the north of New Aberdour (Maps 2; 3; 4). For the most part the deposits form flat or gently undulating ground within 5 km or so of the coast (Peacock, 1971). The sediments consist mainly of ochreous to dark olive brown, laminated to thin-bedded silts and clays up to about 10 m in thickness. Bands of ferruginous nodules occur locally. Pebbles are not common, but where they do occur, they include clasts of white decalcified Cretaceous limestone. The sediments show signs of settlement and minor movement, including slumping, but there is no evidence that they have been affected by a significant postdepositional glacial advance from inland as deduced by Read (1923), or that they are widely covered by solifluction deposits as described by Synge (1956). They have been deformed apparently by ice moving from the sea at Troup Head (Peacock and Merritt, 2000; Map 3).

The deposits of the Kirk Burn Silt Formation were laid down during deglaciation of the Main Late Devensian ice sheet. East Grampian ice had retreated inland but an active ice stream remained in the Moray Firth, forming a barrier a short distance to the north of the present coastline and causing lakes to form against the high ground to the south (Figure 42). It seems that lakes first formed to the north of New Aberdour, draining eastwards along the coast. A larger lake then formed between Gardenstown and Pennan, draining via the Afforsk Spillway into the valley of the Deveron via the valley of the Idoch Water. Much larger lakes formed subsequently between Portknockie and Macduff.

The deposits of the Kirk Burn Silt Formation are prone to landslipping (Chapter 2; Plate 3).

GLACIOMARINE DEPOSITS

Glaciomarine deposits of the Banffshire Coast Drift Group were laid down during the overall retreat of the Moray Firth ice stream at two distinct localities within the district. The older of the two deposits, the *St Fergus Silts Formation*, crops out around the Loch of Strathbeg on Sheet 96 and in the vicinity of the St Fergus gas terminal on Sheet 87E. The younger deposit, the *Spynie Clay* Formation, is associated with Late-glacial raised beaches in the Elgin area.

St Fergus Silt Formation

At the type locality, the St Fergus Silt Formation underlies alluvium of a former freshwater loch that lies behind the belt of coastal sand dunes forming St Fergus Links [NK 124 434], to the north of Peterhead (Map 7). The type section is taken to be BGS Borehole NK15SW1 [NK 1030 5344], in which some 16 m of laminated silty clay and clayey silt with sand laminae and shell fragments were recorded, resting on till (McMillan and Aitken, 1981; Peacock, 1999). The deposits vary from dark grey to dark greyish brown, and are both unbedded and laminated, calcareous and contain dispersed clasts up to cobble size that are probably dropstones. The lithology and fauna suggests that the St Fergus Silt Formation is glaciomarine (Hall and Jarvis, 1989; Peacock, 1999), and two adjusted radiocarbon dates of about 14.9 BP and 14.3 BP (Hall and Jarvis, 1989) on marine bivalves (*Hiatella arctica*) indicate that it was laid down in a period prior to the Windermere Interstadial. The general surface level of the St Fergus Silt Formation and of an associated raised beach imply a marine transgression to at least 12 m OD (see Appendix 1).

Spynie Clay Formation

Late-glacial raised marine beds composed mainly of silty clay, are associated with raised beaches between Elgin and Lossiemouth on Sheet 95 Elgin (Chapter 7). These beds, named the Spynie Clay Formation by Peacock (1999), underlie low-lying ground around Loch Spynie [NJ 237 667] and Lossiemouth Airfield (Map 1). The maximum recorded thickness is 12.5 m (Peacock et al.,

1968). North of the Spynie basin the formation attains a minimum level of 10 m above OD, and possibly reaches over 20 m OD. On the south side of the basin the mapped level is only a little above OD, probably because the sea was excluded by stagnant ice from this area when sedimentation had begun elsewhere.

The type locality of the Spynie Clay Formation is the former Spynie Claypit [NJ 232 672] (Peacock, 1999), where up to 6 m of sand is underlain by thin beds of shell debris and peat, resting on some 9.5 m of 'blue', yellowish red and black clay with scattered boulders, locally interbedded with sand and diamicton (Buchan, 1935; Eyles et al., 1946). Fossils, chiefly the brittle star *Ophiocten sericium* (formerly *Ophiolepis gracilis*) (Plate 25), numerous '*Leda pygmaea*' (probably *Yoldiella lenticula*) and rare *Portlandia arctica*, have been found between 5 and 6.5 m below the surface of the clay (Buchan, 1935). All are decomposed. The regular lamination, the presence of well-dispersed, probably ice-rafted boulders, and the apparent absence of bioturbation support the view that much of the sediment was deposited rapidly in a glaciomarine environment. The low-diversity, high-arctic fauna is similar to that in the Errol Clay Formation of eastern Scotland, which has yielded radiocarbon ages between 12.8 and 14.3 ka BP (Peacock and Browne, 1998; Peacock, 1999).

MEARNS DRIFT GROUP

The Mearns Drift Group comprises interbedded diamictons, glaciolacustrine silts and clays, and glaciofluvial sands and gravels that are all typically vivid reddish brown in colour and contain clasts that are derived mostly from the andesitic volcanic rocks and red Devonian conglomerates, sandstones and siltstones of Strathmore. The group broadly equates with the southern outcrop of the 'Red Series' of Bremner (1916), Sutherland (1984a), Sutherland and Gordon (1993), and Hall and Connell (1991). The deposits were laid down by, or at the margins of, ice that flowed north-eastwards into the North Sea basin from Strathmore (Figure 4). The formations in the group are depicted on Sheet 66E Banchory and Sheet 67 Stonehaven. Deposits occur at least as far north as Nigg Bay (Appendix 1).

Three formations have been identified within the Mearns Drift Group to include the known glacial, glaciofluvial and glaciolacustrine deposits. They are the *Mill of Forest Till, Drumlithie Sand and Gravel, and Ury Silts* formations, respectively. The deposits are the product of the Main Late Devensian glaciation. Locally, some older units may be present at depth, but unless some intervening materials of a different origin and colour occur, it is difficult to distinguish them. No examples of older units of red till are known, but two distinct phases of movement of the Strathmore ice stream have been deduced by Armstrong et al. (1985) from the Dundee district, an earlier movement directed towards the south-east and a later one towards the east-north-east. This reorientation of ice flow is presumed to have occurred during the Main Late Devensian glaciation, either as the result of an expansion of the Southern Upland ice cap as the glaciation progressed (Sutherland, 1984a), or as a response to Scandinavian ice pushing into the North Sea basin. The presence of red till at the mouth of most of the valleys that drain the southern flank of the eastern Grampian Highlands, such as Glen Clova and Glen Esk (Bremner, 1934b, 1936; Synge, 1956), indicate that in general the Strathmore ice either forcibly pushed back ice occupying the valleys to the north, or that the ice within them had already begun to retreat enabling the Strathmore ice to advance towards the mountains. A more complicated sequence of events appears to have occurred at the Balnakettle site, near Fettercairn (Appendix 1).

There are several notable occurrences of dark bluish grey shelly diamicton (*Benholm Clay Formation*) underlying red till along the coast to the south of Aberdeen (Campbell, 1934; Auton et al., 2000), for example at the Burn of Benholm site (Appendix 1). It is interpreted as having been deposited by ice moving onshore from the

Plate 26   Glacitectonised sequence at Balnakettle, north-east of Fettercairn. Till and gravel of the Mearns Drift Group are tectonically intercalated with older (pre-Late Devensian) head gravel composed of angular clasts derived from local Dalradian pelitic and semipelitic bedrock. The top of the tectonised sequence is truncated by a till of the East Grampians Drift Group, formed during a local re-advance (P100707).

North Sea basin, and hence the Benholm Clay has been assigned tentatively to the Logie-Buchan Drift Group (see below).

Mill of Forest Till Formation

Reddish brown tills laid down by the Strathmore ice stream are assigned to the Mill of Forest Till Formation. These typically cohesive, silty and clayey diamictons contain a mixture of clasts derived from Old Red Sandstone strata and Devonian volcanic rocks, together with some clasts of granitic and Dalradian metasedimentary rocks. Well-rounded clasts of igneous and metamorphic rocks are mostly recycled from Old Red Sandstone conglomerates, whereas more angular clasts are derived from outcrops of Dalradian metamorphic and Caledonian igneous rocks forming the adjacent Grampian Highlands. Some of this 'Highland' material may have also been recycled from pre-existing glacigenic sequences during the Main Late Devensian glaciation.

The type section of the Mill of Forest Till Formation occurs in a river cliff [NO 8630 8538], 150 m downstream of Mill of Forest Farm, Stonehaven, where 6 to 7 m of stiff, reddish brown, matrix-supported, silty sandy diamicton crops out beneath 3 m of cobble gravel of the Drumlithie Sand and Gravel Formation (Map 11). The till contains rounded pebbles and cobbles of quartzite, psammite, andesite and feldspathic microgranite, derived mostly from adjacent outcrops of conglomerate (Plate 8b). During the revision survey of Sheet 67 in 1926, Campbell observed a metre-thick unit of grey till at the base of the river cliff.

The Mill of Forest Till Formation is generally less than 5 m thick. Grossly overconsolidated lodgement tills are commonly developed in its basal part whereas friable, sandy to gravelly diamictons (flow tills) up to about 2 m thick commonly occur in its upper part, and locally interdigitate with the overlying Drumlithie Sand and Gravel Formation. One such flow till was recorded in a reference section (BGS trial pit NO77NW1) near Drumelzie [NO 711 790], on Sheet 66E. In another reference section on that sheet (BGS trial pit NO77NE11), at East Mondynes [NO 780 797], a flow till rests on red-brown lodgement till (Auton et al., 1990).

Drumlithie Sand and Gravel Formation

Deposits of the Drumlithie Sand and Gravel Formation were laid down principally as coarsening-upward sequences of sand and gravel at, or near, the margin of the actively retreating Strathmore ice stream, which impinged locally on the coast between the mouth of the River Dee and Stonehaven. Mounds, plateaux and ridges were formed at the retreating ice margin as fans and deltas that were subjected to minor re-advances as the ice withdrew. Eskers are common. Deposits commonly display evidence of slumping, small-scale normal faulting and cryoturbation, and are capped by up to a metre of red-brown sandy flow till.

The sands and gravels contain a mixture of clasts derived from the Silurian and Devonian clastic, volcaniclastic and volcanic rocks of Strathmore together with metamorphic and granitic clasts from the Grampian Highlands. Some of the clasts of crystalline rock have been recycled from Old Red Sandstone conglomerates. For example, the most characteristic durable clasts are well-rounded boulders and pebbles of quartzite from conglomerates. Clasts of soft red-brown sandstone, siltstone, mudstone and decomposed purple andesite are also common, but their relative proportions decrease rapidly away from the Old Red Sandstone outcrops from which they were derived. In its type area, between Auchenblae and Temple of Fiddes [NO 817 818], on Sheet 66E, the Drumlithie Sand and Gravel is typically 5 to 10 m thick. Thickly bedded sands with lenses of gravel are exposed in the partial type section in the Meikle Fiddes esker at Kaim of Clearymuir [NO 798 815] (Map 10).

Ury Silts Formation

The glaciolacustrine Ury Silts Formation consists of characteristic red-brown interlaminated micaceous fine-grained sand, silt and clay. These sediments were laid down in ice-marginal lakes that developed as the Strathmore ice stream retreated south-westwards. Exposures in the Ury Silts Formation are sparse. A partial type section is taken from 14.5 to 18.5 m depth (unbottomed) in BGS Borehole NO88NE4, drilled near the Houff of Ury (Auton et al., 1988), on Sheet 67 (Map 11). This reddish brown silty deposit, which contains sparse isolated pebbles (interpreted as dropstones), is overlain by 1.7 m of moderate reddish brown, friable silty diamicton (flow till). The flow till is overlain by 12.4 m of Drumlithie Sand and Gravel Formation. More typically, there is a gradational contact between the silt and clay unit and the overlying sand and gravel, as in a reference section taken in BGS Borehole NO88NW13, drilled in Craigies Wood [NO 8389 8514], near Stonehaven. The base of the Ury Silts Formation is generally in sharp contact with underlying diamictons of the Mill of Forest Till Formation, or with bedrock.

RELATIONSHIP BETWEEN MEARNS AND EAST GRAMPIAN DRIFT GROUPS

The Mill of Forest Till Formation generally rests directly on bedrock, but in some coastal localities, notably at Nigg Bay (Appendix 1), Cove (Jamieson, 1882b) and Findon (Synge, 1963) red till rests on grey till (Maps 9, 11). Both Jamieson and Synge observed gradational contacts between the tills and concluded that they were laid down during the same (Late Devensian) glaciation. The relationship suggests that the direction of ice flow changed from east-north-east to north-north-east as the Strathmore ice stream became dominant. The grey tills contain clasts derived from the west and are correlated here with the Banchory Till Formation of the East Grampian Drift Group (Table 7). At the Burn of Benholm site (Appendix 1), the Mill of Forest Till Formation rests with a sheared (glacitectonic) contact on the Benholm Clay Formation of the Logie-Buchan Drift Group.

Figure 50
Distribution of
the Logie-Buchan
Drift Group and
related features.

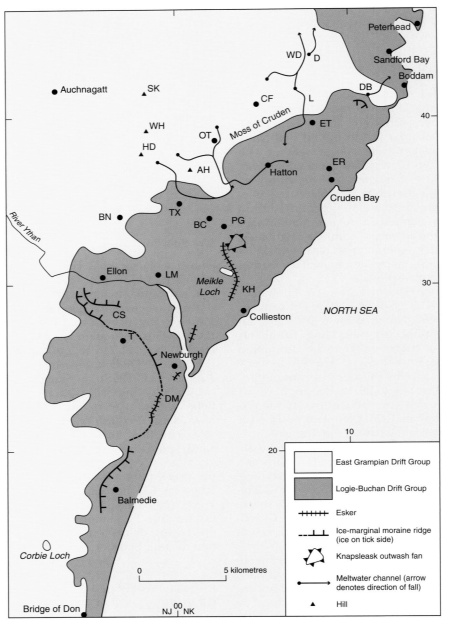

AH   Hill of Auchleuchries
BC   Bellscamphie
BN   Bearnie
CF   Camp Fauld
CS   Cross Stone moraines
D   Dens channel
DB   Den of Boddam
DM   Drums esker/moraine
ER   Errollston
ET   East Teuchan
HD   Hill of Dudwick
KH   Kippet Hills
L   Laeca channel
LM   Lintmill
OT   Old Town
PG   Pitlurg
SK   Skelmuir Hill
T   Tipperty
TX   Tillybrex
WD   West Dens channel
WH   Whitestones Hill

East Grampian Drift Group

Logie-Buchan Drift Group

Esker

Ice-marginal moraine ridge
(ice on tick side)

Knapsleask outwash fan

Meltwater channel (arrow
denotes direction of fall)

Hill

Temporary sections at Ury Home Farm [NO 858 881], on Sheet 67, revealed up to 2.0 m of red-brown clayey lodgement till (Mill of Forest Till Formation) overlying up to 2.2 m of moderate yellowish brown sandy bouldery diamicton with clasts of Dalradian metasedimentary rocks and pink granite (Banchory Till Formation). The relationship between the tills suggests that East Grampian ice advanced into the lower reaches of the valley of the Cowie Water prior to advance of the Strathmore ice (Figure 4). In contrast, at Balnakettle (Map 10), near Fettercairn (Appendix 1) and Cantlayhills [NO 874 905], north of Stonehaven (Map 11), sandy diamictons assigned to the Banchory Till Formation overlie the Mill of Forest Till Formation. At both sites, the upper unit is interpreted as flow till, indicating that

retreat of the East Grampian ice sheet and Strathmore ice stream was contemporaneous.

Deposits of the Drumlithie Sand and Gravel Formation are locally concealed beneath thin diamictons of the East Grampian Drift Group along parts of the Highland boundary. For example, at Balnakettle (Appendix 1) and Cantlayhills, sheared and contorted beds of reddish brown sand and gravel occur beneath yellow-brown sandy diamicton. At Cantlayhills, the diamicton was formed either from a debris flow directly from the East Grampian ice sheet, or as a landslide following deglaciation. In contrast, at Balnakettle (Plate 26) the sand and gravel has been disturbed by a local re-advance of the East Grampian ice sheet following the withdrawal of the Strathmore ice stream in the area (Appendix 1).

LOGIE-BUCHAN DRIFT GROUP

It has been widely believed for over a century that during the last (Main Late Devensian) glaciation ice flowing from Strathmore turned northwards between Stonehaven and Aberdeen and then north-westwards on to the coast of Logie-Buchan, between Aberdeen and Peterhead (Figure 4; Jamieson, 1906; Bremner, 1916; Synge, 1956; Clapperton and Sugden, 1977; Merritt, 1981; Hall, 1984; Munro, 1986; Hall and Connell, 1991). Glacial striae provide evidence of the northerly ice flow (Jamieson, 1882a, 1906; Bremner, 1916; Map 7). The onshore movement is confirmed by the clay mineralogy of the red deposits (Glentworth et al., 1964) together with the unique suite of glacial erratics derived from the floor of the North Sea basin immediately off the coast of Aberdeenshire. The cause of the onshore movement is controversial, but the simplest explanation is that it was constrained to do so by Scandinavian–North Sea ice lying offshore (Chapter 5). The diversion led to the deposition of a distinctive suite of typically vivid reddish brown, interbedded materials that form the hummocky topography and prime agricultural land to the north of Aberdeen on Sheet 77 Aberdeen, Sheet 87W Ellon and Sheet 87E Peterhead. The *Logie-Buchan Drift Group* has been established here to include these glacigenic deposits. It correlates in part with the seismostratigraphical *Wee Bankie Formation* (Stoker et al., 1985; Gatliff et al., 1994), which extends for some 25 to 40 km offshore, where it is bounded to the south of Stonehaven by the prominent Wee Bankie Moraine (Figure 44).

The Logie-Buchan Drift Group (Figure 50) is composed mainly of a complex, interbedded sequence of clayey diamicton, clay, silt, mud, sand and gravel in which individual units can rarely be traced laterally for more than a few tens of metres. Most deposits are vivid reddish brown in colour and calcareous. The sands are typically fine to medium grained, silty and micaceous. Laminae and thin beds of yellowish brown, medium-grained sand and fine gravel commonly contain shell fragments. In addition to locally occurring rock types such as amphibolite, feldspathic psammite, quartzite and meta-greywacke, the deposits contain appreciable proportions of rocks derived from the sea bed to the east, including limestone, dolomite, calcareous siltstone, white and red sandstone of Devonian, Permo–Triassic, possible Jurassic, Cretaceous and Tertiary age (Figure 2). The group underlies the coastal lowlands to the north of Aberdeen, east of Ellon and both south and north of Peterhead, where it forms distinctive, fresh-looking, hummocky topography comprising kettleholes, mounds, plateaux, esker ridges and narrow, winding, steep-sided valleys.

The deposits of the Logie-Buchan Drift Group were first described by Jamieson (1858, 1882a, 1906) as part of his Red Clay Series. They have been described as 'red drift' by Merritt (1981) and the Red Series by Hall (1984) and Hall and Connell (1991). The type area is the parish of Slains, between Collieston and Cruden Bay, on Sheet 87E (Map 7). Typical sediments include silty and sandy diamictons that appear to have formed as flow tills and subaqueous debris flows, and muds formed in a glaciolacustrine, or possibly glacio-estuarine environment. These deposits, interbedded with glaciofluvial sands, fill hollows in the bedrock surface and collectively reach over 25 m in thickness (Merritt, 1981). Topography bears little relationship to the subdrift surface. The sequence is locally dominated by stiff, stony clayey diamicton, especially over bedrock 'highs'. The uppermost metre or so generally consists of firm to stiff, pebbly, silty, clayey diamicton that has probably been formed by solifluction and cryogenic mixing. Thick diamicton-dominated sequences probably include deformation tills, as at Sandford Bay and Errollston (Appendix 1). Laminated silts and clays overlying buff-coloured, pebbly sands with shell fragments commonly occur at the base of the sequence within hollows. Mappable units of glaciofluvial sand and gravel, till and glaciolacustrine deposits are assigned to the *Hatton Till, Kippet Hills Sand and Gravel* and *Ugie Clay* formations, respectively.

Hatton Till Formation

Hall and Jarvis (1995) describe several sites around Bellscamphie where a stiff, reddish brown, relatively stony diamicton that they named the Hatton Till rests on the dark grey Pitlurg Till (of the Banffshire Coast Drift Group). Elsewhere to the north and east of Ellon similar red tills rest on grey and yellowish brown sandy tills of the East Grampian Drift Group. The Hatton Till, which is raised to formational status here, is typical of the more stony red diamictons in the Logie-Buchan Drift Group described above (Merritt, 1981). Its type section is at Bellscamphie (Appendix 1).

Thick units of reddish brown deformation till equivalent to the Hatton Till Formation are exposed at Sandford Bay, near Peterhead, where they have been derived from the south or south-east, and locally incorporate rafts of blue-grey diamicton (Appendix 1). To the north of Peterhead, red tills interdigitate with the Essie Till Formation of the Banffshire Coast Drift Group (see above). In contrast, red tills are not known to interdigitate with yellowish brown, sandy tills of the East Grampian Drift Group to the north of the River Ythan. Furthermore, at Cross Stone, to the south of Ellon, the Hatton Till Formation abuts two morainic ridges indicating that Logie-Buchan ice was free to expand into the Ythan valley following the retreat of East Grampian ice (see below).

Kippet Hills Sand and Gravel Formation

In addition to the thin, laterally impersistent beds of sand and gravel that form part of the sequence, several larger bodies have been identified on Sheet 87E. For example, a sinuous ridge, the Kippet Hills Esker, extends northwards from Cotehill Loch [NK 028 294], near Collieston, past Meikle Loch to Ladies's Brig [NK 029 318], where it widens north-eastwards into a flat-topped, fan-shaped mound at Whitehills (Figure 50; Appendix 1 Kippet Hills). At the partial type section near Knapsleask [NK 0327 3206], the gravel is distinctive in that it contains 40 per cent or more of calcareous material including Palaeozoic dolomite, Mesozoic limestone and calcareous siltstone, Pliocene shelly sandstone (Crag) and shell fragments of early Quaternary and younger age derived from the offshore Aberdeen Ground Formation. Hall and Jarvis (1995) referred to this distinctive glaciofluvial deposit as the Kippet Hills Gravels and Sands and it is named here formally as the *Kippet Hills Sand and Gravel Formation* of the Logie-Buchan Drift Group (Table 7). Hall and Jarvis describe similar material in an abandoned railway cutting at Bellscamphie [NK 0184 3369], which is taken here as another partial type section (Appendix 1 Ellon).

Other deposits of sand and gravel

A distinct suite of pebbly sand deposits lies along the boundary between the Logie-Buchan and East Grampian drift groups to the south of the Buchan Ridge, on Sheet 87W Ellon and Sheet 87E Peterhead. In the vicinity of the Hill of Auchleuchries [NK 006 365] (Map 6), the moundy deposits are capped by yellowish brown till of presumed western derivation, whereas some of the mounds to the north and west of Hatton (Map 7) are capped by red till of the Logie-Buchan Drift Group (Merritt, 1981). The sands contain clasts of a variety of lithologies, but include conspicuous amounts of pink (possibly Peterhead) granite. Surprisingly, they contain very little flint and quartzite from the Buchan Gravels Formation, which crops out just 2 km to the north. The sands overlie dark grey till with shell fragments, which is correlated with the Pitlurg Till of the Banffshire Coast Drift Group at Bellscamphie (Appendix 1). This till unit appears to have been deposited by ice moving onshore from the east-north-east (Bremner, 1928; Hall and Jarvis, 1995). As the pink granite clasts in the pebbly sands are likely also to have been derived from this direction, these glaciofluvial deposits probably relate to the same body of ice that laid down the Pitlurg Till. This association is supported by the presence in the pebbly sands of seams of fissile, chocolate-brown to olive-grey clay (Merritt, 1981). Hall and Jarvis (1995) conclude that the Pitlurg Till was laid down in the Early or Middle Devensian (OIS 4–3) and it follows that the pebbly sands are probably of similar age. However, if the Pitlurg Till correlates with the Whitehills Glacigenic Formation, the sands would have been laid down early in OIS 2 (Table 7).

It is proposed here that the pebbly sands described above be named as the *Auchleuchries Sand and Gravel Formation*, the type section being BGS Borehole NK03NW1 (Merritt, 1981) on the Hill of Auchleuchries [NK 0057 3649]. The Bellscamphie site (Appendix 1 Figure A1.20) provides a reference section. The deposits of the Auchleuchries Sand and Gravel Formation have been placed in the East Grampian Drift Group on Sheet 87W Ellon, because they do not contain a significant proportion of clasts derived from offshore. However, the distribution of the deposits at, or within, the western boundary of the Logie-Buchan Drift Group and the rich shell fauna that was recovered from the Hill of Auchleuchries by Jamieson (1882b, p.172) both suggest that the unit is better placed in the Logie-Buchan Drift Group.

Ugie Clay Formation

Following the withdrawal of the East Grampian ice sheet, the coastal ice that laid down the Logie-Buchan Drift Group dammed the lower reaches of the valleys of the Ythan and Ugie causing extensive lakes to form (Hall, 1984; Hall and Connell, 1991; Figure 42). The fine-grained deposits laid down in these lakes are assigned here to the Ugie Clay Formation. Laminated silts and clays of the formation, mostly reddish brown in colour, occur extensively beneath the floodplains and glaciofluvial terraces of the North and South Ugie waters on Sheet 87E (McMillan and Aitken, 1981) and beneath similar features in the valley of the River Ythan on Sheet 87E (Merritt, 1981; Chapter 6 Glaciolacustrine deposits). Lakes also formed in the Don valley upstream of the Mill of Dyce (Appendix 1), but as the sediments there are dominated by materials derived from the retreating East Grampian ice sheet, they have been assigned to the Glen Dye Silts Formation of the East Grampian Drift Group (see below).

The type section of the Ugie Clay Formation is a stream section [NK 0050 4732] near Baluss Bridge, south of Mintlaw, where over 2 m of reddish brown and dark grey plastic clay is thinly interbedded with yellowish brown sand, gravel and silt. A BGS Borehole NK04NW2 drilled 190 m to the south-south-west of the section proved at least 4 m of the sequence overlying orange-brown, very sandy till, resting in turn on weathered psammitic bedrock. Till exposed nearby has a macrofabric indicating that it was deposited by the East Grampian ice sheet flowing eastwards (Hall and Connell, 1991). The Ugie Clay Formation is capped locally by terraced glaciofluvial sand and gravel. Organic muds sampled from temporary sections in the formation at Baluss Bridge itself, yielded rich Palaeozoic and Mesozoic palynomorph assemblages and unreliable radiocarbon dates (Appendix 1 Ugie valley). Similar laminated organic deposits occur at Errollston clay pit [NK 088 368], near Cruden Bay (Appendix 1).

Plate 27 Sheared lens of the Burn of Benholm Peat Bed within olive grey shelly clay and diamicton (Benholm Clay Formation) in trial pit BBP4, at the Burn of Benholm site north of Johnshaven (Appendix A, 26). The shelly deposits are overlain by red-brown diamicton of the Mill of Forest Till Formation (D4879).

Several deposits of red and brown silt and clay in the Aberdeen area are named here as the *Tullos Clay Member* and assigned tentatively to the Ugie Clay Formation (rather than the Ury Silts Formation). They include those worked in clay pits at Tipperty [NJ 970 268] (Bremner, 1943; Munro, 1986) and Blackdog [NJ 962 139] (Jamieson, 1906; Bremner, 1916; Peacock, 1975). They also occur within the valley between Torry and Tullos Hill in Aberdeen (Simpson, 1948), and have been exposed at the Nigg Bay site (Appendix 1). The deposits are locally fossiliferous and those occurring at lower levels may be glaciomarine (up to about +30 m OD) (Chapter 6   Raised marine deposits)

RELATIONSHIP BETWEEN LOGIE-BUCHAN AND EAST GRAMPIAN DRIFT GROUPS

Apart from at the Hill of Auchleuchries (see above), deposits of the Logie-Buchan Drift Group generally occur at the top of the glacial sequence to the north of Aberdeen (Figure 50). This suggests that they were the youngest glacigenic deposits to be laid down in the area. The interdigitation of bluish grey and red units to the north of Peterhead indicates that the Moray Firth ice stream laid down deposits of the Banffshire Coast Drift Group contemporaneously (see below). To the north of Ellon, red tills generally overlie tills of the East Grampian Drift Group.

In the Aberdeen area, the stratigraphical relationship between deposits of the Logie-Buchan Drift Group and the brown and grey tills derived from the East Grampian ice sheet is more complex. In the north-eastern part of Sheet 77, red-brown till is generally about 1 m thick and underlain by grey/brown till resting on bedrock (Munro, 1986). Isolated, ill-defined rafts of red till, up to 1.5 m in diameter, occur within the grey-brown till beyond the western margin of the Logie-Buchan Drift Group near Ardo House [NJ 929 208] (Map 9). Near Tipperty [NJ 969 277], isolated patches of grey-brown till are incorporated in red-brown till, and *vice versa*. Where a discrete unit of red till overlies grey-brown till, as in a pipeline trench in the vicinity of Mill of Ardo (Munro, 1986, fig. 30 A-B), the boundary between the two is generally regular and sharp. Where there are more complex field relationships between the two till units, or where the grey-brown till overlies the red, contacts are commonly ill-defined and gradational. Similar stratigraphical relationships also occur in several BGS boreholes and trial pits north of Aberdeen (Auton and Crofts, 1986). In most instances (e.g. Borehole NJ91NE14, near Blackdog Rifle Ranges), red-brown silts and clays, overlie greyish brown till, but rarely, in Borehole NJ81SW2, near Little Clinterty, brown sandy till overlies red-brown clayey till.

The local evidence of interdigitation of red tills and grey tills of inland provenance led Clapperton and Sugden (1977) to conclude that the ice responsible for laying down the Logie-Buchan Drift Group was in contact with ice flowing westwards from the East Grampian ice sheet. More commonly, red tills overlie tills of the East

Grampian Drift Group (Bremner, 1916; Auton and Crofts, 1986) suggesting that East Grampian ice expanded to the present position off the coast, if not beyond, before retreating sufficiently to allow the encroachment of ice from offshore (Hall and Connell, 1991). Clapperton and Sugden (1977) concluded that the early expansion of East Grampian ice was the result of fluctuating flow strength between 'Cairngorm–Grampian' and 'Strathmore' ice in the Late Devensian. In contrast, Connell and Hall (1987) concluded that the tills were laid down in separate glaciations, in the Early and Late Devensian, respectively. This was based mainly on evidence from the area north of the River Ythan, where diamictons of the two groups are not seen to interdigitate but locally are observed to be separated by the Ugie Clay Formation. However, as noted above to the north of Ellon, yellowish brown till, derived from inland, caps shelly deposits of the Auchleuchries Sand and Gravel Formation. This indicates the former close proximity of Logie-Buchan Drift Group and East Grampian Drift Group ice at this location, a situation that apparently has not been previously described north of the Ythan (see Jamieson, 1906, p.26–27; Bremner, 1916, p.337).

A pair of crescentic, asymmetrical end moraines have been identified during recent fieldwork on Sheet 87W in the vicinity of Cross Stone [NJ 954 278], near Ellon, at the western boundary of the Logie-Buchan Drift Group (Figure 50; Map 6). These features indicate quite clearly that a lobe of 'Logie-Buchan' ice advanced directly onshore up the valley of the River Ythan following retreat of the East Grampian ice sheet. The latter retreated sufficiently to allow a substantial glacial lake to form in the Ythan valley upstream of Ellon (Merritt, 1981; Hall, 1984). The inland margin of the Logie-Buchan Drift Group can be traced north-eastwards towards the Buchan Ridge, where it is quite distinct in the vicinity of the Den of Boddam [NK 102 411]. It reaches about 80 m above OD at the Den before descending to about 50 m OD to the west of Peterhead (Map 7).

RELATIONSHIP BETWEEN LOGIE-BUCHAN AND BANFFSHIRE COAST DRIFT GROUPS

Red tills overlie yellowish brown and grey tills derived from the west between Ellon and Peterhead (Wilson, 1886; Jamieson, 1906). Immediately to the south of the River Ugie, they overlie greenish grey till that was probably also laid down by East Grampian ice (McMillan and Aitken, 1981). West of Peterhead, at Downiehills [NK 089 471] (Map 7), Jamieson (1858) recorded the occurrence of large blocks of Peterhead Granite in reddish brown till together with possible shell fragments. Later in the same area, he noted the occurrence of dark blue clay intermingled with deposits of the 'Red Series' (Jamieson, 1906). The latter description refers to exposures in the former claypit at Ednie [NK 085 500], where the red deposits were seen to enclose irregular masses described as 'dark grey till' and 'dark blue clay and sand' (Wilson, 1886; BGS records). These observations have caused confusion. On one hand they have been taken to support the view that the deposits of the 'Blue Grey' and

'Red' series are intermingled and are of approximately the same age (Jamieson, 1906; Hall and Connell, 1991). The alternative view is that older 'Blue Grey Series' (Banffshire Coast Drift Group) sediments have been incorporated into the red glaciotectonically by a later, northward movement of ice (Sutherland, 1984a; see Appendix 1 Sandford Bay).

The situation is reversed to the north of the River Ugie, however, where red sediments are locally overlain by the dark grey, shelly Essie Till Formation, which contains erratics of pink granite, quartz and metasedimentary rocks. This diamicton, assigned to the Banffshire Coast Drift Group above, was probably laid down by a late re-advance of the Moray Firth ice stream. Red deposits of the Logie-Buchan Drift Group locally overlie blue-grey tills in the vicinity of Ellon and Aberdeen, but the age of the older tills is uncertain (see Banffshire Coast Drift Group above).

RELATIONSHIP BETWEEN THE LOGIE-BUCHAN AND MEARNS DRIFT GROUPS AND THE BENHOLM CLAY FORMATION

Deposits of the two groups do not appear to abut one another at the surface on land (Figure 4), and offshore no contact has been recognised within the Wee Bankie Formation. However, there are several notable occurrences of dark bluish grey, shelly diamicton underlying red till along the coast to the south of Aberdeen (Campbell, 1934). The best known locality is at the Burn of Benholm (Map 11; Appendix 1). These shelly diamictons have been assigned to the *Benholm Clay Formation* (Auton et al., 2000) and they were almost certainly laid down by ice moving onshore from the North Sea basin during a pre-Devensian glaciation (Table 7). Despite its colour, the Benholm Clay Formation is placed tentatively in the Logie-Buchan Drift Group because of its derivation by ice from the North Sea basin (Plate 27).

EAST GRAMPIAN DRIFT GROUP

This group is broadly equivalent to the sediments that were previously referred informally to the 'Inland Series' (Hall, 1984a; Sutherland and Gordon, 1993). The deposits contain clasts that have been carried by ice flowing from the eastern Grampian Highlands during several glaciations (Figure 4). Although erratics from farther afield do occur (Figure 47), the colour and clast composition of the tills closely reflect the nature of underlying bedrock (commonly deeply weathered), or of rocks cropping out within a few kilometres to the west. The tills are generally sandy, thin (less than 2 m) and patchy, especially across central Buchan, where they are normally pale yellowish brown in colour. Thicker and more widespread tills occur in the valleys of the Dee and Don, where colour ranges from brown to grey. Glaciofluvial and glaciolacustrine deposits are relatively uncommon across large tracts of countryside, especially in central Buchan. The uppermost metre or so of all

materials is commonly severely disturbed by periglacial activity (Connell and Hall, 1987).

As explained in Chapter 5, the sandy nature and pale colour of the tills occurring in central Buchan have led some authors to conclude that they have been weathered since their deposition. These attributes, together with inferred relationships to incorrectly dated organic sediments (Appendix 1 Crossbrae), led some to conclude that they were laid down either during an early Devensian glaciation (Hall, 1984; Sutherland, 1984a; Hall and Connell, 1991) or during a pre-Devensian one (Charlesworth, 1956; Synge, 1956, Fitzpatrick, 1958, 1972; Galloway, 1961a, b, c). The sandiness and colour are equally likely, however, to result from the incorporation of significant proportions of deeply weathered bedrock or previously weathered deposits, in which case the tills need be no older than Late Devensian in age (Clapperton and Sugden, 1977; Hall and Bent, 1990). For example, pedological studies of a till (possibly Hythie Till Formation) near Mintlaw suggest that the soil-forming processes occurred during the Holocene (Van Amerongen, 1976).

Glacigenic deposits of inland provenance are known to occur at several stratigraphical levels (Table 7) and there are possibly representatives of glaciations going back to OIS 8, if not the Anglian (Appendix 1 Kirkhill). The older deposits have been identified through the discovery of intervening units of peat, soil and gelifluctate, but in the absence of firm dating control, their chronostratigraphical context is not known for certain.

Most deposits of inland provenance that have been named in the literature occur at widely dispersed sites where relatively long sequences have been described, for example at the Teindland and Kirkhill localities (Appendix 1). Although formal lithostratigraphical units have been established locally at Boyne Bay and Gardenstown (Appendix 1), only on Sheet 66E Banchory and Sheet 67 Stonehaven have units of this group been mapped out systematically. On these sheets, the glaciofluvial, glaciolacustrine and glacial deposits associated with the former East Grampian ice sheet are placed in the *Lochton Sand and Gravel*, the *Glen Dye Silts* and the *Banchory Till* formations, respectively. These Late Devensian deposits extend northwards onto Sheets 76E and 77, if not farther north, and they form part of the important sequence established at Nigg Bay (Table 7; Appendix 1).

TILL

Banchory Till Formation

Glacial sediments assigned to the Late Devensian Banchory Till Formation are typically sandy, gravelly or bouldery diamictons containing clasts of Dalradian psammite, pelite and semipelite, and Caledonian granitic rocks (Plate 8a). Like most tills related to ice emanating from the East Grampian sheet, their composition closely reflects the nature of the underlying bedrock. For example, a lime-rich fertile soil has developed on the till on the flanks of Rhinbuckie Wood [NO 716 933], where calcareous strata crop out (Map 10). The Banchory Till is generally less than 5 m thick, free-draining and notably overconsolidated. Typically it is pale to yellowish brown, although darker colours occur beneath the water table, ranging from greyish brown to olive-grey. Red and brown colours occur where the tills are developed on decomposed (reddened) granite bedrock.

The type area of the Banchory Till Formation is the upland on the southern flank of the Dee valley, between Banchory and Strachan, on Sheet 66E. A reference section occurs [NO 607 920] at a 5 m-high river cliff of the Burn of Granney (a tributary of the Water of Feugh), just to the west of Sheet 66E. The till is sandy, moderate brown in colour and contains abundant angular clasts of the underlying granite, together with psammite and microgranite. A further reference section [NO 6073 9199] occurs in a small working 400 m west-south-west of Finzean, also just to the west of the Sheet 66E. Bouldery subglacial diamictons of the Banchory Till Formation commonly occur at the base of the East Grampian Drift Group. More sandy and gravelly diamictons are locally interstratified with the overlying Lochton Sand and Gravel Formation and, in places, along the contact between the East Grampian and Mearns drift groups, they overlie the latter deposits (see above).

Three probably Late Devensian members of the Banchory Till Formation have been established in the Aberdeen area (Map 9), but only one is visible at the Nigg Bay site (Table 7; Appendix 1). The *Nigg Till Member* is dark greyish brown, contains clasts derived locally from the west to west-south-west and is typical of surficial tills in the lower Dee valley and to the south. The brown *Kingswells Till Member* is probably a local variant, having a west-south-west to east-north-east fabric and containing much 'Hill of Fare' granite. It is equivalent to Till B of Murdoch (1977) and its type area is the Kingswells–Culter–Aberdeen city area. A third unit, the dark grey *Den Burn Member* with locally derived clasts underlies the above-mentioned till in the vicinity of the Kingswells Roundabout [NJ 878 061]. It has a north-west to south-east fabric and is equivalent to Till A of Murdoch (1977). It is possibly equivalent to the 'grey boulder clay' with clasts derived from the north-west, which Bremner (1934a, 1943) recorded as overlying 'dark shelly boulder clay' in excavations for the Anderson Drive Ring Road. The shelly unit is assigned here to the Banffshire Coast Drift Group (see above).

A fourth member of the Banchory Till Formation occurs in the Ellon–Bellscamphie area. This till, named here as the *Bearnie Till Member*, contains clasts of local provenance but is typically dark grey and contains a rich Jurassic palynoflora (Appendix 1 Site 15). The last two attributes are probably the result of glacial reworking of the older Pitlurg Till Formation (Banffshire Coast Drift Group). The Bearnie Till is overlain by the red, Hatton Till Formation (Table 7; Figure A1.20).

Other units

Most surficial tills of the Late Devensian East Grampian ice sheet occurring to the north and west of Sheet 76E Inverurie and Sheet 77 Aberdeen have not yet been assigned to lithostratigraphical formations. However, the *Hythie Till Formation* (Kirkhill Upper Till) has been identified at several sites in central Buchan, including Oldmill (Appendix 1 Site 8). The Hythie Till probably correlates with the *Byth Till Formation* at the Howe of Byth site and the *Crovie Till Formation* at Gardenstown (Table 7; Appendix 1 Sites 6, 3). The relationship of the Hythie Till Formation to the *Aldie Till Formation* at the Camp Fauld site is uncertain, but the latter unit may be of an earlier Devensian age (Table 7; Appendix 1 Site 14 Moss of Cruden). The granite-rich till at the base of the Sandford Bay sequence is named here as the *Sandford Bay Till Member* of the Hythie Till Formation (Appendix 1).

Pre-Devensian tills derived from the west, north-west and possibly also north have been identified at several localities. They include the *Rottenhill Till Formation* (Kirkhill Lower Till) at Kirkhill assigned to OIS 6, together with two other units of possibly similar age, the *Crossbrae Till Formation* at Crossbrae and the *Bellscamphie Till Formation* in the Ellon area (Table 7; Appendix 1 Site 15). The group also includes the oldest known till in the district, the *Leys Till Formation* of the Kirkhill sequence, which is probably at least as old as OIS 8 (Table 7). The Leys Till is also the oldest Pleistocene glacial deposit known onshore in Scotland.

SANDS AND GRAVELS

Lochton Sand and Gravel Formation

Coarse-grained glaciofluvial deposits of the Lochton Sand and Gravel Formation are characterised by clasts of Dalradian metamorphic rocks and Caledonian granitic rocks, both of relatively local provenance. Sandier units are typically pale yellowish brown in colour. Coarse-, medium- and fine-grained quartzo-feldspathic sands are common, normally forming upward-coarsening units. The formation includes moundy (ice-contact) and terraced (sheet) deposits that both commonly pass downwards into laminated silt and clay of the Glen Dye Silts Formation. Sharp erosional contacts on to the underlying Banchory Till Formation are evident, especially at the base of terraced spreads.

The type locality of the Lochton Sand and Gravel Formation is Lochton Pit [NO 749 926] in the valley of the Burn of Sheeoch, on Sheet 66E (Auton et al., 1988; Brown, 1994) (Map 10). Here, a coarsening-upward deltaic sequence of sand and gravel up to 5.4 m thick overlies interlaminated clay and silt of the Glen Dye Silts Formation. Reference sections occur between 0.4 and 5.2 m depth in BGS Borehole NO69SE15, sited on a kame terrace in the valley of the Water of Feugh, near Heugh Head, and in a trial pit (NO69SW5), dug into an esker ridge south-west of Easter Clune (Auton et al., 1990).

The 'lower sands and gravels' of the Nigg Bay section contain clasts that are predominantly derived from the Dee valley to the west (Appendix 1). Importantly however, they contain very sparse Scandinavian erratics and flint (Chapter 7 Erratics; Figure 47). The unit is named here as the *Ness Sand and Gravel Member* of the Lochton Sand and Gravel Formation.

Other units

Several other units of sand and gravel have been assigned to the East Grampian Drift Group, most predating the Late Devensian (Table 7). The *Pishlinn Burn Gravel Bed*, which underlies the Whitehills Glacigenic Formation at Gardenstown, is tentatively placed within the group, although its age and origin is problematic (Peacock and Merritt, 1997; Appendix 1 Site 3). At the former Tillybrex gravel pit [NK 001 347], 6 km northeast of Ellon (Map 6), over 11 m of weathered gravels with no morphological expression are overlain by brown till of westerly derivation, capped in turn by red diamicton and clay of the Logie-Buchan Drift Group (Merritt, 1981). The gravels at Tillybrex are here named the *Tillybrex Gravel Formation*. Although no contact has been recorded, the gravels appear to occur stratigraphically below patchy deposits of dark grey till correlated with the Pitlurg Till Formation of the Banffshire Coast Drift Group (Hall and Jarvis, 1995; Appendix 1 Ellon). The gravels are underlain by yellowish brown sandy till containing clasts of psammite from the west, together with some well-rounded quartz, quartzite and flint pebbles derived from the Buchan Gravels Formation. This till, correlated with the pre-Devensian Bellscamphie Till by Hall and Jarvis (1995), overlies weathered psammite bedrock. Like the Bellscamphie Till, the Tillybrex Gravel most probably predates the Devensian (Table 7).

The westerly palaeocurrent of the Tillybrex Gravel Formation is broadly similar to the gravels at Oldmill [NK 023 440] and the *Denend Gravel Formation* (Leys Lower Gravel) of the Kirkhill sequence. It suggests that ice advanced into eastern Buchan during a pre-Devensian cold stage(s), perhaps as old as OIS 8 (Appendix 1 Oldmill, Kirkhill and Ellon). Other pre-Devensian units at Kirkhill assigned to the East Grampian Drift Group include the *Pitscow Sand and Gravel Formation* (Kirkhill Lower Sand and Gravel) and the *Swineden Sand Bed* (Table 7; Appendix 1). The *Birnie Gravel Formation*, which underlies the Benholm Clay at the Burn of Benholm site (Appendix 1), is also assigned to the East Grampian Drift Group.

GLACIOLACUSTRINE DEPOSITS

Glen Dye Silts Formation

Thinly interlaminated sandy silts and clays, typically pale olive to greenish grey in colour, were laid down in ice-marginal lakes as the East Grampian ice sheet decayed and retreated in the Late Devensian. They are formally assigned here to the Glen Dye Silts Formation. In the type area of Glen Dye, on Sheet 66E (Map 10), these deposits occupy flat-lying ground near the confluence of the Mill Burn and the Water of Dye, at Miller's Bog [NO 636 861]. The formation is exposed in several sections in the southern bank of the Mill Burn, 600 to 800 m upstream of its confluence with the main river. Here a unit, 0.8 to 3.0 m thick, of micaceous sandy silt and clay with well developed horizontal lamination overlies clayey diamicton of the Banchory Till Formation. The exposed silts and clays show pronounced iron staining and are typically mottled orange and olive-grey in colour.

Deposits of the Glen Dye Silts Formation typically rest with a sharp planar contact on the underlying till or bedrock, but they pass upwards gradationally into deposits of the Lochton Sand and Gravel Formation. This upward transition is recorded in a trial pit (NO68NW2) sited on the floor of a small gravel working at Miller's Bog and a trial pit at Lochton (NO79SE1). The silts are locally interbedded with deposits of the Lochton Sand and Gravel Formation. As well as underlying the Lochton Sand and Gravel Formation in Glen Dye, the Glen Dye Silts Formation is widely developed beneath floodplains, alluvial flats and basin peat elsewhere on Sheet 66E Banchory. Laminated silts and clays are also common in similar settings on Sheet 76E Inverurie and Sheet 77 Aberdeen. They are particularly thick and extensive beneath the floodplain and glaciofluvial terraces of the River Don, between Dyce and Inverurie (Map 8), and beneath similar features within the valley of the River Urie upstream of Inverurie (Auton and Crofts, 1986; Auton et al., 1988; see also Appendix 1 Mill of Dyce and Nigg Bay).

CENTRAL GRAMPIAN DRIFT GROUP

Deposits of the Central Grampian Drift Group occur on Sheet 95 Elgin and Sheet 96W Portsoy. They were laid down by ice that radiated outwards from a centre over Rannoch Moor in the western Highlands, carrying rock fragments from the Central Highland Migmatite Complex and Caledonian igneous rocks as well as the local, dominantly psammitic Grampian Group rocks (Figures 45; 47). The ice that entered the district did so by two routes (Figure 4). Some ice flowed down the Spey valley and merged with ice from the East Grampians. Other ice flowed northwards towards the Moray Firth where it abutted, and merged with a more powerful stream emanating from the north-west Highlands. These combined ice streams flowed into the Moray Firth and along its southern shores. The relative power of the merging ice streams varied through time, resulting in the interdigitation of deposits of the three drift groups. As in the other groups, some formations have been identified that predate the Devensian.

The interplay between ice streams has resulted in complicated glacigenic sequences in the Inverness area, on Sheet 84W (Fletcher et al., 1996). There, psammite-rich tills laid down by central Grampian ice overlie tills dominated by Old Red Sandstone lithologies. The latter were laid down by ice that flowed from the Central Highlands via the Great Glen towards Inverness, where it crossed Old Red Sandstone strata. The more powerful flow of ice from the north-west Highlands forced the ice from the Great Glen eastwards across high ground towards the Elgin area, where it laid down sandstone-rich tills and associated glaciofluvial deposits. Some Mesozoic erratics occur in the surficial tills around Burghead and Lossiemouth, indicating that some ice crossed the southern margin of the Moray Firth (Figure 45). Deposits have been assigned to the Central Grampian Drift Group on Sheet 96W Portsoy, but Sheet 95 Elgin and Sheet 84W Fortrose were published before the group was formally established.

TILL

These tills have been described in general terms in Chapter 6. No formations have been mapped out individually, but a local stratigraphy has been established in the vicinity of the Teindland site (Hall et al., 1995; Table 7; Appendix 1). In the Elgin area, brown sandy tills of the Central Grampian Drift Group contain many well-rounded clasts derived from Old Red Sandstone conglomerates cropping out to the west. They overlie dark grey clayey tills containing shell fragments and Mesozoic erratics derived from the bed of the Moray Firth to the north-west (Peacock et al., 1968; Aitken et al., 1979). Hall et al. (1995a) name the former unit the Tofthead Till and the latter the Altonside Till (Banffshire Coast Drift Group). They also recognise a younger sandy diamicton, the Waterworks Till. They conclude that all three tills, together with locally intervening units of sand and gravel, were laid down in different phases of the Main Late Devensian glaciation, the Waterworks Till being the result of a minor re-advance. The Tofthead Till probably correlates with the *Old Hythe Till* at the Boyne Limestone Quarry (Peacock and Merritt, 2000).

None of the three Late Devensian tills occurring in the vicinity of Teindland appears to be present at the main site. Instead, another sandy till unit occurs that Hall et al. (1995a) name as the Teindland Till and assign to the Early Devensian. It overlies a package of units that includes the *Teindland Buried Soil*, which probably dates from the Ipswichian (OIS 5e). The soil lies stratigraphically above a unit that Hall et al. (1995a) name as the Teindland Gravel. It is composed predominantly of rounded clasts of quartzite and psammite that are probably largely derived from Old Red Sandstone conglomerates. Nearby, it overlies the Red Burn Till, the oldest known till in this area (Appendix 1). Although reddish brown in colour and containing clasts mostly of quartzite and psammite, this diamicton, like the Altonside Till, contains some Mesozoic erratics and hence it is also assigned here to the Banffshire Coast Drift Group, albeit tentatively.

Another old till unit assigned here to the Central Grampian Drift Group occurs at the Boyne Limestone Quarry (Appendix 1). This very sandy diamicton, the *Craig of Boyne Till Formation*, contains much decomposed, easily weathered calc-silicate rock and therefore appears to be weathered to a greater extent than it is (Peacock and Merritt, 2000). Nonetheless, it contains some clasts that have been considerably weathered in situ and is therefore likely to be pre-Devensian in age, like the Red Burn Till at Teindland.

OTHER DEPOSITS

Although the sand and gravel occurring on Sheet 95 Elgin was mainly laid down at the margin of the Moray Firth ice stream as it retreated (Chapter 6), its composition suggests that much of the ice crossing the area was probably sourced in the central Highlands. Therefore, these glaciofluvial deposits should be assigned to the Central Grampian Drift Group, but no formal lithostratigraphical units have yet been set up. The only current named unit is the pre-Devensian Teindland Gravel.

INFORMATION SOURCES

Further geological information held by the British Geological Survey relevant to the Quaternary geology of north-east Scotland is listed below. It includes published maps, memoirs and reports and open-file maps and reports. Other sources include borehole records, site investigation reports, hydrogeological, geochemical and geophysical data and photographs.

Searches of indexes to some collections relevant to the Quaternary of the district can be made on the Geoscience Data Index system in British Geological Survey libraries. This is a developing computer-based system that carries out searches of indexes to digital and nondigital databases for specified geographical areas. It is based on a geographical information system (GIS) linked to a relational database management system. Results of the searches are displayed on maps on screen. At the present time (2003) the databases indexes that are particularly applicable to the Quaternary geology of northeast Scotland are:

- single Onshore Borehole Index (SOBI)
- topographical backdrop based on 1:250 000 scale maps
- outlines of BGS maps at 1:50 000, 1:10 000, 1:10 560 and County Series maps
- geochemistry (stream sediment) sample locations on land
- Land Survey Records (LSR) Site Investigation (Scotland)
- Land Survey Records (LSR) Archives (Scotland)
- Land Survey Records (LSR) Enquires (Scotland)
- Land Survey Records (LSR) Mineral Resources (Scotland)
- memoirs
- mineral Assessment Reports
- Open File Reports
- photographs
- university theses

For further information on data and other geological information, enquiries should be addressed to the National Geoscience Information Service, BGS, Edinburgh.

MAPS

Geological maps

In addition to the printed publication noted here, many BGS maps are available in digital form, which allows the geological information to be used in GIS applications. These **Digital geological map data** must be licensed for use. Details are available from the Intellectual Property Rights Manager at BGS Keyworth. The main datasets are:

DiGMapGB-625 (1:625 000 scale)
DiGMapGB-250 (1:250 000 scale)
DiGMapGB-50 (1:50 000 scale)
DiGMapGB-10 (1:10 000 scale)

The current availability of these can be checked on the BGS web site at: http://www.bgs.ac.uk/products/digitalmaps/digmapgb.html

1:1 000 000 poster maps
Quaternary geology around the United Kingdom, 1994. North sheet.

Sea bed sediments around the United Kingdom, 1987. North sheet.
Quaternary map of the United Kingdom, First edition, 1977. Overprinted with the sheet lines of the 1:50 000/1:63 360 series. North sheet

1:625 000
United Kingdom (North sheet), Quaternary geology, 1977

1:250 000
Maps of the United Kingdom and continental shelf
57N 04W Moray-Buchan, Sea-bed sediments and Quaternary, 1984
57N 02W Peterhead, Sea-bed sediments, 1984; Quaternary, 1986
61N 02W Miller, Quaternary, 1991

1:50 000
Sheet 66E   Banchory, Solid and Drift, 1996
Sheet 67   Stonehaven, Solid and Drift, 1999.
Sheet 77   Aberdeen, Drift, 1980. New edition, in press
Sheet 76E   Inverurie, Solid and Drift, 2002

Table 9
Geological surveyors.

Ashcroft, WA	WAA
Auton, CA	CAA
Barrow, G	GB
Berridge, NG	NGB
Campbell, R	RC
Carroll, S	SC
Crofts, RG	RGC
Duncan, IG	IGD
Edwards, LT	LTE
Fettes, DJ	DJF
Fletcher, TP	TPF
Golledge, N	NG
Gould, D	DG
Grant-Wilson, JS	JSGW
Harris, AL	ALH
Highton, AJ	AJH
Hinxman, LW	LWH
Horne, J	JH
Irvine, DR	DRI
Kneller, BC	BCK
Leslie, AG	AGL
Linn, J	JL
May, F	FM
McCourt, WJ	WJM
McLean, F	FM
Mendum, JR	JRM
Merritt, JW	JWM
Munro, M	MM
Murdoch, WM	WM
Peacock, JD	JDP
Read, HH	HHR
Redwood, SD	SDR
Ross, DL	DLR
Skae, HM	HMS
Smith, CG	CGS
Stephenson, D	DS
Thomas, CW	CWT

Table 10 Survey history of geological maps.

Sheet No	Name	Scale of Survey	Surveyor	Date
67	Stonehaven	1:10 560	DRI	1876–82
		1:10 560	GB	1893–95
		1:10 560	RC	1913–26
		1:10 000	CAA, SC, RGC, DG, JWM, CWT	1984–95
66E	Banchory	1:63 360	HMS	1878–82
		1:10 560	DRI	1878–82
		1:10 560	GB	1893–95
		1:10 000	CAA, SC, RGC, DG, AJH	1984–94
96E	Banff	1:10 560	JH, JSGW	1879–81
		1:10 560	HHR	1918
		1:10 000	JDP	1987
		1:10 000	TPF	1993–96
86E	Turriff	1:10 560	JH, JSGW, LWH	1880–84
		1:10 560	HHR	1917–19
		1:10 000	WAA, AGL, MM	1978–81
		1:10 000	DG, CWT	1988–92
96W	Portsoy	1:10 560	JH	1880–82
		1:10 560	HHR	1916–18
		1:10 000, 1:10 560	TPF, DG, JRM, SDR, DS	1987–89, 1996–97
95	Elgin	1:10 560	JL, JSGW	1877–80
		1:10 560/1:10 000	NGB, ALH, FM, JDP	1961–63
86	Huntly	1:63 360/1:10 560	JH, JSGW, LWH	1880–84
		1:10 560	HHR	1917–19
97	Fraserburgh	1:63 360/1:10 560	JSGW	1879–80
		1:10 560/1:10 000	DJF, JRM, JDP, CGS, CWT	1981
87W	Ellon	1:10 560	JSGW, DRI	1880–81
		1:10 560	HHR	1930
		1:10 000/1:10 560	WAA, IGD, LTE, BDK, AFL,MM	1980s
		1:10 000	DLR	1977–78
		1:10 000	JWM	1995
77	Aberdeen	1:10 560	DRI, JSGW	1880–82
		1:10 000	FM, WMM	1972–77
		1:10 000	CAA, RGC, NG, JWM, DLR, CGS	1984–2000
87E	Peterhead	1:10 560	JSGW	1880–81
		1:10 560/1:10 000	DJF, JRM, JDP, CGS	1981
		1:10 000	DLR, WJM	1976–78
		1:10 000	JWM	1986
76E	Inverurie	1:63 360/1:10 560	DRI, HMS, GB	1879–95
		1:10 000	CAA, DG, AJH, JWM, DLR, CGS	1985–96

Sheet 86E   Turriff, Solid and Drift, 1995
Sheet 87W   Ellon, Solid and Drift, 2002
Sheet 87E   Peterhead, Solid, Drift, 1992
Sheet 96W   Portsoy, Solid and Drift, 2002
Sheet 96E   Banff, Solid and Drift, 2002
Sheet 97   Fraserburgh, Solid, Drift, 1987

1:63 360 (one inch to one mile)
Sheet 95   Elgin, Drift, 1969

Surveyors are listed in Table 9. The survey history of each 1:50 000 and 1:63 360 scale map is given in Table 10.

1:25 000, 1:10 000 and 1:10 560
The coverage of large-scale Quaternary geological maps of the district is shown graphically on the index map (Figure 51). Copies of the fair-drawn maps have been deposited in the British Geological Survey libraries in Edinburgh and at Keyworth for public reference and may also be inspected in the London Information Office, in the Natural History Museum, South Kensington, London. Copies may be purchased directly from The Sales Desk, BGS, Edinburgh, as scanned colour or black and white images, as available.

Applied geology maps

1:25 000
A set of thematic maps (Environmental Geology: Drift Geology) prepared for land-use planning are available for the Aberdeen area. They accompany British Geological Survey Technical Report WA/HI/86/1.

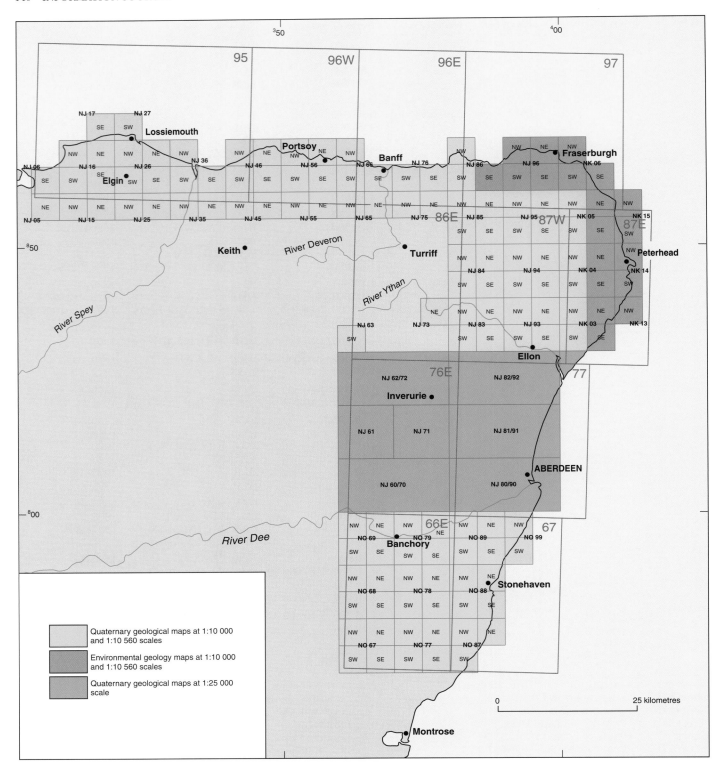

Figure 51 Coverage of large-scale Quaternary geological maps of the district.

1:10 000

Thematic (Environmental Geology) maps of Superficial Deposits, prepared for land-use planning are available for the Peterhead area; the coverage is shown on the graphic index map (Figure 51). These thematic maps accompany *British Geological Survey Technical Report WA/HI/83/1*, but they can be obtained separately as scanned black and white copies.

Mineral assessment maps identifying sand and gravel resources

1:25 000

These colour-printed maps accompany Mineral Assessment Reports (MARs). The asterisk denotes that the sheet number refers only to one of several Ordnance Survey 100 km² sheets covered by the Mineral Resource Sheet (Appendix 2 Figure A2.1, shows the coverage of each sheet).

Mineral Resource Sheet

NJ 36 Garmouth, MAR 41 (Aitken et al., 1979)
NK 04* Peterhead, MAR 58 (McMillan and Aitken, 1981)
NJ 93* Ellon West, MAR 76 (Merritt, 1981)
NJ 03* Ellon East, MAR 76 (Merritt, 1981)
NJ 70 Inverurie and NJ 72 Dunecht, MAR 148 (Auton et al., 1988)
NO 79* Banchory and Stonehaven, MAR 148 (Auton et al., 1988)
NJ 71* Aberdeen sheet 1; Kemnay, MAR 146 (Auton and Crofts, 1986)
NJ 81* Aberdeen, sheet 2; Dyce, MAR 146 (Auton and Crofts, 1986)
NJ 80* Aberdeen, sheet 3; Peterculter, MAR 146 (Auton and Crofts, 1986)
NO 68 Strachan, MAR 149 (Auton et al., 1990)
NO 87/88* Auchenblae and Catterline, MAR 149 (Auton et al., 1990)

Hydrogeology maps

1:625 000
Hydrogeological map of Scotland, 1988
Groundwater vulnerability map of Scotland, 1995

MEMOIRS AND BOOKS

The 1:50 000 and one-inch geological maps are generally accompanied by explanatory sheet memoirs. Those in print and out of print are listed below. The memoirs in print may be obtained either from the BGS, or, with the exception of the title marked with an asterisk* (where the Survey holds the remaining small stock) through The Stationery Office bookshops and approved stockists. Facsimile copies of out of print memoirs can be obtained from BGS; price, on request, is based on the cost of copying. Both in print and out of print memoirs may be consulted at BGS and other libraries.

Memoirs for 1:50 000 and 1:63 360* maps: in print

Sheet 66W   Aboyne   D Gould, 2001
Sheets 76E and 76W   Inverurie and Alford   D Gould, 1997
Sheet 77   Aberdeen   M Munro, 1986
Sheet 95*   Elgin district   J D Peacock, N G Berridge, A L Harris, F May, P J Brand, R W Elliot, and P J Fenning, 1968

Memoirs for 1:50 000 and 1:63 360 maps: out of print

Sheet 76   Central Aberdeenshire   J S G Wilson and L W Hinxman, 1890

Sheets 86 and 96   Banff, Huntly and Turriff   H H Read, 1923
Sheet 87 and 96   North-east Aberdeenshire with detached portions of Banffshire   J S G Wilson, 1886
Sheet 85/95   Geology of Lower Strathspey   L W Hinxman and J S G Wilson, 1902
Sheet 97   Northern Aberdeenshire, eastern Banffshire   J S G Wilson, 1882

Geochemical atlas

The Geochemical Baseline Survey of the Environment (G-BASE) is based on the collection of stream sediment and stream water samples at an average density of one sample per 1.5 km². The fine (minus 150 μm) fractions of stream sediment samples are analysed for a wide range of elements, using automated instrumental methods.

The samples from north-east Scotland were collected in 1977–80. The results are published in atlas form (*Regional geochemistry of the East Grampians area*, British Geological Survey, 1991) covering the east of Scotland, from the Moray Firth to the Firth of Forth. The geochemical data, with location and site information, are available as hard copy for sale or in digital form under licensing agreement. The coloured geochemical atlas is also available in digital form (on CD-ROM or floppy disk) under licensing agreement. BGS offers a client-based service of interactive GIS interrogation of the G-BASE data.

DOCUMENTARY COLLECTIONS

Borehole and site investigation record collection

BGS holds collections of borehole and site investigation records, which may be consulted and copies of most can be purchased from the National Geological Records Centres at BGS Keyworth and BGS Edinburgh. The collections from the district currently include the records of some 300 site investigations undertaken from across the district. These are concentrated mainly around the urban areas on the eastern coastal fringe (Figure 52). There is a digital index, (LSR) Site Investigation (Scotland), and the reports themselves are held on microfiche. The collections also contain locations and logs of approximately 5 500 boreholes. Their distribution, by geological sheet, is given in Table 11. Index

Table 11
Number of boreholes on each geological sheet.

Geological sheet	Number of boreholes
95 Elgin	796
96W Portsoy	28
96E Banff	63
97 Fraserburgh	138
86E Turriff	11
87W Ellon	298
87E Peterhead	641
76E Inverurie	474
77 Aberdeen	1805
66E Banchory	554
67 Stonehaven	686

Figure 52
Distribution of
site investigation
records held by
BGS in 1999.

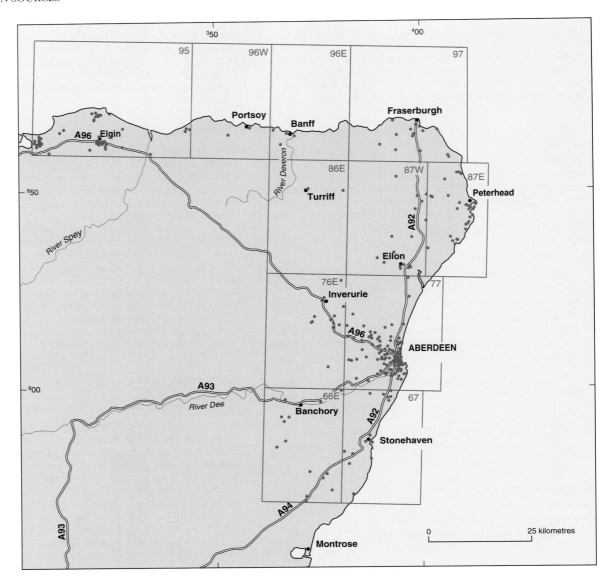

information for these boreholes has been digitised and incorporated within the SOBI database. The logs are either hand-written or typed and many of the older records are driller's logs.

Hydrogeological data

Records of water boreholes are held at BGS, Edinburgh.

Geochemical data

Records of stream-sediment and other analyses are held at BGS, Keyworth.

Seismic data

Records of earthquakes are held at BGS, Edinburgh.

MATERIAL COLLECTIONS

Geological Survey photographs

Some 726 photographs illustrating aspects of the geology of north-east Scotland are deposited for reference in the libraries at BGS Edinburgh, BGS Keyworth and in the BGS Information Office, London. The photographs were taken at various times over the last century and depict details of various rocks as well as Quaternary sediments exposed either naturally or in excavations. The collection also includes many general views, particularly of coastal scenery. There is a digtal index of titles and a listing can be supplied on request. The photographs can be supplied as black and white or colour prints and 35 mm colour transparencies, at a fixed tariff, from the Photographic Department, BGS, Edinburgh.

ADDRESSES FOR DATA SOURCES

British Geological Survey (Headquarters),
Keyworth, Nottingham NG12 5GG
Telephone   0115 936 3100
Fax   0115 936 3200

London Information Office at the Natural History Museum
Earth Galleries, Exhibition Road, South Kensington, London
SW7 2DE.
Telephone   0171 589 4090
Fax   0171 584 8270
e-mail   sales@bgs.ac.uk
www.bgs.ac.uk

British Geological Survey, Murchison House, West Mains Road,
Edinburgh EH 93 LA.
Telephone   0131 667 1000
Fax   0131 668 2683

REFERENCES

Most of the references listed below are held in the Library of the British Geological Survey, Murchison House, Edinburgh, and at Keyworth, Nottingham. Copies of the references can be purchased subject to current copyright legislation.

AALBERSBERG, G, and LITT, T. 1998. Multiproxy climate reconstructions for the Eemian and Early Weichselian. *Journal of Quaternary Science*, Vol. 13, 367–390.

ABER, J S, CROOT, D G, and FENTON, F M. 1989. *Glaciotectonic landforms and structures*. (Dordrecht, The Netherlands: Kluwer Academic Publishers.)

AITKEN, A M. 1983. A preliminary study of the sand and gravel deposits of Strathmore (1:25 000 sheets NN72, 82, 92 and parts of 71, 81 and 91: sheets NO03, 14, 24, 25, 34, 35, 45, 55, 56, 64, 65, 66, 67, 75, 76, 77, 86, 87, 88 and parts of NO02, 04, 12, 13, 23, 36, 44, 46, 54 and 78). *Institute of Geological Sciences, Industrial Mineral Assessment Unit, Open File Report*, WF/SC/83/5.

AITKEN, A M, MERRITT, J W, and SHAW, A J. 1979. The sand and gravel resources of the country around Garmouth, Grampian Region. Description of 1:25 000 resource Sheet NJ36. *Mineral Assessment Report of the Institute of Geological Sciences*, No. 41.

AITKEN, J F. 1991. Sedimentology and palaeoenvironmental significance of late Devensian to mid Holocene deposits in the Don valley, north-east Scotland. Unpublished PhD thesis, University of Aberdeen.

AITKEN, J F. 1993. A re-appraisal of supposed iceberg 'dump' and 'grounding' structures from Pleistocene glaciolacustrine sediments, Aberdeenshire. *Quaternary Newsletter*, Vol. 71, 1–10.

AITKEN, J F. 1995. Lithofacies and depositional history of a Late Devensian ice-contact deltaic complex, northeast Scotland. *Sedimentary Geology*, Vol. 99, 111–130.

AITKEN, J F. 1998. Sedimentology of Late Devensian glacio-fluvial outwash in the Don Valley, Grampian Region. *Scottish Journal of Geology*, Vol. 34, 97–117.

ALLEY, R B, and MACAYEAL, D R. 1994. Ice-rafted debris associated with binge-purge oscillations of the Laurentide Ice Sheet. *Palaeoceanography*, Vol. 9, 503–511.

ALLEY, R B, and ten others. 1993. Abrupt increase in Greenland snow accumulation at the end of the Younger Dryas event. *Nature*, Vol. 362, 527–529.

ALVERSON, K, and OLDFIELD, F. 2000. Past global changes and their significance for the future: an introduction. *Quaternary Science Reviews*, Vol. 19, 3–7.

ANDERSON, J G C. 1943. Sands and gravels of Scotland. Quarter-inch Sheet 9 Elgin–Banff–Aberdeen. *Wartime Pamphlet of the Geological Survey of Great Britain*, No. 30.

ANDERSON, J G C. 1945. Sands and gravels of Scotland: Quarter-Inch Sheet 12, Stonehaven–Perth–Dundee. *Wartime Pamphlet of the Geological Survey of Great Britain*, No. 30.

ANDERTON, R, BRIDGES, P H, LEEDER, M R, and SELLWOOD, B W. 1979. *A dynamic stratigraphy of the British Isles*. (London: Allen and Unwin.)

ANDREWS, I J, and six others. 1990. *United Kingdom offshore regional report: the geology of the Moray Firth*. (London: HMSO for the British Geological Survey.)

ANSARI, M H. 1992. Stratigraphy and palaeobotany of Middle Pleistocene interglacial deposits in the North Sea. Unpublished PhD thesis, University Wales; Bangor.

ARMSTRONG, M, and PATERSON, I B. 1985. Some recent discoveries of ice-wedge cast networks in north-east Scotland — a comment. *Scottish Journal of Geology*, Vol. 21, 107–108.

ARMSTRONG, M, PATERSON, I B, and BROWNE, M A E. 1985. Geology of the Perth and Dundee district. *Memoir of the British Geological Survey*, Sheets 48W, 48E and 49 (Scotland).

ASHCROFT, W A, and WILSON, C D V. 1976. A geophysical survey of the Turriff basin of the Old Red Sandstone, Aberdeenshire. *Journal of the Geological Society of London*, Vol. 132, 27–43.

ASHCROFT, W A, and MUNRO, M. 1978. The structure of the eastern part of the Insch mafic intrusion, Aberdeenshire. *Scottish Journal of Geology*, Vol. 14, 55–79.

ASHWORTH, J R. 1975. The sillimanite zones of the Huntly–Portsoy area in the north-east Dalradian, Scotland. *Geological Magazine*, Vol. 112, 113–136.

ATKINSON, T C, BRIFFA, K R, and COOPE, G R. 1987. Seasonal temperatures in Britain during the last 22 000 years, reconstructed using beetle remains. *Nature, London*, Vol. 325, 587–593.

AUSTIN, W E N, and KROON, D. 1996. Late glacial sedimentology, foraminifera and stable isotope stratigraphy of the Hebrides continental shelf, northwest Scotland. 187–213 in Late Quaternary Palaeoceanography of the North Atlantic Margins. ANDREWS, J T, and AUSTIN, W E N, BERGSTEN, H, and JENNINGS, A E (editors). *Geological Society, Special Publication*, No. 111.

AUTON, C A. 1992. The utility of conductivity surveying and resistivity sounding in evaluating sand and gravel deposits and mapping drift sequences in northeast Scotland. *Engineering Geology*, Vol. 32, 11–28.

AUTON, C A. 2000. The Buchan Gravels Formation. 79–81 in *The Quaternary of the Banffshire coast and Buchan: Field Guide.* MERRITT, J W, CONNELL, E R, and BRIDGLAND, D R (editors). (London: Quaternary Research Association.)

AUTON, C A, and CROFTS, R G. 1986. The sand and gravel resources of the country around Aberdeen, Grampian Region. Description of 1:25 000 resource sheets NJ71, 80, 81 and 91, with parts of NJ61, 90 and 92, and with parts of NO89 and 99. *Mineral Assessment Report, British Geological Survey*, No. 146.

AUTON, C A, MERRITT, J W, and ROSS, D L. 1988. The sand and gravel resources of the country around Inverurie and Dunecht, and between Banchory and Stonehaven, Grampian Region. Mineral Assessment Report 148. *British Geological Survey Technical Report*, WF/88/1.

AUTON, C A, THOMAS, C W, and MERRITT, J W. 1990. The sand and gravel resources of the country around Strachan and between Auchenblae and Catterline, Grampian Region. Mineral Assessment Report 149. *British Geological Survey Technical Report*, WF/90/7.

AUTON, C A, GORDON, J E, MERRITT, J W, and WALKER, M J C. 2000. The glacial and interstadial sediments at the Burn of Benholm,

Kincardineshire: evidence for onshore pre-Devensian ice movement in northeast Scotland. *Journal of Quaternary Science*, Vol. 15, 141–156.

BAKER, A, SMART, P L, and EDWARDS, R L. 1995. Paleoclimate implications of mass spectrometric dating of a British flowstone. *Geology*, Vol. 23, 309–312.

BALLANTYNE, C K, and HARRIS, C. 1994. *The periglaciation of Great Britain.* (Cambridge: Cambridge University Press.)

Ballantyne, C K, and five others. 1998. High-resolution reconstruction of the last ice sheet in NW Scotland. *Terra Nova*, Vol. 10, 63–67.

BARD, E, and five others. 1987. Retreat velocity of the North Atlantic polar front during the last deglaciation determined by ^{14}C accelerator mass spectrometry. *Nature, London*, Vol. 328, 791–794.

BARNE, J H, ROBSON, C F, KAZNAKOVA, S S, DOODY, J P, and DAVIDSON, N C (editors). 1996. *Coasts and seas of the United Kingdom: North-east Scotland: Cape Wrath to St. Cyrus.* (Peterborough: Nature Conservation Committee.)

BASHAM, I R. 1968. Deeply weathered rock and associated soils of the Insch and Boganclough masses, Aberdeen. Unpublished PhD thesis, University of Aberdeen.

BASHAM, I R. 1974. Mineralogical changes associated with deep weathering of gabbro in Aberdeenshire. *Clay Minerals*, Vol. 10, 189–202.

BAUMANN, K H, and six others. 1995. Reflection of Scandinavian ice sheet fluctuations in Norwegian sea sediments during the past 150 years. *Quaternary Research*, Vol. 43, 185–197.

BEHRE, K E. 1989. Biostratigraphy of the last glacial period in Europe. *Quaternary Science Reviews*, Vol. 8, 25–44.

BELL, D. 1895. The shelly clays and gravels of Aberdeenshire considered in relation to the question of submergence. *Quaterly Journal of the Geological Society of London*, Vol. 51, 472–479.

BENN, D I, and EVANS, D J A. 1996. The interpretation and classification of subglacially-deformed sediments. *Quaternary Science Reviews*, Vol. 15, 23–52.

BENN, D I, and EVANS, D J A. 1998. *Glaciers and glaciation.* (London: Arnold.)

BENN, D I, LOWE, J J, and WALKER, M J C. 1992. Glacier response to climatic change during the Loch Lomond Stadial and early Flandrian: geomorphological and palynological evidence from the Isle of Skye, Scotland. *Journal of Quaternary Science*, Vol. 7, 125–144.

BENT, A J A. 1986. Aspects of Pleistocene glaciomarine sequences in the northern North Sea. Unpublished PhD thesis, University of Edinburgh.

BERSTAD, S, and DYPVIK, H. 1982. Sedimentological evolution and natural radioactivity of Tertiary sediments from the central North Sea. *Journal of Sedimentary Petrology*, Vol. 5, 77–88.

BIRKS, H J B. 1977. The Flandrian forest history of Scotland: a preliminary synthesis. 119–135 in *British Quaternary studies: recent advances.* SHOTTON, F W (editor). (Oxford: Clarendon Press.)

BIRKS, H J B, and RANSOM, M E. 1969. An interglacial peat at Fugla Ness, Scotland. *New Phytologist*, Vol. 68, 777–796.

BISHOP, W W, and COOPE, G R. 1977. Stratigraphical and faunal evidence for Lateglacial and early Flandrian environments in south-west Scotland. 61–88 in *Studies in the Scottish Lateglacial.* GRAY, J M, and LOWE, J J (editors). (Oxford and New York: Pergamon.)

BOND, G, and LOTTI, R. 1995. Iceberg discharges into the North Atlantic on millennial time scales during the last glaciation. *Science*, Vol. 267, 1005–1010.

BOND, G, and six others. 1993. Correlation between climate records from North Atlantic sediments and Greenland ice. *Nature*, Vol. 365, 143–147.

BOULTON, G S. 1968. Flow tills and related deposits on some Vestspitsbergen glaciers. *Journal of Glaciology*, Vol. 7, No. 51, 391–412.

BOULTON, G S. 1970. On the deposition of subglacial and melt-out tills at the margins of certain Svalbard glaciers. *Journal of Glaciology*, Vol. 9, 231–245.

BOULTON, G S. 1979. Processes of glacier erosion on different substrata. *Journal of Glaciology*, Vol. 23, 15–38.

BOULTON, G S. 1987. A theory of drumlin formation by subglacial deformation. 25–80 in *Drumlin Symposium.* MENZIES, J, and ROSE, J (editors). (Rotterdam: Balkema.)

BOULTON, G S, and DEYNOUX, M. 1981. Sedimentation in glacial environments and the identification of tills and tillites in ancient sedimentary sequences. *Precambrian Research*, Vol. 15, 397–422.

BOULTON, G S, and DOBBIE, K E. 1993. Consolidation of sediments by glaciers: relations between sediment geotechnics, soft-bed glacier dynamics and subglacial groundwater flow. *Journal of Glaciology*, Vol. 39, 26–44.

BOULTON, G S, and PAUL, M A. 1976. The influence of genetic processes on some geotechnical properties of glacial tills. *Quarterly Journal of Engineering Geology*, Vol. 9, 159–193.

BOULTON, G S, DENT, D L, and MORRIS, E M. 1974. Subglacial shearing and crushing, and the role of water pressure in tills from SE Iceland. *Geografiska Annaler, Series A Physical Geography*, Vol. 56A, No. 314, 135–145.

BOULTON, G S, JONES, A S, CLAYTON, K M, and KENNING, M J. 1977. A British ice-sheet model and patterns of glacial erosion and deposition in Britain 231–246 in *British Quaternary studies: recent advances.* SHOTTON, F W (editor). (Oxford: Claredon Press.)

BOULTON, G S, SMITH, G D, JONES, A S, and NEWSOME, J. 1985. Glacial geology and glaciology of the last mid-latitude ice sheets. *Journal of the Geological Society of London*, Vol. 142, 447–474.

BOULTON, G S, PEACOCK, J D, and SUTHERLAND, D G. 1991. Quaternary. 503–543 in *Geology of Scotland.* (Third edition). CRAIG, G Y (editor). (London: The Geological Society.)

BOWEN, D Q. 1989. The last interglacial-glacial cycle in the British Isles. *Quaternary International*, Vol. 3, Pt 4, 41–47.

BOWEN, D Q, (editor). 1999. A revised correlation of Quaternary deposits in the British Isles. *Geological Society Special Report*, No. 23.

BOWEN, D Q, ROSE, J, McCABE, A M, and SUTHERLAND, D G. 1986. Correlation of Quaternary glaciations in England, Ireland, Scotland and Wales. 299–340 *in* Quaternary Glaciations in the Northern Hemisphere. SIBRAVA, V, BOWEN, D Q, and RICHMOND, G M (editors). *Quaternary Science Reviews*, Vol. 5.

BOWEN, D Q, and SYKES, G A. 1988. Correlation of marine events and glaciations on the northeast Atlantic margin. *Philosophical Transactions of the Royal Society of London*, Vol. B, 318, 619–635.

BRADY, G S, CROSSKEY, H W, and ROBERSTON, D. 1874. A monograph of the post-Tertiary Entomostraca of Scotland. *Palaeontographical Society Monograph*, 1–232.

BREMNER, A. 1912. *The physical geology of the Dee valley.* (Aberdeen: The University Press.)

BREMNER, A. 1915. The capture of the Geldie by the Feshie. *Scottish Geographical Magazine*, Vol. 31, 589–596.

BREMNER, A. 1916. Problems in the glacial geology of north-east Scotland and some fresh facts bearing on them. *Transactions of the Edinburgh Geological Society*, Vol. 10, 334–347.

BREMNER, A. 1917. Low-level kettle-holes in and near Aberdeen. *Transactions of the Edinburgh Geological Society*, Vol. 11, 23–24.

BREMNER, A. 1919. A geographical study of the high plateau of the south-eastern Highlands. *Scottish Geographical Magazine*, Vol. 35, 331–351.

BREMNER, A. 1920a. The glacial geology of the Stonehaven District. *Transactions of the Edinburgh Geological Society*, Vol. 11, 25–41.

BREMNER, A. 1920b. Limits of valley glaciation in the basin of the Dee. *Transactions of the Edinburgh Geological Society*, Vol. 11, 61–68.

BREMNER, A. 1921. *The physical geology of the Don basin.* (Aberdeen: The University Press.)

BREMNER, A. 1922. Deeside as a field for the study of geology. *The Deeside Field*, Vol. 1, 33–36.

BREMNER, A. 1928. Further problems in the glacial geology of north-eastern Scotland. *Transactions of the Edinburgh Geological Society*, Vol. 12, 147–164.

BREMNER, A. 1931. The valley glaciation in the district round Dinnet, Cambus o'May and Ballater. *The Deeside Field*, Vol. 5, 15–24.

BREMNER, A. 1934a. The glaciation of Moray and ice movements in the north of Scotland. *Transactions of the Edinburgh Geological Society*, Vol. 13, 17–56.

BREMNER, A. 1934b. Meltwater drainage channels and other glacial phenomena of the Highland Border Belt from Cortachy to the Bervie Water. *Transactions of the Edinburgh Geological Society*, Vol. 13, 174–175.

BREMNER, A. 1936. The glaciation of Glenesk. *Transactions of the Edinburgh Geological Society*, Vol. 13, 378–382.

BREMNER, A. 1938. The glacial epoch in north-east Scotland. *The Deeside Field*, Vol. 8, 64–68.

BREMNER, A. 1939. Notes on the glacial geology of east Aberdeenshire. *Transactions of the Edinburgh Geological Society*, Vol. 13, 474–475.

BREMNER, A. 1942. The origins of the Scottish river system. *Scottish Geographical Magazine*, Vol. 58, 15–20, 54–59, 99–103.

BREMNER, A. 1943. The glacial epoch in the north-east. 10–30 in *The Book of Buchan (Jubilee Volume).* TOCHER, J F (editor). (Aberdeen: P Scrogie Ltd & Aberdeen University Press.)

BRIDGLAND, D R. 2000a. Flint-rich gravels in Aberdeenshire. 96–101 in *The Quaternary of the Banffshire coast and Buchan: Field Guide.* MERRITT, J W, CONNELL, E R, and BRIDGLAND, D R (editors). (London: Quaternary Research Association.)

BRIDGLAND, D R. 2000b. Discussion: the characteristics, variations and likely origin of the Buchan Ridge Gravel. 139–143 in *The Quaternary of the Banffshire coast and Buchan: Field Guide.* MERRITT, J W, CONNELL, E R, and BRIDGLAND, D R (editors). (London: Quaternary Research Association.)

BRIDGLAND, D R, and SAVILLE, A. 2000. Den of Boddam. 102–115 in *The Quaternary of the Banffshire coast and Buchan: Field Guide.* MERRITT, J W, CONNELL, E R, and BRIDGLAND, D R (editors). (London: Quaternary Research Association.)

BRIDGLAND, D R, SAVILLE, A, and SINCLAIR, J M. 1997. New evidence for the origin of the Buchan Ridge Gravel, Aberdeenshire. *Scottish Journal of Geology*, Vol. 33, 43–50.

BRIDGLAND, D R, SAVILLE, A, and SINCLAIR, J. 2000. Skelmuir Hill. 126–138 in *The Quaternary of the Banffshire coast and Buchan: Field Guide.* MERRITT, J W, CONNELL, E R, and BRIDGLAND, D R (editors). (London: Quaternary Research Association.)

BRITISH GEOLOGICAL SURVEY. 1977. Moray–Buchan Sheet 57°N–04°W. Solid geology. 1:250 000. (Southampton: Ordnance Survey for British Geological Survey.)

BRITISH GEOLOGICAL SURVEY. 1992. Peterhead, Scotland Sheet 87E. Solid and Drift. 1:50 000. (Keyworth: British Geological Survey.)

BRITISH GEOLOGICAL SURVEY. 2002. Lexicon of named rocks units [online]. Keyworth, Nottingham: British Geological Survey. Last revised 05/22/2002. Available from http://www.bgs.ac.uk.

BRITISH GEOLOGICAL SURVEY. 1998. *United Kingdom Minerals Yearbook 1997.* (Keyworth, Nottingham: British Geological Survey.)

BRITISH STANDARDS INSTITUTION. 1975. *BS 812: Methods for sampling and testing of mineral aggregates, sands and fillers.* (London: British Standards Institution.)

BROECKER, W S. 1984. Terminations. 687–698 in *Milankovitch and Climate. Part 2.* BERGER, A, IMBRIE, J, HAYS, J, KUKLA, G, and SALTZMAN, B (editors). (Norwell, Mass: D Reidel.)

BROECKER, W S. 1994. Massive iceberg discharges as triggers for global climate change. *Nature*, Vol. 372, 421–424.

BROECKER, W S, and DENTON, G H. 1990. What drives glacial cycles? *Scientific American*, Vol. 262, 43–50.

BROWN, A G. 1987. Long-term sediment storage in the Severn and Wye catchments. 307–332 in *Palaeohydrology in practice.* GREGORY, K J, LEWIN, J, and THORNES, J B (editors). (Chichester: Wiley.)

BROWN, I M. 1993. Pattern of deglaciation of the last (Late Devensian) Scottish ice sheet: evidence from ice-marginal deposits in the Dee valley, north-east Scotland. *Journal of Quaternary Science*, Vol. 8, 235–250.

BROWN, I M. 1994. Former glacial lakes in the Dee valley: origin, drainage and significance. *Scottish Journal of Geology*, Vol. 30, 147–158.

BUCHAN, A. 1935. Investigations of the glacial and post-glacial deposits of Loch Spynie. Unpublished PhD thesis, University of Aberdeen.

BUCHARDT, B. 1978. Oxygen isotope palaeotemperatures from the Tertiary period in the North Sea area. *Nature, London*, Vol. 275, 121–123.

BUILDING RESEARCH ESTABLISHMENT. 1968. Shrinkage of natural aggregates in concrete. *Digest Building Research Establishment*, No. 35, Series 2.

CABRERA, D E. 1976. A study of the soils and their underlying deposits at Cruden Bay, Grampian. Unpublished MSc thesis, University of Aberdeen.

CAMBRIDGE, P G. 1982. A note on supposed crag shells from the Kippet Hills, Aberdeenshire. *Bulletin of the Geological Society of Norfolk*, Vol. 32, 37–38.

CAMERON, T D J, STOKER, M S, and LONG, D. 1987. The history of Quaternary sedimentation in the UK sector of the North Sea Basin. *Journal of the Geological Society, London*, Vol. 144, Pt 1, 43–58.

CAMERON, T D J, and six others. 1992. *United Kingdom Offshore Regional Report: the Geology of the Southern North Sea.* (London: HMSO for the British Geological Survey, London.)

CAMERON, D G, and seven others. 1998. *Directory of Mines and Quarries, 1998.* (5th edition). (Keyworth, Nottingham: British Geological Survey.)

CAMPBELL, R. 1934. On the occurrence of shelly boulder clay and interglacial deposits in Kincardineshire. *Transactions of the Edinburgh Geological Society*, Vol. 13, 176–182.

CARR, S. 1998. Last glacial maximum in the North Sea Basin. Unpublished PhD thesis, Royal Holloway, University of London.

CARR, S. 1999. The micromorphology of Last Glacial Maximum sediments in the Southern North Sea. *Catena*, Vol. 35, 123–145.

CARROLL, S. 1995a. Geology of the Stonehaven district. 1:10 000 sheets NO88NW, NO88NE, NO88SW and NO88SE (South of the Highland Boundary Fault). *British Geological Survey Technical Report*, WA/94/19.

CARROLL, S. 1995b. Geology of the Inverbervie and Catterline District. 1:10 000 sheets NO87NW, NO87NE and NO87SW. *British Geological Survey Technical Report*, WA/94/20.

CARROLL, S. 1995c. Geology of the Laurencekirk District. 1:10 000 sheets NO77SW and NO77SE. *British Geological Survey Technical Report*, WA/95/81.

CARROLL, S. 1995d. Geology of the Fettercairn District. 1:10 000 sheets NO67NW, NO67NE, NO67SW, NO67SE and NO68SE (South of the Highland Boundary Fault). *British Geological Survey Technical Report*, WA/95/91.

CASELDINE, C J, and EDWARDS, K J. 1982. Interstadial and last interglacial deposits covered by till in Scotland: comments and new evidence. *Boreas*, Vol. 11, 119–123.

CHAMBERS, R. 1857. Geological notes on Banffshire. *Proceedings of the Royal Society of Edinburgh*, Vol. 3, 332–333.

CHAPMAN, M R, SHACKLETON, N J, and DUPLESSY, J-C. 2000. Sea surface temperature variabillity during the last glacial–interglacial cycle: assessing the magnitude and pattern of climate change in the North Atlantic. *Palaeogeography, Palaeoclimatology, Palaeoecology*, Vol. 157, 1–25.

CHARLESWORTH, J K. 1926. The re-advance, marginal kame-moraine of the south of Scotland, and some later stages of retreat. *Transactions of the Royal Society of Edinburgh*, Vol. 55, 25–50.

CHARLESWORTH, J K. 1956. The Late-glacial history of the Highlands and Islands of Scotland. *Transactions of the Royal Society of Edinburgh*, Vol. 62, 769–928.

CLAPPERTON, C M. 1977. Buchan. 37–39 in *INQUA X Congress 1977. Guidebook for Excursions A10 and C10. The Northern Highlands of Scotland*. CLAPPERTON, C M (editor). (Norwich: Geo Abstracts.)

CLAPPERTON, C M. 1997. Greenland ice cores and North Atlantic sediments: implications for the last glaciation in Scotland. 45–58 in *Reflections on the ice age in Scotland*. GORDON, J E (editor). (Glasgow: Scottish Association of Geography Teachers and Scottish Natural Heritage.)

CLAPPERTON, C M, and GEMMELL, A M D. 1998. Windy Hills SSSI, Aberdeenshire: site survey and management phase 1. *Scottish Natural Heritage Commissioned Report*, F98LF10 (Unpublished report).

CLAPPERTON, C M, and SUGDEN, D E. 1975. The glaciation of Buchan — a reappraisal. 19–22 in *Quaternary studies in north east Scotland*. A M D GEMMELL (editor). (Aberdeen: Department of Geography, University of Aberdeen.)

CLAPPERTON, C M, and SUGDEN, D E. 1977. The Late Devensian glaciation of north-east Scotland. 1–13 in *Studies in the Scottish Lateglacial environment*. GRAY, J M, and LOWE, J J (editors). (Oxford: Pergamon Press.)

CLAYTON, K M. 1974. Zones of glacial erosion. *Institute of British Geographers Special Publication*, Vol. 7, 163–176.

CLIMAP PROJECT MEMBERS. 1976. The surface of the ice age Earth. *Science*, Vol. 191, 1131–1136.

CONNELL, E R. 1984a. Kirkhill Quarry. Site. 59 in *Buchan field guide*. HALL, A M (editor). (Cambridge: Quaternary Research Association.)

CONNELL, E R. 1984b. Kirkhill Quarry. Deposits between the lower and upper palaeosols. 62–64 in *Buchan field guide*. HALL, A M (editor). (Cambridge: Quaternary Research Association.)

CONNELL, E R. 1984c. Kirkhill Quarry. Deposits above the upper palaeosol. 64–69 in *Buchan field guide*. HALL, A M (editor). (Cambridge: Quaternary Research Association.)

CONNELL, E R. 1984d. Sandhole (Denhead) Sand Pit. 83–86 in *Buchan field guide*. HALL, A M (editor). (Cambridge: Quaternary Research Association).

CONNELL, E R. 2000a. Tofthead, Fochabers. 38–39 in *The Quaternary of the Banffshire coast and Buchan: field guide*. MERRITT, J W, CONNELL, E R, and BRIDGLAND, D R (editors). (London: Quaternary Research Association.)

CONNELL, E R. 2000b. Sandford Bay, Peterhead. 57–61 in *The Quaternary of the Banffshire coast and Buchan: field guide*. MERRITT, J W, CONNELL, E R, and BRIDGLAND, D R (editors). (London: Quaternary Research Association.)

CONNELL, E R, and HALL, A M. 1984a. Kirkhill Quarry. Deposits beneath the lower palaeosol. 60–62 in *Buchan field guide*. HALL, A M (editor). (Cambridge: Quaternary Research Association).

CONNELL, E R, and HALL, A M. 1984b. Boyne Bay Quarry. 97–101 in *Buchan field guide*. HALL, A M (editor). (Cambridge: Quaternary Research Association).

CONNELL, E R, and HALL, A M. 1984c. Sandford Bay. 106–107 in *Buchan field guide*. HALL, A M (editor). (Cambridge: Quaternary Research Association).

CONNELL, E R, and HALL, A M. 1987. The periglacial stratigraphy of Buchan. 277–285 in *Periglacial processes and landforms in Britain and Ireland*. BOARDMAN, J (editor). (Cambridge: Cambridge University Press).

CONNELL, E R, and ROMANS, J C C. 1984. Kirkhill Quarry. Palaeosols. 70–76 in *Buchan field guide*. HALL, A M (editor). (Cambridge: Quaternary Research Association.)

CONNELL, E R, EDWARDS, K J, and HALL, A M. 1982. Evidence for two pre-Flandrian palaeosols in Buchan, Scotland. *Nature, London*, Vol. 297, 570–572.

CONNELL, E R, HALL, A M, SHAW, D, and RILEY, L A. 1985. Palynology and significance of radiocarbon dated organic materials from Cruden Bay Brick Pit, Grampian Region, Scotland. *Quaternary Newsletter*, Vol. 47, 19–25.

CRAIG, G Y (editor). 1991. *Geology of Scotland*. (Third edition). (London: The Geological Society.)

CRAMPTON, C B, and CARRUTHERS, R G. 1914. The geology of Caithness. *Memoir of the Geological Survey of Great Britain*, Sheets 110, 116 with parts of 109, 115 and 117 (Scotland).

CROFTS, R S. 1975. Sea bed topography off north east Scotland. *Scottish Geographical Magazine*, Vol. 91, 52–64.

CULLINGFORD, R A, and SMITH, D E. 1980. Late Devensian raised shorelines in Angus and Kincardineshire, Scotland. *Boreas*, Vol. 9, 21–38.

CULLINGFORD, R A, SMITH, D E, and FIRTH, C R. 1991. The altitude and age of the Main Postglacial Shoreline in eastern Scotland. *Quaternary International*, Vol. 9, 39–52.

CUMMING, G A, and BATE, P A. 1933. The Lower Cretaceous erratics of the Fraserburgh district, Aberdeenshire. *Geological Magazine*, Vol. 70, 397–413.

DALE, B. 1985. Dinoflagellate cyst analysis of Upper Quaternary sediments in core GIK 15530–4 from the Skagerrak. *Norsk Geologisk Tidsskrift*, Vol. 65, 97–102.

DANSGAARD, W, and ten others. 1993.   Evidence for general instability of past climate from a 250-kyr ice-core record. *Nature*, Vol. 364, 218–220.

DAWSON, A G, LONG, D, and SMITH D E. 1988.   The Storegga Slides: evidence from eastern Scotland of a possible tsunami. *Marine Geology*, Vol. 82, 271–276.

DEPARTMENT OF AGRICULTURE AND FISHERIES FOR SCOTLAND. 1962. *Scottish Peat*. (Second Report of the Scottish Peat Committee). (Edinburgh: HMSO.)

DEPARTMENT OF AGRICULTURE AND FISHERIES FOR SCOTLAND. 1964. *Scottish Peat Surveys*. (South West Scotland). (Edinburgh: HMSO.)

DEPARTMENT OF AGRICULTURE AND FISHERIES FOR SCOTLAND. 1965a.   *Scottish Peat Surveys*. (Central Scotland). (Edinburgh: HMSO.)

DEPARTMENT OF AGRICULTURE AND FISHERIES FOR SCOTLAND. 1965b. *Scottish Peat Surveys*. (Western Highlands and Islands). (Edinburgh: HMSO.)

DEPARTMENT OF AGRICULTURE AND FISHERIES FOR SCOTLAND. 1968. *Scottish Peat Surveys*. (Caithness, Shetland and Orkney). (Edinburgh: HMSO.)

DONNER, J J. 1957.   The geology and vegetation of Late-glacial retreat stages in Scotland.   *Transactions of the Royal Society of Edinburgh*, Vol. 63, 221–264.

DONNER, J J. 1960.   Pollen analysis of the Burn of Benholm peat bed, Kincardineshire.   *Societas Scientiarum Fennica, Commentationes Biologicae*, Vol. 22, 1–13.

DONNER, J J. 1963.   The Late- and Post-Glacial raised beaches in Scotland.   *Annales Academiae Scientiarum Fennicae*, Vol. 68, Pt II, Series A III Geologica-Geographica, 1–13.

DONNER, J J. 1979.   The Early or Middle Devensian peat at Burn of Benholm, Kincardineshire.   *Scottish Journal of Geology*, Vol. 15, 247–250.

DOWDESWELL, J A, ELVERHØI, A, ANDREWS, J T, and HEBBELN, D. 1999.   Asynchronous deposition of ice-rafted layers in the Nordic seas and North Atlantic Ocean.   *Nature, London*, Vol. 400, 348–351.

DREIMANIS, A, and VAGNERS, U J. 1971.   The effect of lithology on the texture of tills.   66–82 in *Research Methods in Pleistocene Geomorphology*. YATSU, and FALCONER (editors). (Norwich: Geo Abstracts.)

DROOP, G T R, and CHARNLEY, N. 1985.   Comparative geobarometry of pelitic hornfelses associated with the newer gabbros: a preliminary study.   *Journal of the Geological Society of London*, Vol. 142, 53–62.

DRYBURGH, P M. 1978.   *Scotland's peat resources: an introduction to their potential*. (Edinburgh: School of Engineering, University of Edinburgh.)

DULLER, G A T, WINTLE, A F, and HALL, A M. 1995.   Luminescence dating and its application to key pre-Late Devensian sites in Scotland.   *Quaternary Science Reviews*, Vol. 14, 495–519.

DURNO, S E. 1956.   Pollen analysis of peat deposits in Scotland.   *Scottish Geographical Magazine*, Vol. 72, 177–187.

DURNO, S E. 1957.   Certain aspects of vegetational history in north-east Scotland.   *Scottish Geographical Magazine*, Vol. 73, 176–184.

DURNO, S E. 1970.   Pollen diagrams from three buried peats in the Aberdeen area.   *Transactions of the Botanical Society of Edinburgh*, Vol. 41, 43–50.

EDMOND, J M, and GRAHAM, J D. 1977.   Peterhead power station cooling water intake tunnel: an engineering case study. *Quarterly Journal of Engineering Geology*, Vol. 10, 281–301.

EDWARDS, A G. 1970.   Scottish aggregates: their suitability for concrete with regard to rock constituents.   *Current Paper of the Building Research Station*, No. 28/70.

EDWARDS, K J. 1978.   Palaeoenvironmental and archaeological investigations in the Howe of Cromar, Grampian Region, Scotland. Unpublished PhD thesis, University of Aberdeen.

EDWARDS, K J. 1979.   Environmental impact in the prehistoric period.   27–42 in *Early Man in the Scottish Landscape. Scottish Archaeological Forum*. THOMS, L M (editor). Vol. 9. (Edinburgh: Edinburgh University Press.)

EDWARDS, K J, and CONNELL, E R. 1981.   Interglacial and interstadial sites in north-east Scotland.   *Quaternary Newsletter*, Vol. 33, 22–28.

EDWARDS, K J, and ROWNTREE, K M. 1980.   Radiocarbon and palaeoenvironmental evidence for changing rates of erosion at a Flandrian stage site in Scotland.   207–223 in *Timescales in Geomorphology*. CULLINGFORD, R A, DAVIDSON, D A, and LEWIN, J (editors). (Chichester and New York: Wiley.)

EDWARDS, K J, CASELDINE, C J, and CHESTER, D K. 1976.   Possible interstadial and interglacial pollen floras from Teindland, Scotland.   *Nature*, Vol. 264, 742–744.

EHLERS, J. 1988.   Skandinavische Geschiebe in Großbritannien.   *Der Geschiebesammler*, Vol. 22, 49–64.

EHLER, J, and WINGFIELD, R. 1991.   The extension of the Late Weichselian/Late Devensian ice sheets in the North Sea Basin. *Journal of Quaternary Science*, Vol. 6, 313–326.

EHLERS, J, GIBBARD, P L, and ROSE, J. 1991.   Glacial deposits of Britain and Europe: general overview.   493–501 in *Glacial deposits in Great Britain and Ireland*. EHLERS, J, GIBBARD, P L, and ROSE, J (editors). (Rotterdam: Balkema.)

EMILIANI, C. 1954.   Depth habitats of some species of pelagic foraminifera as indicated by oxygen isotope ratios.   *American Journal of Science*, Vol. 252, 149.

ENGLAND, J. 1986.   Glacial erosion of a high Arctic valley. *Journal of Glaciology*, Vol. 32, 60–64.

EVANS, D, CHESHER, J A, DEEGAN, C E, and FANNIN, N G T. 1981. The offshore geology of Scotland in relation to the IGS shallow drilling programme, 1970–1978.   *Report of the Institute of Geological Sciences*, No. 81/12.

EVANS, D, MORTON, A C, WILSON, S, JOLLEY, D, and BARREIRO, B A. 1997.   Palaeoenvironmental significance of marine and terrestrial Tertiary sediments on the NW Scottish Shelf in BGS borehole 77/7.   *Scottish Journal of Geology*, Vol. 33, 31–42.

EYLES, N, McCABE, A M, and BOWEN, D Q. 1994.   The stratigraphic and sedimentological significance of Late Devensian Ice Sheet surging in Holderness, Yorkshire, UK.   *Quaternary Science Reviews*, Vol. 13, 727–759.

EYLES, V A, ANDERSON, J G C, BONNELL, D G R, and BUTTERWORTH, A R I C. 1946.   Brick clays of north-east Scotland. Part I —Description of occurrences; Part II — Report on analyses and physical tests.   *Geological Survey of Great Britain: Wartime Pamphlet*, No. 47.

FIRTH, C R, and HAGGART, B A. 1989.   Loch Lomond Stadial and Flandrian shorelines in the inner Moray Firth area, Scotland.   *Journal of Quaternary Science*, Vol. 4, 37–50.

FITZPATRICK , E A. 1956.   An indurated soil horizon formed by permafrost.   *Soil Science*, Vol. 7, 248–254.

FITZPATRICK, E A. 1958.   An introduction to the periglacial geomorphology of Scotland.   *Scottish Geographical Magazine*, Vol. 74, 28–36.

FITZPATRICK, E A. 1963.   Deeply weathered rock in Scotland, its occurrence, age and contribution to the soils.   *Journal of Soil Science*, Vol. 14, 33–43.

FITZPATRICK, E A. 1965. An interglacial soil at Teindland, Morayshire. *Nature*, Vol. 207, 621–622.

FITZPATRICK, E A. 1969. Some aspects of soil evolution in north-east Scotland. *Soil Science*, Vol. 107, 403–408.

FITZPATRICK, E A. 1972. The principal Tertiary and Pleistocene events in north-east Scotland. 1–4 in *North-East Scotland Geographical Essays*. CLAPPERTON, C M (editor). (Aberdeen: Department of Geography, University of Aberdeen.)

FITZPATRICK, E A. 1975a. Errollston Farm. 19–21 in *Aberdeen Field Excursion Guide*. (Informal Publication: Quaternary Research Association.)

FITZPATRICK, E A. 1975b. Particle size distribution and stone orientation patterns in some soils of north-east Scotland. 49–60 in *Quaternary Studies in North East Scotland*. GEMMELL, A M D (editor). (Aberdeen: Department of Geography, University of Aberdeen.)

FITZPATRICK, E A. 1975c. Windy Hills. 23–25 in *Aberdeen Field Excursion Guide*. (Informal Publication: Quaternary Research Association.)

FITZPATRICK, E A. 1987. Periglacial features in the soils of north east Scotland. 153–162 in *Periglacial processes and landforms in Britain and Ireland*. BOARDMAN, J (editor). (Cambridge: Cambridge University Press.)

FLETCHER, T P, and five others. 1996. Geology of the Fortrose and eastern Inverness district. *Memoir of the British Geological Survey*, Sheet 84W (Scotland).

FLETT, J S, and READ, H H. 1921. Tertiary gravels of the Buchan district of Aberdeenshire. *Geological Magazine*, Vol. 58, 215–225.

FRASER, G K. 1943. Peat deposits of Scotland. Part I — General account. *Wartime Pamphlet of the Geological Survey of Great Britain: Scotland*, No. 36.

FRASER, G K. 1948. Peat deposits of Scotland. Part II — Peat mosses of Aberdeenshire, Banffshire and Morayshire (area of quarter-inch sheet 9). *Wartime Pamphlet of the Geological Survey of Great Britain: Scotland*, No. 36.

FRIIS, H. 1976. Weathering of a Neogene fluviatile fining-upwards sequence at Voervadsbro, Denmark. *Bulletin of the Geological Society of Denmark*, Vol. 25, 99–105.

FRONVAL, T, JANSEN, E, BLOEMENDAL, J, and JOHNSEN, S. 1995. Oceanic evidence for coherent fluctuations in Fennoscandian and Laurentide ice sheets on millennium timescales. *Nature*, Vol. 374, 443–446.

FUNNELL, B M. 1995. Global sea-level and the (pen-)insularity of late Cenozoic Britain. 3–13 *in* Island Britain: a Quaternary Perspective. PREECE, R C (editor). *Geological Society of London, Special Publication*, No. 96.

GALLOWAY, R W. 1958. Periglacial phenomena in Scotland. Unpublished PhD thesis, University of Edinburgh.

GALLOWAY, R W. 1961a. Solifluction in Scotland. *Scottish Geographical Magazine*, Vol. 77, 75–87.

GALLOWAY, R W. 1961b. Periglacial phenomena in Scotland. *Geografiska Annaler*, Vol. 43, 348–353.

GALLOWAY, R W. 1961c. Ice wedges and involutions in Scotland. *Biuletyn Periglacjalny*, Vol. 10, 169–193.

GATLIFF, R W, and eleven others. 1994. *United Kingdom offshore regional report: the geology of the central North Sea*. (London: HMSO for the British Geological Survey.)

GEIKIE, A. 1878. On the Old Red Sandstone of western Europe. *Transactions of the Royal Society of Edinburgh*, Vol. 28, 345–452.

GEMMELL, A M D. 1975. The Kippet Hills. 14–19 in *Aberdeen Field Excursion Guide*. (Informal Publication: Quaternary Research Association.)

GEMMELL, A M D, and AUTON, C A. 2000. Windy Hills. 81–92 in *The Quaternary of the Banffshire coast and Buchan: field guide*. MERRITT, J W, CONNELL, E R, and BRIDGLAND, D R (editors). (London: Quaternary Research Association.)

GEMMELL, A M D, and KESEL, R H. 1979. Developments in the study of the Buchan flint deposits. 66–77 in *Early Man in the Scottish landscape*. THOMS, L M (editor). (Edinburgh: Edinburgh University Press.)

GEMMELL, A M D, and KESEL, R H. 1982. Letters to the editors. The 'Pliocene' Gravels of Buchan: a reappraisal: reply. *Scottish Journal of Geology*, Vol. 18, 333–335.

GEMMELL, A M D, and RALSTON, I B M. 1984. Some recent discoveries of ice-wedge cast networks in north-east Scotland. *Scottish Journal of Geology*, Vol. 20, 115–118.

GEMMELL, A M D, and RALSTON, I B M. 1985. Ice wedge polygons in north-east Scotland: a reply. *Scottish Journal of Geology*, Vol. 21, 109–111.

GEMMELL, A M D, and STOVE, G C. 1999. Windy Hills SSSI, Aberdeenshire: site survey and management phase 2. *Scottish Natural Heritage Commissioned Report*, F98LF11 (Unpublished report).

GEMMELL, S, HANSOM, J, and HOEY, T B. 1996. Geomorphology conservation and management of the River Spey and Spey Bay SSSIs, Moray. *Scottish Natural Heritage Commissioned Report*.

GLASSER, N F, and HALL, A M. 1997. Calculating Quaternary erosion rates in North East Scotland. *Geomorphology*, Vol. 20, 29–48.

GLENTWORTH, R. 1954. *The soils of the country around Banff, Huntly and Turriff (Sheets 86 and 96)*. (Memoir of the Soil Survey of Great Britain: Scotland). (Edinburgh: Her Majesty's Stationery Office for Department of Agriculture and Fisheries for Scotland.)

GLENTWORTH, R, and MUIR, J W. 1963. *The soils of the country around Aberdeen, Inverurie and Fraserburgh. Sheets 77, 76 and 87/97*. (Memoir of the Soil Survey of Great Britain: Scotland). (Edinburgh: Her Majesty's Stationery Office for Department of Agriculture and Fisheries for Scotland.)

GLENTWORTH, R, MITCHELL, W A, and MITCHELL, B D. 1964. The red glacial drift deposits of north-east Scotland. *Clay Minerals Bulletin*, Vol. 5, 373–381.

GODWIN, H, and WILLIS, E H. 1959. Radiocarbon dating of the Late-glacial Period in Britain. *Proceedings of the Royal Society of London, Series B*, Vol. 150, 199–215.

GOODWIN, I D. 1998. Did changes in Antarctic ice volume influence Late Holocene sea-level lowering? *Quaternary Science Reviews*, Vol. 17, 319–332.

GORDON, D, and six others. 1989. Dating of Late Pleistocene Interglacial and Interstadial Periods in the United Kingdom from Speleothem Growth Frequency. *Quaternary Research*, Vol. 31, 14–26.

GORDON, J E. 1993a. Boyne Quarry. 233–236 *in* Quaternary of Scotland. GORDON, J E, and SUTHERLAND, D G (editors). *Geological Conservation Review Series*, No. 6 (London: Chapman and Hall.)

GORDON, J E. 1993b. Kippet Hills. 242–245 *in* Quaternary of Scotland. GORDON, J E, and SUTHERLAND, D G (editors). *Geological Conservation Review Series*, No. 6 (London: Chapman and Hall.)

GORDON, J E. 1993c. Nigg Bay. 479–482 *in* Quaternary of Scotland. GORDON, J E, and SUTHERLAND, D G (editors). *Geological Conservation Review Series*, No. 6 (London: Chapman and Hall.)

GORDON, J E. 1993d. Burn of Benholm. 482–485 in Quaternary of Scotland. GORDON, J E, and SUTHERLAND, D G (editors). *Geological Conservation Review Series*, No. 6 (London: Chapman and Hall.)

GORDON, J E, (editor). 1997. *Reflections on the Ice Age in Scotland: an update on Quaternary studies.* (Glasgow: Scottish Association of Geography Teachers and Scottish Natural Heritage).

GORDON, J E, and SUTHERLAND, D G (editors). 1993a. Quaternary of Scotland. *Geological Conservation Review Series*, No. 6 (London: Chapman and Hall.)

GORDON, J E, and SUTHERLAND, D G. 1993b. Windy Hills. 216–218 in Quaternary of Scotland. GORDON, J E, and SUTHERLAND, D G. *Geological Conservation Review Series*, No. 6 (London: Chapman and Hall.)

GORNITZ, V. 1995. Sea-level rise: a review of recent past and near-future trends. *Earth Surface Processes and Landforms*, Vol. 20, 7–20.

GOULD, D. 1996. Geology of the Findochty and Bin of Cullen areas. 1:10 000 sheets NJ46NE, NJ46SE and part of NJ45NE. *British Geological Survey Technical Report*, WA/96/68.

GOULD, D. 1997. Geology of the country around Inverurie and Alford. *Memoir of the British Geological Survey*, Sheet 76W and 76E (Scotland).

GOULD, D. 2001. Geology of the Aboyne district. *Memoir of the British Geological Survey*, Sheet 66W (Scotland).

GRAHAM, D K, HARLAND, R, GREGORY, D M, LONG, D, and MORTON, A C. 1990. The biostratigraphy and chronostratigraphy of BGS borehole 78/4, north Minch. *Scottish Journal of Geology*, Vol. 26, 65–75.

GRANT, R. 1960. The soils of the country round Elgin. *Memoir of the Soil Survey* (Scotland).

GRAY, J M, and LOWE, J J (editors). 1977. *Studies in the Scottish Lateglacial Environment.* (Oxford: Pergamon Press.)

GREENWOOD, P G, and RAINES, M G. 1994. Ground probing radar surveys in Scotland and East Anglia. Part 2 — Ground Probing Radar Sections. *British Geological Survey Technical Report*, WN/94/12.

GREENWOOD, P G, AUTON, C A, and RAINES, M G. 1995. Ground probing radar surveys in Scotland and East Anglia. Part 1 — Ground Probing Radar Report. *British Geological Survey Technical Report*, WN/95/6.

GREGORY, D M, and BRIDGE, V A. 1979. On the Quaternary foraminiferal species *Elphidium ?ustulatum* Todd 1957: its stratigraphic and palaeontological implications. *Journal of Foraminiferal Research*, Vol. 9, 70–75.

GROUSSET, F E, PUJOL, C, LABEYRIE, L, AUFFRET, G, and BOELAERT, A. 2000. Were the North Atlantic Heinrich events triggered by the behavior of the European ice sheets? *Geology*, Vol. 28, 123–126.

GUNSON, A R. 1975. The vegetation history of north-east Scotland. 61–72 in *Quaternary studies in north-east Scotland*. GEMMELL, A M D (editor). (Aberdeen: Department of Geography, University of Aberdeen.)

HALL, A M. 1982. The Pliocene gravels of Buchan: a reappraisal: discussion. *Scottish Journal of Geology*, Vol. 18, 336–338.

HALL, A M. 1983. Deep weathering and landform evolution in north-east Scotland. Unpublished PhD thesis, University of St Andrews.

HALL, A M. 1984a. *Buchan field guide.* (Cambridge: Quaternary Research Association.)

HALL, A M. 1984b. Oldmill. 106 in *Buchan Field Guide*. HALL, A M (editor). (Cambridge: Quaternary Research Association).

HALL, A M. 1985. Cenozoic weathering covers in Buchan, Scotland and their significance. *Nature*, Vol. 315, 392–395.

HALL, A M. 1986. Deep weathering patterns in north-east Scotland and their geomorphological significance. *Zeitschrift für Geomorphologie*, Vol. NF 30, 407–422.

HALL, A M. 1987. Weathering and relief development in Buchan, Scotland. Part II. 991–1005 in *International Geomorphology* 1986. GARDINER, V (editor). (Chichester: John Wiley.)

HALL, A M. 1991. Pre-Quaternary landscape evolution in the Scottish Highlands. *Transactions of the Royal Society of Edinburgh: Earth Sciences*, Vol. 8, 1–26.

HALL, A M. 1993a. Moss of Cruden. 218–221 in Quaternary of Scotland. GORDON, J E, and SUTHERLAND, D G (editors). *Geological Conservation Review Series*, No. 6 (London: Chapman and Hall.)

HALL, A M. 1993b. Pittodrie. 221–223 in Quaternary of Scotland. GORDON, J E, and SUTHERLAND, D G (editors). *Geological Conservation Review Series*, No 6 (London: Chapman and Hall.)

HALL, A M. 1993c. Hill of Longhaven. 223–225 in Quaternary of Scotland. GORDON, J E, and SUTHERLAND, D G (editors). *Geological Conservation Review Series*, No. 6 (London: Chapman and Hall.)

HALL, A M. 1995. Was all of Lewis glaciated in the Late Devensian? *Quaternary Newsletter*, Vol. 76, 1–7.

HALL, A M. 1997. Quaternary stratigraphy: the terrestrial record. 59–71 in *Reflections on the Ice Age in Scotland*. GORDON, J E (editor). (Glasgow: Scottish Association of Geography Teachers and Scottish Natural Heritage).

HALL, A M. 2000. The Teindland Interglacial site. 34–37 in *The Quaternary of the Banffshire coast and Buchan: field guide*. MERRITT, J W, CONNELL, E R, and BRIDGLAND, D R (editors). (London: Quaternary Research Association.)

HALL, A M, and BENT, A J A. 1990. The limits of the last British Ice Sheet in northern Scotland and the adjacent shelf. *Quaternary Newsletter*, Vol. 61, 2–12.

HALL, A M, and CONNELL, E R. 1982. Recent excavations at the Greensand locality of Moreseat, Grampian Region. *Scottish Journal of Geology*, Vol. 18, 291–296.

HALL, A M, and CONNELL, E R. 1986. A preliminary report on the Quaternary sediments at Leys gravel pit, Buchan. *Quaternary Newsletter*, Vol. 48, 17–28.

HALL, A M, and CONNELL, E R. 1991. The glacial deposits of Buchan, northeast Scotland. 129–136 in *Glacial Deposits in Great Britain and Ireland*. EHLERS, J, GIBBARD, P L, and ROSE, J (editors). (Rotterdam: Balkema.)

HALL, A M, and CONNELL, E R. 2000. Howe of Byth Quarry. 72–78 in *The Quaternary of the Banffshire coast and Buchan: field guide*. MERRITT, J W, CONNELL, E R, and BRIDGLAND, D R (editors). (London: Quaternary Research Association.)

HALL, A M, and JARVIS, J. 1989. A preliminary report on the Late Devensian glaciomarine deposits around St Fergus, Grampian Region. *Quaternary Newsletter*, Vol. 59, 5–7.

HALL, A M, and JARVIS, J. 1993a. Kirkhill. 225–230 in Quaternary of Scotland. GORDON, J E, and SUTHERLAND, D G (editors). *Geological Conservation Review Series*, No. 6 (London: Chapman and Hall.)

HALL, A M. and JARVIS, J. 1993b. Bellscamphie. 230–233 in Quaternary of Scotland. GORDON, J E, and SUTHERLAND, D G (editors). *Geological Conservation Review Series*, No. 6 (London: Chapman and Hall.)

HALL, A M, and JARVIS, J. 1994. A concealed Lower Cretaceous outlier at Moss of Cruden, Grampian Region. *Scottish Journal of Geology*, Vol. 30, 163–166.

HALL, A M, and JARVIS, J. 1995. A multiple till sequence near Ellon, Grampian Region: T. F. Jamieson's 'indigo boulder clay' re-examined. *Scottish Journal of Geology*, Vol. 31, 53–59.

HALL, A M, and PEACOCK, J D. 2000. King Edward. 43–45 in *The Quaternary of the Banffshire coast and Buchan: field guide*. MERRITT, J W, CONNELL, E R, and BRIDGLAND, D R (editors). (London: Quaternary Research Association.)

HALL, A M, and SUGDEN, D E. 1987. Limited modification of mid-latitude landscapes by ice sheets. *Earth Surface Processes and Landforms*, Vol. 12, 531–542.

HALL, A M, and WHITTINGTON, G. 1989. Late Devensian glaciation of southern Caithness. *Scottish Journal of Geology*, Vol. 25, 307–324.

HALL, A M, MELLOR, T, and WILSON, J. 1989. The clay mineralogy and age of deeply weathered rock in north-east Scotland. *Zeitschrift der Geomorphologie Supplement Bund*, Vol. 72, 97–108.

HALL, A M, WHITTINGTON, G, DULLER, G A T, and JARVIS, J. 1995a. Late Pleistocene environments in lower Strathspey, Scotland. *Transactions of the Royal Society of Edinburgh: Earth Sciences*, Vol. 85, 253–273.

HALL, A M, DULLER, G, JARVIS, J, and WINTLE, A G. 1995b. Middle Devensian ice-proximal gravels at Howe of Byth, Grampian Region. *Scottish Journal of Geology*, Vol. 31, 61–64.

HALL, A R. 1980. Late Pleistocene deposits at Wing, Rutland. *Philosophical Transactions of the Royal Society of London*, Vol. B289, 135–164.

HALLAM, A, and SELLWOOD, B. 1976. Middle Mesozoic sedimentation in relation to tectonics in the British area. *Journal of Geology*, Vol. 84, 302–321.

HAMBLIN, R J O. 1973. The Haldon Gravels of south Devon. *Proceedings of the Geologists' Association*, Vol. 84, 459–476.

HANCOCK, J M. 1975. The petrography of the Chalk. *Proceedings of the Geologists' Association*, Vol. 86, 499–536.

HARDE, K W. 1981. *A field guide in colour to beetles*. (London: Octopus Books.)

HARKNESS, D D, and WILSON, H W. 1979. Scottish Universities Research and Reactor Centre radiocarbon measurements III. *Radiocarbon*, Vol. 21, 203–256.

HARKNESS, D D, and WILSON, H W. 1981. Scottish Universities Research and Reactor Centre radiocarbon measurements IV. *Radiocarbon*, Vol. 23, 252–304.

HARLAND, R. 1983. Distribution maps of recent dinoflagellate cysts in bottom sediments from the North Atlantic Ocean and adjacent seas. *Palaeontology*, Vol. 26, 321–387.

HARLAND, R. 1988. Quaternary dinoflagellate cyst biostratigraphy of the North Sea. *Palaeontology*, Vol. 31, 877–903.

HARLAND, R. 1992. Dinoflagellate cysts of the Quaternary System. 253–266 in *A stratigraphic index of dinoflagellate cysts*. POWELL, A J (editor). (London: Chapman and Hall.)

HARLAND, R. 1993. Dinoflagellate cyst analysis of the Quaternary 'shelly till' at the Burn of Benholm, Scottish Sheet 57A. *British Geological Survey Technical Report*, WH93/173R.

HARRIS, A L, HASELOCK, P J, KENNEDY, M J, and MENDUM, J R. 1994. The Dalradian Supergroup in Scotland, Shetland and Ireland. 33–53 *in* A revised correlation of Precambrian rocks in the British Isles. GIBBONS, W, and HARRIS, A L (editors). *Special Report of the Geological Society of London*, No. 22.

HART, J K, and BOULTON, G S. 1991. The interrelation of glacitectonic and glaciodepositional processes within the glacial environment. *Quaternary Science Reviews*, Vol. 10, 335–350.

HART, R. 1941. Soil studies in relation to geology in an area in north-east Scotland. Part 1. The mineralogy of the soil parent materials. *Journal of Agricultural Science*, Vol. 31, 438–447.

HEDBERG, H D (editor). 1976. International stratigraphic guide. (New York: John Wiley & Sons)

HEINRICH, H. 1988. Origin and consequences of cyclic ice rafting in the north-east Atlantic Ocean during the past 130 000 years. *Quaternary Research*, Vol. 29, 142–152.

HICOCK, S R, LIAN, O B, and MATHEWES, R W. 1999. 'Bond cycles' recorded in terrestrial Pleistocene sediments of southwestern British Columbia, Canada. *Journal of Quaternary Science*, Vol. 14, 443–449.

HILL, M O, and eight others. 1999. Climate changes and Scotland's natural heritage: an environmental audit. *Scottish Natural Heritage Research and Monitoring Report*, No. 132.

HINXMAN, L W, and TEALL, J J H. 1896. The geology of West Aberdeenshire, Banffshire, parts of Elgin and Inverness. *Memoir of the Geological Survey* (Scotland).

HINXMAN, L W, and WILSON, J S G. 1902. The geology of Lower Strathspey. *Memoir of the Geological Survey, Scotland*, Sheet 85 (Scotland).

HOLMES, R. 1977. Quaternary deposits of the central North Sea, 5. The Quaternary geology of the UK sector of the North Sea between 56° and 58°N. *Report of the Institute of Geological Sciences*, No. 77/14.

HOLMES, R. 1991. Foula Sheet 60°N–04°W. Quaternary Geology. 1:250 000 Map Series. British Geological Survey, Edinburgh.

HOLMES, R. 1997. Quaternary stratigraphy: the offshore record. 72–94 in *Reflections on the ice age in Scotland*. GORDON, J E (editor). (Glasgow: Scottish Association of Geography Teachers and Scottish Natural Heritage.)

HOLMES, R, JEFFREY, D H, RUCKLEY, N A R, and WINGFIELD, R. 1993. Quaternary sediments around the United Kingdom, North Sheet, 1:1 000 000. British Geological Survey, Edinburgh.

HUIJZER, B, and VANDENBERGHE, J. 1998. Climatic reconstruction of the Weichselian pleniglacial in northwestern and central Europe. *Journal of Quaternary Science*, Vol. 13, 391–417.

HULL, E. 1895. The glacial deposits of Aberdeenshire. *Geological Magazine*, Vol. 2, 450–452.

HYDRAULICS RESEARCH. 1997. Coastal cells in Scotland. *Scottish Natural Heritage Research, Survey and Monitoring Report*, No. 56.

IMBRIE, J, and IMBRIE, K P. 1979. *Ice Ages: solving the mystery*. (London: MacMillan.)

IMBRIE, J, and eight others. 1984. The orbital theory of Pleistocene climate: support from a revised chronology of the marine ^{18}O record. 269–305 in *Milankovitch and climate: understanding the response to astronomical forcing*. BERGER, A L, and four others (editors). (Dordrecht: Reidel).

JAMIESON, T F. 1858. On the Pleistocene deposits of Aberdeenshire. *Quarterly Journal of the Geological Society of London*, Vol. 14, 509–532.

JAMIESON, T F. 1859. On an outlier of Lias in Aberdeenshire. *Quarterly Journal of the Geological Society of London*, Vol. 15, 131–133.

JAMIESON, T F. 1860a.   On the occurrence of crag strata beneath boulder clay in Aberdeenshire.   *Quarterly Journal of the Geological Society of London*, Vol. 16, 371–373.

JAMIESON, T F. 1860b.   On the drift and rolled gravel of the north of Scotland.   *Quarterly Journal of the Geological Society of London*, Vol. 16, 347–371.

JAMIESON, T F. 1862.   On the ice-worn rocks of Scotland.   *Quarterly Journal of the Geological Society of London*, Vol. 18, 164–184.

JAMIESON, T F. 1865.   On the history of the last geological changes in Scotland.   *Quarterly Journal of the Geological Society of London*, Vol. 21, 161–203.

JAMIESON, T F. 1866.   On the glacial phenomena of Caithness.   *Quarterly Journal of the Geological Society of London*, Vol. 22, 261–281.

JAMIESON, T F. 1874.   On the last stage of the glacial period in North Britain.   *Quarterly Journal of the Geological Society of London*, Vol. 30, 317–338.

JAMIESON, T F. 1882a.   On the crag shells of Aberdeenshire and the gravel beds containing them.   *Quarterly Journal of the Geological Society*, Vol. 38, 145–159.

JAMIESON, T F. 1882b.   On the Red Clay of the Aberdeenshire coast and the direction of ice-movement in that quarter.   *Quaterly Journal of the Geological Society of London*, Vol. 38, 160–177.

JAMIESON, T F. 1906.   The glacial period in Aberdeenshire and the southern border of the Moray Firth.   *Quarterly Journal of the Geological Society of London*, Vol. 62, 13–39.

JAMIESON, T F. 1910.   On the surface geology of Buchan. 19–25 in *The Book of Buchan*. TOCHER, J F, (editor). (Aberdeen and Peterhead: Aberdeen University Press and P. Scrogie Ltd).

JAMIESON, T F, JUKES-BROWNE, A J, and MILNE, J. 1897.   Cretaceous fossils in Aberdeenshire.   *Report of the British Association for the Advancement of Science*, 333–342.

JANSEN, J H F. 1976.   Late Pleistocene and Holocene history of the northern North Sea, based on acoustic reflection records.   *Netherlands Journal of Sea Research*, Vol. 10, 1–43.

JAPSEN, P. 1997.   Regional Neogene exhumation of Britain and the western North Sea.   *Journal of the Geological Society of London*, Vol. 154, 239–247.

JAPSEN, P. 1998.   Regional velocity-depth anomalies, North Sea Chalk: a record of overpressure and Neogene uplift and erosion.   *American Association of Petroleum Geologists*, Vol. 82, 2031–2074.

JARDINE, W G, and five others. 1988.   A late Middle Devensian interstadial site at Sourlie, near Irvine, Strathclyde.   *Scottish Journal of Geology*, Vol. 24, 288–295.

JEFFERY, D H, and LONG, D. 1989.   Early Pleistocene sedimentation and geographic change in the UK sector of the North Sea.   *Terra Abstracts*, Vol. 1, 425.

JENSEN, K A, and KNUDSEN, K L. 1988.   Quaternary foraminiferal stratigraphy in boring 81/29 from the central North Sea.   *Boreas*, Vol. 17, 273–287.

JOHNSON, H, RICHARDS, P C, LONG, D, and GRAHAM, C C. 1993.   *United Kingdom Offshore Regional Report: the Geology of the Northern North Sea*. (London: HMSO for the British Geological Survey.)

JONES, L H P, and MILNE, A A. 1956.   Birnessite, a new manganese oxide mineral from Aberdeenshire, Scotland.   *Mineralogical Magazine*, Vol. 31, 283–288.

JORDT, H, FALEIDE, J I, BJØLYKKE, K, and IBRAHAM, M T. 1998.   Cenozoic sequence stratigraphy of the central and northern North Sea basin: tectonic development, sediment distribution and provenance areas.   *Marine and Petroleum Geology*, Vol. 12, 845–879.

KARLLSON, W, VOLLSET, J, BJORLYKKE, K, and JORGENSEN, P. 1979.   Changes in the mineralogical composition of Tertiary sediments from North Sea wells.   281–289 in *Proceedings of the 6th International Clay Conference*. Development in sedimentology, No. 27. MORTLAND, M M, and FARMER, V C (editors). (Amsterdam, Netherlands: Elsevier.)

KESEL, R H, and GEMMELL, A M D. 1981.   The 'Pliocene' gravels of Buchan: a reappraisal.   *Scottish Journal of Geology*, Vol. 17, 185–203.

KNUDSEN, K L, and SEJRUP, H P. 1993.   Pleistocene stratigraphy in the Devils Hole area, central North Sea: foraminiferal and amino-acid evidence.   *Journal of Quaternary Science*, Vol. 8, 1–14.

KOPPI, A J. 1977.   Weathering of Tertiary gravels, a schist and a metasediment in north-east Scotland. Unpublished PhD thesis, University of Aberdeen.

KOPPI, A J, and FITZPATRICK, E A. 1980.   Weathering in Tertiary gravels in NE Scotland.   *Journal of Soil Science*, Vol. 31, 525–532.

KOTILAINEN, A T, and SHACKLETON, N J. 1995.   Rapid climate variability in the North Pacific Ocean during the past 95 000 years.   *Nature, London*, Vol. 377, 323–326.

KROON, D, AUSTIN, W E N, CHAPMAN, M R, and GANSSEN, G M. 1997.   Deglacial surface circulation changes in the northeastern Atlantic: temperature and salinity records off NW Scotland on a century scale.   *Palaeoceanography*, Vol. 12, 755–763.

LAGERKLINT, I M, and WRIGHT, J D. 1999.   Late glacial warming prior to Heinrich event 1: the influence of ice rafting and large ice sheets on the timing of initial warming.   *Geology*, Vol. 27, 1099–1102.

LAMB, H H. 1977.   *Climate: present, past and future*. Vol. 2. Climatic history and the future. (London: Methuen.)

LAMBECK, K. 1991.   A model for Devensian and Flandrian glacial rebound and sea-level change in Scotland.   33–62 in *Glacial isostacy, sea level and mantle rheology*. SABADINI, R, LAMBECK, K, and BOSCHI, E, (editors). (Dordrecht: Kluwer Academic Publishers.)

LAMBECK, K. 1993a.   Glacial rebound of the British Isles, I; preliminary model results.   *Geophysical Journal International*, Vol. 115, 941–959.

LAMBECK, K. 1993b.   Glacial rebound of the British Isles, II; a high resolution, high precision model.   *Geophysical Journal International*, Vol. 115, 960–990.

LAMBECK, K. 1995.   Late Devensian and Holocene shorelines of the British Isles and North Sea from models of glacio-hydro-isostatic rebound.   *Journal of the Geological Society of London*, Vol. 152, 437–448.

LARSEN, E, and five others. 1987.   Cave stratigraphy in western Norway; multiple Weichselian glaciations and interstadial vertebrate fauna.   *Boreas*, Vol. 16, 267–292.

LAW, G R. 1962.   The sub-surface geology of the Bay of Nigg– Tullos area. Unpublished MSc thesis, University of Aberdeen.

LAWSON, T J. 1984.   Reindeer in the Scottish Quaternary.   *Quaternary Newsletter*, Vol. 42, 1–7.

LAWSON, T J, and ATKINSON, T C. 1995.   Quaternary chronology. 12–18 in *The Quaternary of Assynt and Coigach: field guide*. LAWSON, T J (editor). (Cambridge: Quaternary Research Association.)

LAXTON, J L. 1992.   A particle-size classification of sand and gravel deposits as a basis for end-use assessment.   *Engineering Geology*, Vol. 32, 29–37.

LEHMAN, S. 1996.   True grit spells double trouble.   *Nature, London,* Vol. 382, 25–26.

LESLIE, A G. 1984.   Field relations in the north-eastern part of the Insch mafic igneous mass, Aberdeenshire.   *Scottish Journal of Geology,* Vol. 20, 225–235.

LEWIN, J, and WEIR, M J C. 1977.   Morphology and recent history of the Lower Spey.   *Scottish Geographical Magazine,* Vol. 93, 45–51.

LIDMAR-BERGSTROM, K. 1982.   Pre-Quaternary geomorphological evolution in southern Fennnoscandia.   *Sveriges Geologiska Undersökning,* Vol. 785, 1–202.

LINDROTH, C H. 1985.   The Carabidae (Coleoptera) of Fennoscandia and Denmark. In *Fauna Entomologica Scandinavica.* BRILL, E J (editor). Part 1. Vol. 15. (Copenhagen: Scandanavian Science Press Ltd.)

LINTON, D L. 1951.   Problems of Scottish scenery.   *Scottish Geographical Magazine,* Vol. 67, 65–85.

LINTON, D L. 1963.   The forms of glacial erosion.   *Transactions of the Institute of British Geographers,* Vol. 33, 1–28.

LONG, D, SMITH, D E, and DAWSON, A G. 1989.   A Holocene tsunami deposit in eastern Scotland.   *Journal of Quaternary Science,* Vol. 4, 61–66.

LOWE, J J. 1984.   A critical evaluation of pollen-stratigraphic investigations of pre-Late Devensian sites in Scotland.   *Quaternary Science Reviews,* Vol. 3, 405–432.

LOWE, J J, and eight others. 1999.   The chronology of palaeoenvironmental changes during the last Glacial-Holocene transition: towards an event stratigraphy for the British Isles.   *Journal of the Geological Society of London,* Vol. 156, 397–410.

LUBINSKY, I. 1980.   Marine bivalve molluscs of the Canadian central and eastern Arctic: faunal composition and zoogeography.   *Canadian Bulletin of Fisheries and Aquatic Sciences,* Vol. 207, 1–111.

LUMSDEN, G I. 1958.   The glacial and post-glacial history of the lower reaches of the River Don, Aberdeenshire.   *Transactions of the Geological Society of Edinburgh,* Vol. 17, 141–155.

MACKIE, W. 1905.   Some notes on the distribution of erratics over eastern Moray.   *Transactions of the Edinburgh Geological Society,* Vol. 8, 91–97.

MACKIE, W. 1923.   The principles that regulate the distribution of particles of heavy minerals in sedimentary rocks, as illustrated by the sandstones of the north-east of Scotland.   *Transactions of the Edinburgh Geological Society,* Vol. 11, 138–164.

MACPHERSON, E. 1971.   *The marine molluscs of arctic Canada.* (Publications in Biological Oceanography, No 3). (Ottawa: National museums of Canada.)

MADGETT, P A, and CATT, J A. 1978.   Petrography, stratigraphy and weathering of Late Pleistocene tills in East Yorkshire, Lincolnshire and north Norfolk.   *Proceedings of the Yorkshire Geological Society,* Vol. 42, 55–108.

MAIZELS, J K, and AITKEN, J F. 1991.   Palaeohydrological change during deglaciation in upland Britain: a case study from northeast Scotland.   105–145 in *Temperate Palaeohydrology.* STARKEL, L, GREGORY, K J, and THORNES, J B (editors). (Chichester: John Wiley.)

MANGERUD, J, and six others. 1981.   A Middle Weichselian ice-free period in Western Norway: the Ålesund interstadial.   *Boreas,* Vol. 10, 447–462.

MANGERUD, J, JANSEN, E, and LANDVIK, J.Y. 1996.   Late Cenozoic history of the Scandinavian and Barents Sea ice sheets.   *Global and Planetary Change,* Vol. 12, 11–26.

MASLIN, M A, and five others. 1995.   Northwest Pacific Site 882: the initiation of Northern Hemisphere glaciation. Ocean Drilling Program, College Station, TX.   *Proceedings of the Ocean Drilling Program, Scientific Reports,* Vol. 145, 315–329.

MASLIN, M A, LI, X S, LOUTRE, M-F, and BERGER, A. 1998.   The contribution of orbital forcing to the progressive intensification of Northern Hemisphere glaciation.   *Quaternary Science Reviews,* Vol. 17, 411–426.

MAUZ, B. 1998.   The onset of the Quaternary: a review of new findings in the Pliocene–Pleistocene chronostratigraphy.   *Quaternary Science Reviews,* Vol. 17, 357–364.

MAYLE, F E, and nine others. 1999.   Climate variations in Britain during the last Glacial-Holocene transition (15.0–11.5 cal ka BP): comparison with the GRIP ice-core record.   *Journal of the Geological Society of London,* Vol. 156, 411–423.

MCCABE, A M. 1996.   Dating and rhythmicity from the last deglacial cycle in the British Isles.   *Journal of the Geological Society of London,* Vol. 153, 499–502.

MCCABE, A M, and CLARK, P U. 1998.   Ice-sheet variability around the North Atlantic Ocean during the last deglaciation.   *Nature, London,* Vol. 392, 373–377.

MCCABE, A M, KNIGHT, J, and MCCARRON, S. 1998.   Evidence for Heinrich event 1 in the British Isles.   *Journal of Quaternary Science,* Vol. 13, 549–568.

MCDONALD, B C, and SHILTS, W W. 1975.   Interpretation of faults in glaciofluvial sediments. 123–131 *in* Glaciofluvial and glaciolacustrine sedimentation. JOPLING, A V, and MCDONALD, B C (editors).   *Society for Economic Paleontologists and Mineralogists (Society for Sedimentary Geology) Special Publication,* No. 23.

MCEWEN, L J. 1997.   Geomorphological change and fluvial landscape evolution in Scotland during the Holocene. 116–129 in *Reflections on the Ice Age in Scotland.* GORDON, J E (editor). (Glasgow: Scottish Association of Geography Teachers and Scottish Natural Heritage.)

MCGOWN, A. 1971.   The classification for engineering purposes of tills from moraines and associated landforms.   *Journal of Engineering Geology,* Vol. 4, 115–130.

MCGOWN, A, and DERBYSHIRE, E. 1977.   Genetic influences on the properties of tills.   *Quarterly Journal of Engineering Geology,* Vol. 10, 389–410.

MCGREGOR, D M, and WILSON, J S G. 1967.   Gravity and magnetic surveys of the younger gabbros of Aberdeenshire.   *Quarterly Journal of the Geological Society of London,* Vol. 123, 99–123

MCLEAN, F. 1977.   The glacial sediments of a part of Aberdeenshire. Unpublished PhD thesis, University of Aberdeen.

MCMILLAN, A A, and AITKEN, A M. 1981.   The sand and gravel resources of the country west of Peterhead, Grampian Region. Description of 1:25 000 sheet NK04 and parts of NJ94, 95 and NK 05, 14 and 15.   *Mineral Assessment Report of the Institute of Geological Sciences,* No. 58.

MCMILLAN, A A, and MERRITT, J W. 1980.   A reappraisal of the 'Tertiary' deposits of Buchan, Grampian Region.   *Report of the Institute of Geological Sciences,* No. 80/1, 18–25.

MCMILLAN, A A, AND HAMBLIN, R J O. 2000.   A mapping related lithostratigraphical framework for the Quaternary of the UK.   *Quaternary Newsletter,* No. 92, 21–34.

MENKE, B, and BEHRE, K E. 1973.   History of vegetation and biostratigraphy.   *Eiszeitalter und Gegenwart,* Vol. 23, 251–267.

MERRITT, J W. 1981.   The sand and gravel resources of the country around Ellon, Grampian Region. Description of 1:25 000 resource sheets NJ93 with parts of NJ82, 83 and 92, and NK03 and parts of NK02 and 13.   *Mineral Assessment Report of the Institute of Geological Sciences*, No. 76.

MERRITT, J W. 1992a.   A critical review of methods used in the appraisal of onshore sand and gravel resources in Britain. *Engineering Geology*, Vol. 32, 1–9.

MERRITT, J W. 1992b.   The high-level marine shell-bearing deposits of Clava, Inverness-shire, and their origin as glacial rafts. *Quaternary Science Reviews*, Vol. 11, 759–779.

MERRITT, J W, and CONNELL, E R. 2000.   Oldmill Quarry. 68–71 in *The Quaternary of the Banffshire coast and Buchan: field guide*. MERRITT, J W, CONNELL, E R, and BRIDGLAND, D R (editors). (London: Quaternary Research Association.)

MERRITT, J W, and MCMILLAN, A A. 1982.   Letters to the editors. The 'Pliocene' gravels of Buchan: a reappraisal. *Scottish Journal of Geology*, Vol. 18, 329–332.

MERRITT, J W, and PEACOCK, J D. 1983a. A preliminary study of the sand and gravel deposits around Aberdeen (1:25 000 sheets NJ71, 80, 81, 90, 91 and parts of 61 and 92; NO69 and 79 and parts of 89).   *Institute of Geological Sciences, Industrial Mineral Assessment Unit, Open File Report*, WA/HI/83/4.

MERRITT, J W, and PEACOCK, J D. 1983b.   A preliminary study of the sand and gravel deposits around Inverness and Nairn, Highland Region and Forres and Elgin, Grampian Region (1:25 000 sheets NH63, 64, 73, 75, 84, 85, 94, 95 and 96; NJ05, 06, 15, 16, 17, 25, 26, 27, 35 and part of 04).   *Institute of Geological Sciences, Industrial Mineral Assessment Unit, Open File Report*, WA/HI/83/5.

MERRITT, J W, and PEACOCK, J D. 2000.   Glacial meltwater channels in Banffshire and Buchan with particular reference to the area around the Tore of Troup.   46–48 in *The Quaternary of the Banffshire coast and Buchan: field guide*. MERRITT, J W, CONNELL, E R, and BRIDGLAND, D R (editors). (London: Quaternary Research Association.)

MERRITT, J W, AUTON, C A, and ROSS, D L. 1988.   Summary assessment of the sand and gravel resources of northeast Scotland.   *British Geological Survey Technical Report*, WF/88/2.

MERRITT, J W, AUTON, C A, and FIRTH, C R. 1995.   Ice-proximal glaciomarine sedimentation and sea-level change in the Inverness area, Scotland: a review of the deglaciation of a major ice stream of the British Late Devensian ice sheet.   *Quaternary Science Reviews*, Vol. 14, 289–329.

MERRITT, J W, CONNELL, E R, and BRIDGLAND, D R.   2000. *The Quaternary of the Banffshire coast and Buchan: field guide*. (London: Quaternary Research Association.)

MERRITT, J W, CONNELL, E R, HALL, A M, and PEACOCK, J D. 2000. An introduction to the Cainozoic geology of coastal Morayshire, Banffshire and Buchan.   1–20 in *The Quaternary of the Banffshire coast and Buchan: field guide*. MERRITT, J W, CONNELL, E R, and BRIDGLAND, D R (editors). (London: Quaternary Research Association.)

MILLER, G H. 1990.   Additional dating methods.   405–413 in *Geomorphological Techniques*. GOUDIE, A (edited for the British Geomorphological Research Group). (London: Unwin Hyman.)

MILLER, G H, and MANGERUD, J. 1985.   Aminostratigraphy of European marine interglacial deposits. *Quaternary Science Reviews*, Vol. 4, 215–278.

MILLER, G H, BRIGHAM, J K, and CLARK, P. 1982.   Alteration of the Total aIle/Ile ratio by different methods of sample preparation.   9–20 *in* Amino acid geochronology laboratory. January 1981 through May 1982. Report of current activities. MILLER, G H (editor).   *Institute of Arctic and Alpine Research and Department of Geological Sciences, University of Colorado.*

MILLER, G H, SEJRUP, H P, MANGERUD, J, and ANDERSEN, B G. 1983.   Amino acid ratios in Quaternary molluscs and foraminifera from western Norway: aminostratigraphy and paleotemperature estimates.   *Boreas*, Vol. 12, 107–124.

MILLER, G H, and seven others. 1987.   Racemization-derived late Devensian temperature reduction in Scotland.   *Nature, London*, Vol. 326, 593–595.

MILLER, H. 1859.   *Sketchbook of popular geology.* (Edinburgh: Constable.)

MILNE, J. 1892a.   Geology of Buchan.   *Transactions of the Buchan Field Club*, Vol. 2, 23–38.

MILNE, J. 1892b.   Drift rocks in Buchan. *Transactions of the Buchan Field Club*, Vol. 2, 181–198.

MITCHELL, G F, PENNY, L F, SHOTTON, F W, and WEST, R G. 1973. A correlation of Quaternary deposits in the British Isles. *Geological Society of London Special Report*, No. 4.

MUNRO, M. 1970. A re-asssessment of the 'younger' basic igneous rocks between Huntly and Portsoy based on new borehole evidence.   *Scottish Journal of Geology*, Vol. 6, 41–52.

MUNRO, M. 1986.   Geology of the country around Aberdeen. *Memoir of the British Geological Survey*, Sheet 77 (Scotland).

MUNRO, M, and GALLACHER, J W. 1984.   Cumulate relations in the 'Younger Basic' masses of the Huntly–Portsoy area, Grampian Region.   *Scottish Journal of Geology*, Vol. 20, 361–382.

MURDOCH, W M. 1975.   The geomorphology and glacial deposits of the area around Aberdeen.   14–18 in *Quaternary Studies in North East Scotland*. GEMMELL, A M D (editor). (Aberdeen: University of Aberdeen.)

MURDOCH, W M. 1977.   The glaciation and deglaciation of south east Aberdeenshire. Unpublished PhD thesis, University of Aberdeen.

MUSSON, R M W. 1994.   A catalogue of British earthquakes. *British Geological Survey Technical Report*, WL/94/04.

MUSSON, R M W. 1996.   The seismicity of the British Isles. *Annali di Geofisica*, Vol. 39, 463–469.

MUSSON, R M W, NEILSON, G, and BURTON, P W. 1986. Macroseismic reports on historical British earthquakes X11: 1927 January 24 North Sea.   *British Geological Survey Global Seismology Unit Report*, WL/GS/86/281.

MYKURA, W. 1983. Old Red Sandstone.   205–251 in *Geology of Scotland*. (second edition). CRAIG, G Y (editor). (Edinburgh: Scottish Academic Press.)

NATURAL ENVIRONMENT RESEARCH COUNCIL. 1998.   UKDMAP CD-ROM. *United Kingdom Digital Marine Atlas*. (Third edition). (Bidston Observatory, Birkenhead: British Oceanographic Data Centre.)

NESJE, A, and SEJRUP, H P. 1988.   Late Weichselian/Late Devensian ice sheets in the North Sea and adjacent areas. *Boreas*, Vol. 17, 371–384.

NICOL, J. 1860.   On the geological structure of the vicinity of Aberdeen and the north-east of Scotland.   *Report of the British Association for the Advancement of Science, 29th meeting*, 116–119.

NORTON, P E P, and SPAINK, G. 1973.   The earliest occurrence of *Macoma balthica* (L.) as a fossil in the North Sea deposits. *Malacologia*, Vol. 14, 33–37.

OCKELMANN, W K. 1954.   On the interrelationship and the zoogeography of northern species of Yoldia Møller, S. Str. (*Mollusca, fam. Ledidae*).   *Meddelelser om Grønland*, Vol. 107, Nr. 7.

OGILVIE, A G. 1923. The physiography of the Moray Firth coast. *Transactions of the Royal Society of Edinburgh*, Vol. 53, 377–404.

OLSEN, L. 1997. Rapid shifts in glacial extension characterise a new concepted model for glacial variations during the Mid and Late Weichselian in Norway. *Norges Geologiste Undersokelse, Bulletin*, 433, 54–55.

OSBORNE, P J. 1988. A late Bronze Age insect fauna from the River Avon, Warwickshire, England: its implications for the terrestrial and fluvial environment and for climate. *Journal of Archaeological Science*, Vol. 15, 715–727.

OWEN, D E. 1987. Commentary: usage of stratigraphic terminology in papers, illustrations, and talks. *Journal of Sedimentary Petrology*, Vol. 57, 363–372.

OWENS, B, and MARSHALL, J, (compilers). 1978. Micro-palaeontological biostratigraphy of samples from around the coasts of Scotland. *Report of the Institute of Geological Sciences*, No. 78/20.

PATERSON, I B. 1974. The supposed Perth Re-advance in the Perth district. *Scottish Journal of Geology*, Vol. 10, 53–66.

PAUL, M A, and EYLES, N. 1990. Constraints on the preservation of diamict facies (melt-out tills) at the margins of stagnant glaciers. *Quaternary Science Reviews*, Vol. 9, 51–69.

PAUL, M A, and LITTLE, J A. 1991. Geotechnical properties of glacial deposits in lowland Britain. 389–403 in *Glacial deposits in Great Britain and Ireland*. EHLERS, J, GIBBARD, P L, and ROSE, J (editors). (Rotterdam: Balkema.)

PEACOCK, J D. 1966. Note on the drift sequence near Portsoy, Banffshire. *Scottish Journal of Geology*, Vol. 2, 35–37.

PEACOCK, J D. 1971. A re-interpretation of the coastal deposits of Banffshire and their place in the late-Glacial history of NE Scotland. *Bulletin of the Geological Survey of Great Britain*, Vol. 37, 81–89.

PEACOCK, J D. 1975. Scottish Late- and Post-glacial marine deposits. 45–48 in *Quaternary studies in north east Scotland*. GEMMELL, A M D (editor). (Aberdeen: University of Aberdeen.)

PEACOCK, J D. 1979. Mill of Dyce. 203–256 in Scottish Universities Research and Reactor Centre radiocarbon measurements III. HARKNESS, D D, and WILSON, H W (editors). *Radiocarbon*, Vol. 21.

PEACOCK, J D. 1980. An overlooked record of interglacial or interstadial sites in north-east Scotland. *Quaternary Newsletter*, Vol. 32, 14–15.

PEACOCK, J D. 1983. Planning for development: Peterhead project. *British Geological Survey Technical Report*, WA/HI/83/1.

PEACOCK, J D. 1984. Errolston. 108–109 in *Buchan field guide*. HALL, A M (editor). (Cambridge: Quaternary Research Association.)

PEACOCK, J D. 1985. Comments on the Quaternary deposits and landforms of Scotland and the neighbouring shelves: a review. *Quaternary Science Reviews*, Vol. 4, i–ii.

PEACOCK, J D. 1995. Late Devensian to early Holocene palaeoenvironmental changes in the Viking Bank area, northern North Sea. *Quaternary Science Reviews*, Vol. 14, 1029–1042.

PEACOCK, J D. 1996. Marine molluscan proxy data applied to Scottish late glacial and Flandrian sites: strengths and limitations. 215–228 in Late Quaternary Palaeoceanography of the North Atlantic Margins. ANDREWS, J T, AUSTIN, W E N, BERGSTEN, H, and JENNINGS, A E (editors). *Special Publication of the Geological Society of London*, No. 111.

PEACOCK, J D. 1997. Was there a re-advance of the British ice sheet into the North Sea between 15 ka and 14 ka BP? *Quaternary Newsletter*, Vol. 81, 1–8.

PEACOCK, J D. 1999. The pre-Windermere Interstadial (Late Devensian) raised marine strata of eastern Scotland and their macrofauna: a review. *Quaternary Science Reviews*, Vol. 18, 1655–1679.

PEACOCK, J D. 2000. A glaciated pavement and crossing striations between Fraserburgh and Rattray Head. 56 in *The Quaternary of the Banffshire coast and Buchan: field guide*. MERRITT, J W, CONNELL, E R, and BRIDGLAND, D R (editors). (London: Quaternary Research Association.)

PEACOCK, J D, and BROWNE, M A E. 1998. Radiocarbon dates from the Errol Beds (pre-Windermere Interstadial raised marine deposits) in eastern Scotland. *Quaternary Newsletter*, Vol. 86, 1–7.

PEACOCK, J D, and CONNELL, E R. 2000. Glacial lakes in the valleys of the North Ugie and South Ugie. 49–52 in *The Quaternary of the Banffshire coast and Buchan: field guide*. MERRITT, J W, CONNELL, E R, and BRIDGLAND, D R (editors). (London: Quaternary Research Association.)

PEACOCK, J D, and HARKNESS, D D. 1990. Radiocarbon ages and the full-glacial to Holocene transition in seas adjacent to Scotland and southern Scandinavia: a review. *Transactions of the Royal Society of Edinburgh: Earth Sciences*, Vol. 81, 385–396.

PEACOCK, J D, and MERRITT, J W. 1997. Glacigenic rafting at Castle Hill, Gardenstown, and its significance for the glacial history of northern Banffshire, Scotland. *Journal of Quaternary Science*, Vol. 12, 283–294.

PEACOCK, J D, and MERRITT, J W. 2000a. Glacial deposits at the Boyne Limestone Quarry, Portsoy, and the late-Quaternary history of coastal Banffshire. *Journal of Quaternary Science*, Vol. 15, 543–555.

PEACOCK, J D, and MERRITT, J W. 2000b. Boyne Bay Limestone Quarry. 25–32 in *The Quaternary of the Banffshire coast and Buchan: field guide*. MERRITT, J W, CONNELL, E R, and BRIDGLAND, D R (editors). (London: Quaternary Research Association.)

PEACOCK, J D, and MERRITT, J W. 2000c. So-called 'pre-Glacial' raised beach deposits deposits on the Banffshire coast. 33 in *The Quaternary of the Banffshire coast and Buchan: Field Guide*. (London: Quaternary Research Association.)

PEACOCK, J D, and MERRITT, J W. 2000d. Castle Hill, Garden-stown. 40–42 in *The Quaternary of the Banffshire coast and Buchan: field guide*. MERRITT, J W, CONNELL, E R, and BRIDGLAND, D R (editors). (London: Quaternary Research Association.)

PEACOCK, J D, and MERRITT, J W. 2000e. Exposures in the Blackhills Sand and Gravel Formation and related deposits. 53–55 in *The Quaternary of the Banffshire coast and Buchan: field guide*. MERRITT, J W, CONNELL, E R, and BRIDGLAND, D R (editors). (London: Quaternary Research Assocation.)

PEACOCK, J D, BERRIDGE, N G, HARRIS, A L, and MAY, F. 1968. The geology of the Elgin district. *Memoir of the Geological Survey*, Sheet 95 (Scotland).

PEACOCK, J D, and seven others. 1977. Sand and gravel resources of the Grampian region. *Report of the Institute of Geological Sciences*, No. 77/2.

PEACOCK, J D, GRAHAM, D K, and WILKINSON, I P. 1978. Late-glacial and post-glacial marine environments at Ardyne, Scotland, and their significance in the interpretation of the history of the Clyde sea area. *Report of the Institute of Geological Sciences*, No. 78/17.

PEACOCK, J D, and five others. 1992. Late Devensian and Flandrian palaeoenvironmental changes in the Scottish continental shelf west of the Outer Hebrides. *Journal of Quaternary Science*, Vol. 7, 145–161.

PEACOCK, J D, HALL, A M, and CONNELL, E R. 2000. St Fergus. 62–64 in *The Quaternary of the Banffshire coast and Buchan: field guide*. (London: Quaternary Research Association.)

PEARS, N V. 1968. Post-glacial tree-lines of the Cairngorm Mountains. *Transactions of the Botanical Society of Edinburgh*, 40, 361–394.

PEARS, N V. 1970. Post-glacial tree-lines of the Cairngorm Mountains; some modifications based on radiocarbon dating. *Transactions of the Botanical Society of Edinburgh*, Vol. 40, 536–544.

PEARSON, D G, EMELEUS, C H, and KELLEY, S P. 1996. Precise $^{40}Ar/^{39}Ar$ age for the initiation of Palaeogene volcanism in the Inner Hebrides and its regional significance. *Journal of the Geological Society of London*, Vol. 153, 815–818.

PENNEY, D. 1990. Quaternary ostracod chronology of the central North Sea: the record from BH 81/29. *Courier Forschungsinstitut Senckenberg*, Vol. 123, 97–109.

PHILLIPS, L. 1976. Pleistocene vegetational history and geology in Norfolk. *Philosophical Transactions of the Royal Society of London*, Vol. B275, 215–286.

PIRAZZOLI, P.A. 1993. Global sea-level changes and their measurement. *Global and Planetary Change*, Vol. 8, 135–148.

PONS, A, GUIOT, J L, BEAULIEU, J L, DE, and REILLE, M. 1992. Recent contributions to the climatology of the last glacial–interglacial cycle based on French pollen sequences. *Quaternary Science Reviews*, Vol. 11, 439–448.

POPE, R D. 1977. Revised version of 'A check list of British insects' by G S Kloet, and W D Hinks. Second edition, part 3. *Royal Entomological Society of London, Handbooks for the identification of British Insects*, Volume 11, Part 3.

POWELL, A J. 1992. Dinoflagellate cysts of the Tertiary System. 253–266 in *A stratigraphic index of dinoflagellate cysts*. POWELL, A J (editor). (London: Chapman and Hall.)

PRESTWICH, J. 1837. Observations on the Ichthyolites of Gamrie in Banffshire, and on the accompanying red conglomerates and sandstones. *Transactions of the Geological Society of London*, Vol. 5, 139–148.

PRICE, R J. 1983. *Scotland's environment during the last 30 000 years*. (Edinburgh: Scottish Academic Press.)

QUATERNARY RESEARCH ASSOCIATION. 1975. *Field excursion guide: Aberdeen*. (Informal publication: Quaternary Research Association.)

RAMSAY, D M. 1965. Factors influencing aggregate impact value in rock aggregate. *Quarry Manager's Journal*, Vol. 49, 129–134.

RAMSAY, D M, DHIR, R K, and SPENCE, I M. 1973. The reproducibility of results in the aggregate impact test: the influence of non-geological factors. '*Quarry Managers' Journal*, Vol. 57, 179–181.

RAMSAY, D M, DHIR, R K, and SPENCE, I M. 1974. The role of rock and clast fabric in the physical performance of crushed-rock aggregate. *Engineering Geology*, Vol. 8, 267–285.

RAMSAY, K. 1999. Geotechnical properties of glacial tills in the Aberdeen area. Unpublished MSc thesis, Heriot-Watt University, Edinburgh.

RAYMO, M E. 1994. The initiation of Northern Hemisphere glaciation. *Annual Review of Earth and Planetary Sciences*, Vol. 22, 353–383.

RAYMO, M E. 1997. The timing of major climate terminations. *Paleoceanography*, Vol. 12, 577–585.

RAYMO, M E, and RUDDIMAN, W F. 1992. Tectonic forcing of late Cenozoic climate. *Nature, London*, Vol. 359, 117–122.

READ, H H. 1923. The geology of the country around Banff, Huntly and Turriff (Lower Banffshire and north-west Aberdeenshire). *Memoir of the Geological Survey, Scotland, sheets* 86 and 96.

READ, H H, BREMNER, A, CAMPBELL, R, and GIBB, A W. 1923. Records of the occurrence of boulders of Norwegian rocks in Aberdeenshire and Banffshire. *Transactions of the Edinburgh Geological Society*, Vol. 11, 230–231.

RIDGWAY, J M. 1982. Common clay and shale. *Mineral Dossier Mineral Resources Consultative Committee*, No. 22.

RIDING, J B. 1998. A palynological investigation of a Quaternary sample from Boyne Bay, Banffshire (Scottish 1:50 000 Sheet 96). *British Geological Survey: Edinburgh, Technical Report*, WH/98/17R.

RITCHIE, W. 1992. Scottish landform examples — 4. Coastal parabolic dunes of the Sands of Forvie. *Scottish Geographical Magazine*, Vol. 108, 39–44.

RITCHIE, W, SMITH, J S, and ROSE, N. 1978. *Beaches of northeast Scotland*. (Aberdeen: Department of Geography, University of Aberdeen.)

ROBINS, N S. 1990. *Hydrogeology of Scotland*. (London: HMSO for the British Geological Survey.)

ROMANS, J C C. 1977. Stratigraphy of buried soil at Teindland Forest, Scotland. *Nature, London*, Vol. 268, 622–623.

ROSE, J. 1985. The Dimlington Stadial/Dimlington Chronozone: a proposal for the naming of the main glacial episode of the Late Devensian in Britain. *Boreas*, Vol. 14, 225–230.

RUDDIMAN, W F, and KUTZBACH, J E. 1991. Plateau uplift and climatic change. *Scientific American*, Vol. 264, 66–75.

RUDDIMAN, W F, and McINTYRE, A. 1973. Time-transgressive deglacial retreat of polar waters from the North Atlantic. *Quaternary Research*, Vol. 3, 117–130.

RUDDIMAN, W F, and McINTYRE, A. 1976. North-east Atlantic palaeoclimatic changes over the past 600 000 years. 111–146 *in* Investigation of late Quaternary palaeoceanography and palaeoclimatology. CLINE, R M, and HAGS, J D (editors). *Geological Society of America Memoir*, 145.

RUDDIMAN, W F, and McINTYRE, A. 1981. The mode and mechanism of the last deglaciation: oceanic evidence. *Quaternary Research*, Vol. 16, 125–134.

RUDDIMAN, W F, and RAYMO, M E. 1988. Northern Hemisphere climate régimes during the past 3 Ma: possible tectonic connections. *Philosophical Transactions of the Royal Society of London*, Vol. B318, 411–430.

RUDDIMAN, W F, SANCETTA, C D, and McINTYRE, A. 1977. Glacial/interglacial response rate of subpolar North Atlantic waters to climatic change: the record in oceanic sediments. *Philosophical Transactions of the Royal Society of London*, Vol. B280, 119–142.

RUDDIMAN, W F, RAYMO, M E, MARTINSON, D G, CLEMENT, B M, and BACKMAN, J. 1989. Pleistocene evolution: northern hemisphere ice sheets and North Atlantic Deep-Sea Core V23-82: correlation with the terrestrial record. *Palaeoceanography*, Vol. 4, 353–412.

SALTER, J W. 1857. On the Cretaceous fossils of Aberdeenshire. *Quarterly Journal of the Geological Society of London*, Vol. 13, 83–89.

SAVILLE, A. 2000. The NMS Project on flint mining in Scotland. 93–95 in *The Quaternary of the Banffshire coast and Buchan: field guide*. MERRITT, J W, CONNELL, E R, and BRIDGLAND, D R (editors). (London: Quaternary Research Association.)

SCHRAG, D P. 2000. Of ice and elephants. *Nature, London,* Vol. 404, 23–24.

SCOTT, D. 1890. Excursion to Links of St Fergus. *Transactions of the Buchan Field Club,* Vol. 1, 86–92.

SEJRUP, H P, and KNUDSEN, K L. 1993. Paleoenvironments and correlations of interglacial sediments in the North Sea. *Boreas,* Vol. 22, 223–235.

SEJRUP, H P, and nine others. 1987. Quaternary stratigraphy of the Fladen area, central North Sea: a multidisciplinary study. *Journal of Quaternary Science,* Vol. 2, 35–58.

SEJRUP, H P, and six others. 1994. Late Weichselian glaciation history of the northern North Sea. *Boreas,* Vol. 23, 1–13.

SEJRUP, H P, and seven others. 1995. Quaternary of the Norwegian Channel; paleoceanography and glaciation history. *Norsk Geologisk Tidsskrift,* Vol. 75, 65–87.

SEJRUP, H P, and five others. 1998. The Jaeren area: a border zone of the Norwegian Channel Ice Stream. *Quaternary Science Reviews,* Vol. 17, 801–812.

SEJRUP, H P, and five others. 2000. Quaternary glaciations in southern Fennoscandia: evidence from southwestern Norway and the northern North Sea region. *Quaternary Science Reviews,* Vol. 19, 667–685.

SELBY, I. 1989. Quaternary geology of the Hebridean continental margin. Unpublished PhD thesis, University of Nottingham.

SHACKLETON, N J, and OPDYKE, N D. 1973. Oxygen isotope and palaeomagnetic stratigraphy of equatorial Pacific core V28–238: oxygen isotope temperatures and ice volumes on a 10^5 year and 10^6 year scale. *Quaternary Research,* Vol. 3, 39–55.

SHACKLETON, N J, and sixteen others. 1984. Oxygen isotope calibration of the onset of ice-rafting and history of glaciation in the North Atlantic region. *Nature, London,* Vol. 307, 620–623.

SHACKLETON, N J, BERGER, A, and PELTIER, W R. 1990. An alternative astronomical calibration of the lower Pleistocene time based on ODP Site 677. *Transactions of the Royal Society of Edinburgh for Earth Sciences,* Vol. 81, 251–261.

SHAW, J. 1979. Genesis of the Sveg tills and rogen moraines of Central Sweden: a model of basal meltout. *Boreas,* Vol. 8, 409–426.

SHENNAN, I. 1989. Holocene crustal movements and sea-level changes in Britain. *Journal of Quaternary Science,* Vol. 4, 77–89.

SIMPSON, S. 1948. The glacial deposits of Tullos and Bay of Nigg, Aberdeen. *Transactions of the Royal Society of Edinburgh,* Vol. 61, 687–687.

SIMPSON, S. 1955. A re-interpretation of the drifts of north-east Scotland. *Transactions of the Edinburgh Geological Society,* Vol. 16, 189–199.

SISSONS, J B. 1958. Supposed ice-dammed lakes in Britain with particular reference to the Eddleston Valley, southern Scotland. *Geografiska Annaler,* Vol. 40, 159–187.

SISSONS, J B. 1961a. A subglacial drainage system by the Tinto Hills, Lanarkshire. *Transactions of the Edinburgh Geological Society,* Vol. 18, 175–193.

SISSONS, J B. 1961b. Some aspects of glacial drainage channels in Britain. *Scottish Geographical Magazine,* Vol. 77, 15–36.

SISSONS, J B. 1965. Quaternary. 467–503 in *The geology of Scotland.* CRAIG, G Y, (editor). (Edinburgh: Oliver and Boyd).

SISSONS, J B. 1967. *The evolution of Scotland's scenery.* (Edinburgh & London: Oliver & Boyd.)

SISSONS, J B. 1974a. A lateglacial ice-cap in the central Grampians. *Transactions of the Institute of British Geographers,* Vol. 62, 95–114.

SISSONS, J B. 1974b. Lateglacial marine erosion in Scotland. *Boreas,* Vol. 3, 41–48.

SISSONS, J B. 1974c. The Quaternary in Scotland: a review. *Scottish Journal of Geology,* Vol. 10, 311–337.

SISSONS, J B. 1976. *The geomorphology of the British Isles: Scotland.* (London: Methuen.)

SISSONS, J B. 1979. The Loch Lomond Stadial in the British Isles. *Nature, London,* Vol. 280, 199–203.

SISSONS, J B. 1981. The last Scottish ice-sheet: facts and speculative discussion. *Boreas,* Vol. 10, 1–17.

SISSONS, J B. 1982. Interstadial and last interglacial deposits covered by till in Scotland: a reply. *Boreas,* Vol. 11, 123–124.

SISSONS, J B. 1983. Quaternary. 399–424 in *The geology of Scotland.* (second edition). G Y CRAIG (editor). (Edinburgh: Scottish Academic Press.)

SKEMPTON, A W, and NORTHEY, R D. 1952. The sensitivity of clays. *Géotechnique,* Vol. 3, 30–53.

SKINNER, A C, McELVENNEY, E, RUCKLEY, N, RISE, L, and ROKOENGEN, K. 1986. Cormorant Sheet 61°N–00°. Quaternary geology. 1:250 000. British Geological Survey.)

SKINNER, B J, and PORTER, S C. 1995. *The Dynamic Earth.* (Third edition). (New York: Wiley.)

SMITH, C G. 1983. Planning for development: Aberdeen project. *British Geological Survey Technical Report,* WA/HI/86/1.

SMITH, D E. 1984. Lower Ythan Valley. 47–58 in *Buchan field guide.* HALL, A M (editor). (Cambridge: Quaternary Research Association.)

SMITH, D E. 1993. Philorth Valley. 251–254 in Quaternary of Scotland. GORDON, J E, and SUTHERLAND, D G (editors). *Geological Conservation Review Series,* No. 6 (London: Chapman and Hall.)

SMITH, D E, and CULLINGFORD, R A. 1985. Flandrian relative sea-level changes in the Montrose Basin area. *Scottish Geographical Magazine,* Vol. 101, 91–105.

SMITH, D E, and DAWSON, A G. 1990. Tsunami waves in the North Sea. *New Scientist,* Vol. 127, 46–49.

SMITH, D E, CULLINGFORD, R A, and SEYMOUR, W P. 1982. Flandrian relative sea-level changes in the Philorth Valley, north-east Scotland. *Transactions of the Institute of British Geographers (New Series),* Vol. 7, 321–336.

SMITH, D E, CULLINGFORD, R A, and BROOKS, C L. 1983. Flandrian relative sea level changes in the Ythan Valley, north-east Scotland. *Earth Surface Processes and Landforms,* Vol. 8, 423–438.

SMITH, D E, FIRTH, C R, BROOKS, C L, ROBINSON, M, and COLLINS, P E F. 1999. Relative sea-level rise during the Main Post-glacial Transgression in NE Scotland, U.K. *Transactions of the Royal Society of Edinburgh: Earth Sciences,* Vol. 90, 1–27.

SMITH, J, SCOTT, T, and STEEL, J. 1904. The post-drift fossils of the Clyde drainage area at low levels. 528–545 in *The geology and palaeontology of the Clyde drainage area.* MURDOCH, J B (editor). (Glasgow: Geological Society of Glasgow.)

SMITH, J S, MATHER, A S, and GEMMELL, A M D. 1977. *A landform inventory of Grampian Region.* (Aberdeen: Department of Geography, University of Aberdeen.)

SMITH, M R, and COLLIS, L (editors). 1993. Aggregates. Sand, gravel and crushed rock aggregates for construction purposes.

(Second Edition.)   *Special Publication of the Geological Society of London*, No. 9.

SMITH, S M, and HEPPELL, D. 1991.   Check list of British marine Mollusca.   *National Museums of Scotland Information Series (Natural History)*, No. 11.

SOIL SURVEY OF SCOTLAND. 1983.   East of Scotland. Land capability for agriculture. Sheet 5. 1:250 000. (Aberdeen: The Ordnance Survey, Southampton for the Macaulay Institute for Soil Research.)

STACE, C. 1991.   *New Flora of the British Isles*. (Cambridge: Cambridge University Press.)

STEERS, A. 1937.   The Culbin Sands and Burghead Bay. *Geological Journal*, Vol. 90, 498(Second Edition.) 528.

STEPHENSON, D, and GOULD, D. 1995.   *British regional geology: the Grampian Highlands*. (fourth edition). (London: HMSO for the British Geological Survey.)

STEVENSON, A S. 1991.   Miller Sheet 61° N–02°W. Quaternary Geology. 1:250 000 Map Series. British Geological Survey.

STEWART, F S. 1991.   A reconstruction of the eastern margin of the late Weichselian ice sheet in northern Britain. Unpublished PhD thesis, University of Edinburgh.

STEWART, F S, and STOKER, M S. 1990.   Problems associated with seismic facies analysis of diamicton-dominated, shelf glacigenic sequences. *Geo-Marine Letters*, Vol. 10, 151–156.

STOKER, M S. 1988.   Pleistocene ice-proximal glaciomarine sediments in boreholes from the Hebrides shelf and Wyville-Thomson Ridge, NW UK continental shelf.   *Scottish Journal of Geology*, Vol. 24, 249–262.

STOKER, M S. 1990.   Glacially influenced sedimentation on the Hebridean slope, northwestern United Kingdom continental margin.   349–362 *Special Publication of the Geological Society of London*, No. 53.

STOKER, M S, and BENT, A J A. 1985.   Middle Pleistocene glacial and glaciomarine sedimentation in the west central North Sea.   *Boreas*, Vol. 14, 325–332.

STOKER, M S, and BENT, A J A. 1987.   Lower Pleistocene deltaic and marine sediments in boreholes from the central North Sea. *Journal of Quaternary Science*, Vol. 2, 87–96.

STOKER, M S, and GRAHAM, C. 1985.   Pre-Late Weichselian submerged rock platforms off Stonehaven.   *Scottish Journal of Geology*, Vol. 21, 205–208.

STOKER, M S, and HOLMES, R. 1991.   Submarine end-moraines as indicators of Pleistocene ice limits of NW Britain.   *Journal of the Geological Society of London*, Vol. 148, 431–434.

STOKER, M S, SKINNER, A C, FYFE, J A, and LONG, D. 1983. Palaeomagnetic evidence for early Pleistocene in the central and northern North Sea.   *Nature, London*, Vol. 304, 332–334.

STOKER, M S, LONG, D, and FYFE, J A. 1985.   A revised Quaternary stratigraphy for the central North Sea.   *Report of the British Geological Survey*, No. 17/2.

STOKER, M S, HITCHEN, K, and GRAHAM, C C. 1993.   *United Kingdom offshore regional report: the geology of the Hebrides and West Shetland shelves, and adjacent deepwater areas*. (London: HMSO for the British Geological Survey.)

STOKER, M S, and eight others. 1994.   A record of late Cenozoic stratigraphy, sedimentation and climate change from the Hebrides Slope, NE Atlantic Ocean.   *Journal of the Geological Society of London*, Vol. 151, 235–249.

STOKER, M S, HOWE, J A, and STOKER, S J. 1999.   Late Vendian–?Cambrian glacially influenced deep-water sedimentation, Macduff Slate Formation (Dalradian),

N E Scotland.   *Journal of the Geological Society of London*, Vol. 156, 55–61.

STUIVER, M, and REIMER, P J. 1993.   Extended ^{14}C data base and revised calib 3.0 ^{14}C age calibration program. *Radiocarbon*, Vol. 35, 215–230.

SUGDEN, D E. 1986.   Pre-glacial weathered rock in till, Tyrebagger, Aberdeen and its implications.   397–401 in *Essays for Professor R E H Mellor*. RITCHIE, W, STONE, J C, and MATHER, A S (editors). (Aberdeen: University of Aberdeen.)

SUGDEN, D E. 1989.   Modification of old land surfaces by ice sheets.   *Zeitschrift fur Geomorphologie N F Supplemunt-Bunt*, Vol. 72, 163–172.

SUGDEN, D E, and JOHN, B S. 1976.   *Glaciers and landscape: a geomorphological approach*. (London: Arnold.)

SUTHERLAND, D G. 1981.   The high-level marine shell beds of Scotland and the build-up of the last Scottish ice sheet.   *Boreas*, Vol. 10, 247–254.

SUTHERLAND, D G. 1984a.   The Quaternary deposits and landforms of Scotland and the neighbouring shelves: a review. *Quaternary Science Reviews*, Vol. 3, 157–254.

SUTHERLAND, D G. 1984b.   Gardenstown (Gamrie).   89–93 in *Buchan field guide*. HALL, A M (editor). (Cambridge: Quaternary Research Association.)

SUTHERLAND, D G. 1984c.   King Edward.   93–96 in *Buchan field guide*. HALL, A M (editor). (Cambridge: Quaternary Research Association.)

SUTHERLAND, D G. 1993a.   Teindland Quarry.   236–240 in The Quaternary of Scotland. GORDON, J E, and SUTHERLAND, D G (editors).   *Geological Conservation Review Series*, No. 6 (London: Chapman and Hall.)

SUTHERLAND, D G. 1993b.   Castle Hill.   240–242 in The Quaternary of Scotland. GORDON, J E, and SUTHERLAND, D G (editors).   *Geological Conservation Review Series*, No. 6 (London: Chapman and Hall.)

SUTHERLAND, D G. 1999.   Scotland.   99–114 in A revised correlation of Quaternary deposits in the British Isles. BOWEN, D Q. (editor). *Geological Society of London Special Report*, No. 23.

SUTHERLAND, D G, and GORDON, J E. 1993.   The Quaternary in Scotland.   13–47 in *The Quaternary of Scotland*. GORDON, J E, and SUTHERLAND, D G (editors).   *Geological Conservation Review Series*, No. 6 (London: Chapman and Hall.)

SYKES, G. 1991.   Amino acid dating.   161–176 in *Quaternary dating methods — a user's guide. Technical Guide 4*. SMART, P L, and FRANCES, P D (editors). (Cambridge: Quaternary Research Association.)

SYNGE, F M. 1956.   The glaciation of north-east Scotland. *Scottish Geographical Magazine*, Vol. 72, 129–143.

SYNGE, F M. 1963.   The Quaternary succession round Aberdeen, north-east Scotland.   353–361 in *Report on the 6th International Congress on the Quaternary*. (Warsaw, 1961: International Union for Quaternary Research, Lodz.)

THOMAS, G S P. 1984.   Sedimentation of a sub-aqueous esker-delta at Strabathie, Aberdeenshire.   *Scottish Journal of Geology*, Vol. 20, 9–20.

THOMAS, G S P, and CONNELL, R, J. 1985.   Iceberg drop, dump, and grounding structures from Pleistocene glacio-lacustrine sediments, Scotland.   *Journal of Sedimentary Petrology*, Vol. 55, 243–249.

THOMSON, K, and HILLIS, R R. 1995.   Tertiary structuration and erosion of the inner Moray Firth.   249–269 in The tectonics, sedimentation and palaeoceanography of the North Atlantic

Region. SCRUTTON, R A, STOKER, M S, SHIMMIELD, G B, and TUDHOPE, A W (editors). *Geological Society of London Special Publication*, No. 90.

TIPPING, R M, and MILBURN, P. 2000. Palaeoenvironmental evidence from peat beneath the floor of the Den of Boddam. 116–125 in *The Quaternary of the Banffshire coast and Buchan: field guide.* MERRITT, J W, CONNELL, E R, and BRIDGLAND, D R (editors). (London: Quaternary Research Association.)

TRAVERSE, A, and GINSBURG, R N. 1966. Palynology of the surface sediments of the Great Bahama Bank, as related to water movement and sedimentation. *Marine Geology*, Vol. 4, 417–459.

TURNER, C. 1970. The Middle Pleistocene deposits at Marks Tey, Essex. *Philosophical Transactions of the Royal Society of London*, Vol. B257, 373–440.

TURNER, W. 1870. On the bones of a seal found in red clay near Grangemouth, with remarks on the species. *Proceedings of the Royal Society of Edinburgh*, Vol. 7, 105–114.

VALEN, V, LARSEN, E, and MANGERUD, J. 1995. High resolution paleomagnetic correlation of Middle Weichselian ice-dammed lake sediments in two coastal caves, western Norway. *Boreas*, Vol. 24, 141–153.

VAN AMERONGEN, J C. 1976. The study of two soils in an area near Mintlaw, North-East Scotland. Unpublished MSc thesis, University of Aberdeen.

VAN DEN BROEK, J M M, and VAN DER WAALS, L. 1967. The late Tertiary peneplain of southern Limburg. *Geologie en Mijnbouw*, Vol. 46, 318–332.

VASARI, Y. 1977. Radiocarbon dating of the Late Glacial and Early Flandrian vegetational succession in the Scottish Highlands and the Isle of Skye. 143–163 in *Studies in the Scottish Lateglacial environment.* GRAY, J M, and LOWE, J J (editors). (Oxford: Pergamon Press.)

VASARI, Y, and VASARI, A. 1968. Late- and post-glacial macrophytic vegetation in the lochs of northern Scotland. *Acta Botanica Fennica*, Vol. 80, 1–120.

WALKER, A B. 1998. UK earthquake monitoring 1997/1998: British Geological Survey Seismic Monitoring and Information Service Nineth Annual Report. *British Geological Survey Technical Report*, WL/98/03.

WALKER, A D. 1973. The age of the Cuttie's Hillock Sandstone (Permo–Triassic) of the Elgin area. *Scottish Journal of Geology*, Vol. 9, 177–183.

WALKER, A D, and six others. 1982. *Soil and land capability for agriculture: Eastern Scotland.* (Handbook of the Soil Survey of Scotland, No. 5). (Aberdeen: Macauley Institute for Soil Research.)

WALKER, M J C, and six others. 1992. Allt Odhar and Dalcharn: two pre-Late Devensian (Late Weichselian) sites in northern Scotland. *Journal of Quaternary Science*, Vol. 7, 69–86.

WALKER, M J C, and six others. 1999. Isotopic 'events' in the GRIP ice core: a stratotype for the Late Pleistocene. *Quaternary Science Reviews*, Vol. 18, 1143–1150.

WALTON, K. 1956. Rattray: a study in coastal evolution. *Scottish Geographical Magazine*, Vol. 72, 85–96.

WALTON, K. 1959. Ancient elements in the coastline of north-east Scotland. 93–109 in *Geographical essays in memory of Alan G Ogilvy, Edinburgh.* MILLER, R, and WATSON, J W (editors). (London: Nelson).

WHITTAKER, A. and eleven others. 1991. A guide to stratigraphical procedure. *Journal of the Geological Society, London*, Vol. 148, 813–824.

WHITTINGTON, G. 1994. *Bruckenthalia spiculfolia* (Salisb.) Reichenb. (Ericaceae) in the Late Quaternary of western Europe. *Quaternary Science Reviews*, Vol. 13, 761–768.

WHITTINGTON, G, HALL, A M, and JARVIS, J. 1993. A pre-Late Devensian pollen site from Camp Fauld, Buchan, north-east Scotland. *New Phytologist*, Vol. 125, 867–874.

WHITTINGTON, G, and six others. 1998. Devensian organic interstadial deposits and ice sheet extent in Buchan, Scotland. *Journal of Quaternary Science*, Vol. 13, 309–324.

WILKINSON, I P. 1993. Calcareous micropalaeontological analysis of the Quaternary 'shelly till' at Burn of Benholm. *British Geological Survey Technical Report*, WH93/232R.

WILKINSON, I P. 1998. Pleistocene Microfossils from Boyne Bay, Banffshire. *British Geological Survey Technical Report*, WH97/224R.

WILSON, J S G. 1882. Northern Aberdeenshire and eastern Banffshire. *Memoirs of the Geological Survey*, Sheet 97 (Scotland).

WILSON, J S G. 1886. North-east Aberdeenshire with detached portions of Banffshire. *Memoirs of the Geological Survey*, Scotland, Sheet 87.

WILSON, J S G, and HINXMAN, L W. 1890. Geology of central Aberdeenshire. *Memoir of the Geological Survey of Scotland*, Sheet 76 (Scotland).

WILSON, M J, and TAIT, J M. 1977. Halloysite in some soils from north-east Scotland. *Clay Minerals*, Vol. 12, 59–66.

WILSON, M J, BAIN, D C, and MITCHELL, W A. 1968. Saponite from the Dalradian metalimestones of north-east Scotland. *Clay Minerals*, Vol. 7, 343–349.

WILSON, M J, and five others. 1981. A swelling hematite/layer sillicate complex in weathered granite. *Clays and Clay Minerals*, Vol. 16, 261–278.

WILSON, M J, RUSSELL, J D, TAIT, J M, CLARK, D R, and FRASER, A R. 1984. Macaulayite, a new mineral from north-east Scotland. *Mineralogical Magazine*, Vol. 48, 127–129.

WORSLEY, P. 1991. Possible early Devensian glacial deposits in the British Isles. 47–51 in *Glacial deposits in Great Britain and Ireland.* EHLERS, J, GIBBARD, P L, and ROSE, J (editors). (Rotterdam: A A Balkema.)

WRIGHT, J S. 1997. Deep weathering profiles (saprolites) in north east Scotland. *Scottish Geographical Magazine*, Vol. 113, 189–194.

ZENKEVITCH, L. 1963. *Biology of the seas of the U.S.S.R.* (London: Allen and Unwin.)

ZIEGLER, P A. 1981. *Evolution of sedimentary basins of north-west Europe.* (London: Heyden.)

Index

MAPS

Map 1   Glacial and glaciofluvial features and the distribution of tills in the Elgin district

Map 2   Glacial and glaciofluvial features and the distribution of glacigenic deposits on Sheet 96W Portsoy.

Map 3   Glacial and glaciofluvial features and the distribution of glacigenic deposits on Sheet 96E Banff.

Map 4   Glacial and glaciofluvial features and the distribution of glacigenic deposits on Sheet 97 Fraserburgh.

Map 5   Glacial and glaciofluvial features and the distribution of glacigenic deposits on Sheet 86E Turriff.

Map 6   Glacial and glaciofluvial features and the distribution of glacigenic deposits on Sheet 87W Ellon.

Map 7   Glacial and glaciofluvial features and the distribution of glacigenic deposits on Sheet 87E Peterhead.

Map 8   Glacial and glaciofluvial features and the distribution of glacigenic deposits on Sheet 76E Inverurie.

Map 9   Glacial and glaciofluvial features and the distribution of glacigenic deposits on Sheet 77 Aberdeen.

Map 10   Glacial and glaciofluvial features and the distribution of glacigenic deposits on Sheet 66E Banchory.

Map 11   Glacial and glaciofluvial features and the distribution of glacigenic deposits on Sheet 67 Stonehaven.

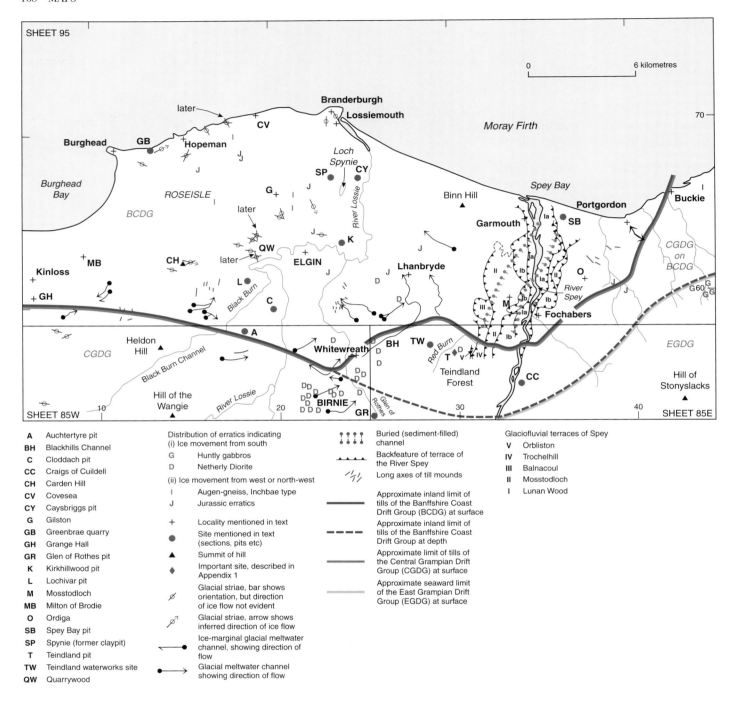

Map 1   Glacial and glaciofluvial features and the distribution of tills in the Elgin district (after Peacock et al., 1968).

Map 2 Glacial and glaciofluvial features and the distribution of glacigenic deposits on Sheet 96W Portsoy.

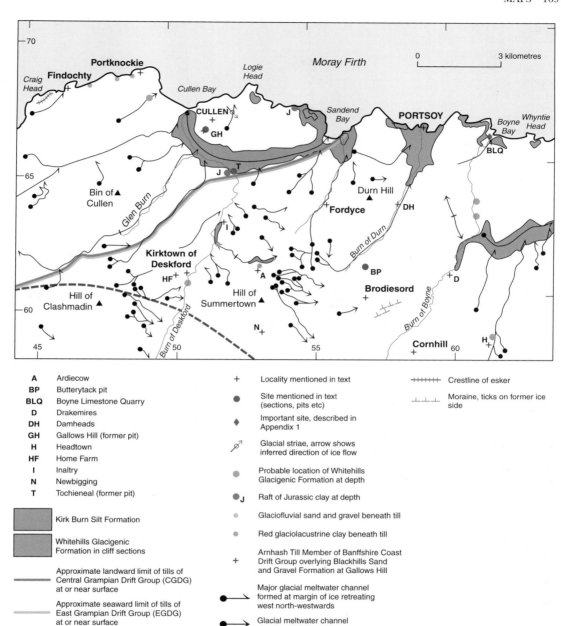

A	Ardiecow	
BP	Butterytack pit	
BLQ	Boyne Limestone Quarry	
D	Drakemires	
DH	Damheads	
GH	Gallows Hill (former pit)	
H	Headtown	
HF	Home Farm	
I	Inaltry	
N	Newbigging	
T	Tochieneal (former pit)	

Kirk Burn Silt Formation

Whitehills Glacigenic Formation in cliff sections

Approximate landward limit of tills of Central Grampian Drift Group (CGDG) at or near surface

Approximate seaward limit of tills of East Grampian Drift Group (EGDG) at or near surface

Approximate inland limit of concealed deposits of Banffshire Coast Drift Group (BCDG) at or near surface

+ Locality mentioned in text

Site mentioned in text (sections, pits etc)

Important site, described in Appendix 1

Glacial striae, arrow shows inferred direction of ice flow

Probable location of Whitehills Glacigenic Formation at depth

J Raft of Jurassic clay at depth

Glaciofluvial sand and gravel beneath till

Red glaciolacustrine clay beneath till

+ Arnhash Till Member of Banffshire Coast Drift Group overlying Blackhills Sand and Gravel Formation at Gallows Hill

Major glacial meltwater channel formed at margin of ice retreating west north-westwards

Glacial meltwater channel showing direction of flow

Subglacial meltwater channel with 'up-and-down' profile

Crestline of esker

Moraine, ticks on former ice side

Map 3   Glacial and
glaciofluvial features
and the distribution
of glacigenic
deposits on
Sheet 96E Banff.

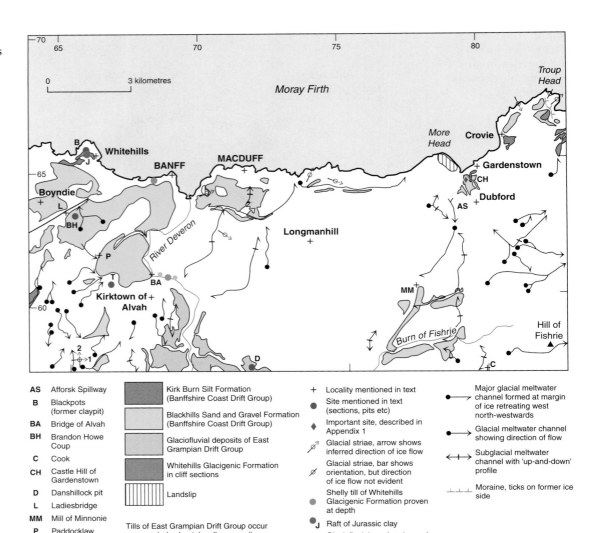

AS	Afforsk Spillway		
B	Blackpots (former claypit)		
BA	Bridge of Alvah		
BH	Brandon Howe Coup		
C	Cook		
CH	Castle Hill of Gardenstown		
D	Danshillock pit		
L	Ladiesbridge		
MM	Mill of Minnonie		
P	Paddocklaw		
T	Tipperty		

Kirk Burn Silt Formation
(Banffshire Coast Drift Group)

Blackhills Sand and Gravel Formation
(Banffshire Coast Drift Group)

Glaciofluvial deposits of East
Grampian Drift Group

Whitehills Glacigenic Formation
in cliff sections

Landslip

Tills of East Grampian Drift Group occur
across whole sheet, locally concealing
deposits of the Banffshire Coast Drift Group

+   Locality mentioned in text

●   Site mentioned in text
(sections, pits etc)

◆   Important site, described in
Appendix 1

⌀   Glacial striae, arrow shows
inferred direction of ice flow

⌀   Glacial striae, bar shows
orientation, but direction
of ice flow not evident

●   Shelly till of Whitehills
Glacigenic Formation proven
at depth

●   Raft of Jurassic clay

●   Glaciofluvial sand and gravel
beneath till

Major glacial meltwater
channel formed at margin
of ice retreating west
north-westwards

Glacial meltwater channel
showing direction of flow

Subglacial meltwater
channel with 'up-and-down'
profile

Moraine, ticks on former ice
side

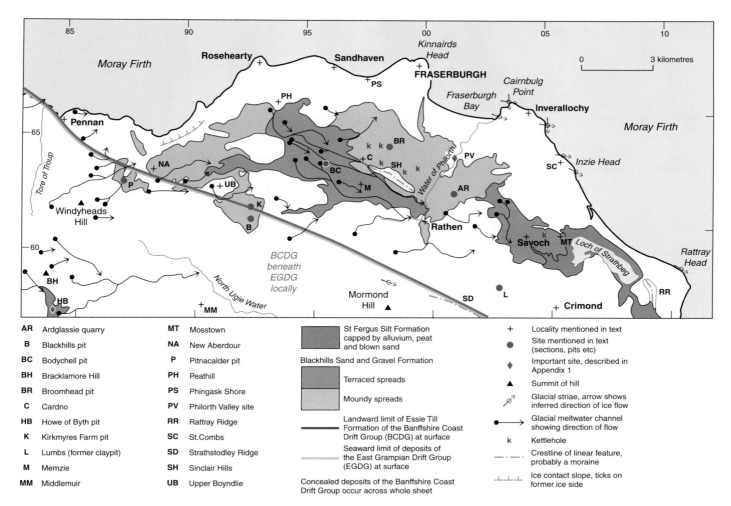

AR	Ardglassie quarry
B	Blackhills pit
BC	Bodychell pit
BH	Bracklamore Hill
BR	Broomhead pit
C	Cardno
HB	Howe of Byth pit
K	Kirkmyres Farm pit
L	Lumbs (former claypit)
M	Memzie
MM	Middlemuir

MT	Mosstown
NA	New Aberdour
P	Pitnacalder pit
PH	Peathill
PS	Phingask Shore
PV	Philorth Valley site
RR	Rattray Ridge
SC	St.Combs
SD	Strathstodley Ridge
SH	Sinclair Hills
UB	Upper Boyndlie

St Fergus Silt Formation capped by alluvium, peat and blown sand

Blackhills Sand and Gravel Formation

Terraced spreads

Moundy spreads

Landward limit of Essie Till Formation of the Banffshire Coast Drift Group (BCDG) at surface

Seaward limit of deposits of the East Grampian Drift Group (EGDG) at surface

Concealed deposits of the Banffshire Coast Drift Group occur across whole sheet

+ Locality mentioned in text

● Site mentioned in text (sections, pits etc)

◆ Important site, described in Appendix 1

▲ Summit of hill

⤢ Glacial striae, arrow shows inferred direction of ice flow

●→ Glacial meltwater channel showing direction of flow

k Kettlehole

—·—·— Crestline of linear feature, probably a moraine

⊥⊥⊥ Ice contact slope, ticks on former ice side

Map 4 Glacial and glaciofluvial features and the distribution of glacigenic deposits on Sheet 97 Fraserburgh.

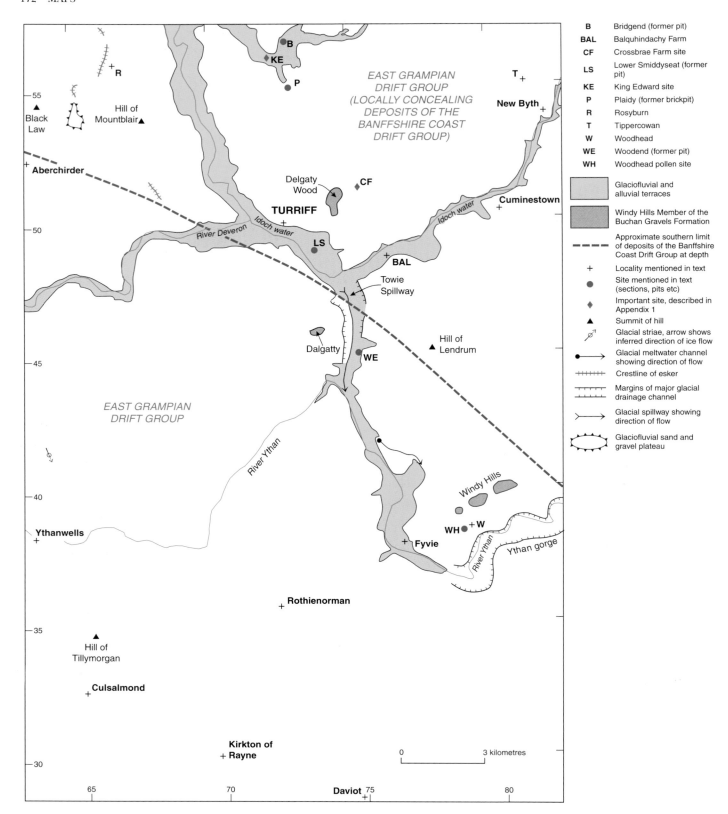

B	Bridgend (former pit)
BAL	Balquhindachy Farm
CF	Crossbrae Farm site
LS	Lower Smiddyseat (former pit)
KE	King Edward site
P	Plaidy (former brickpit)
R	Rosyburn
T	Tippercowan
W	Woodhead
WE	Woodend (former pit)
WH	Woodhead pollen site

Glaciofluvial and alluvial terraces

Windy Hills Member of the Buchan Gravels Formation

Approximate southern limit of deposits of the Banffshire Coast Drift Group at depth

Locality mentioned in text

Site mentioned in text (sections, pits etc)

Important site, described in Appendix 1

Summit of hill

Glacial striae, arrow shows inferred direction of ice flow

Glacial meltwater channel showing direction of flow

Crestline of esker

Margins of major glacial drainage channel

Glacial spillway showing direction of flow

Glaciofluvial sand and gravel plateau

Map 5 Glacial and glaciofluvial features and the distribution of glacigenic deposits on Sheet 86E Turriff.

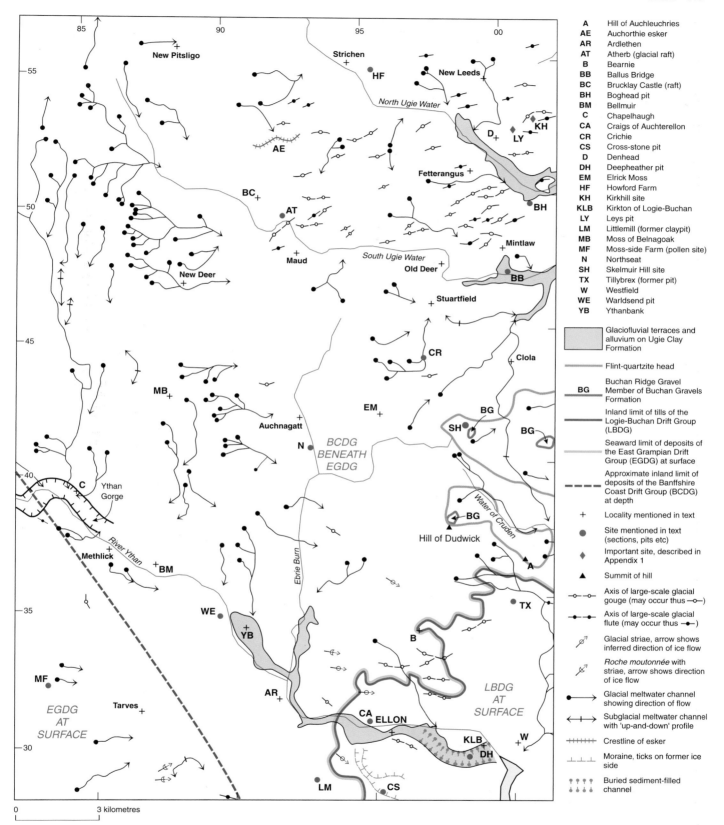

A	Hill of Auchleuchries
AE	Auchorthie esker
AR	Ardlethen
AT	Atherb (glacial raft)
B	Bearnie
BB	Ballus Bridge
BC	Brucklay Castle (raft)
BH	Boghead pit
BM	Bellmuir
C	Chapelhaugh
CA	Craigs of Auchterellon
CR	Crichie
CS	Cross-stone pit
D	Denhead
DH	Deepheather pit
EM	Elrick Moss
HF	Howford Farm
KH	Kirkhill site
KLB	Kirkton of Logie-Buchan
LY	Leys pit
LM	Littlemill (former claypit)
MB	Moss of Belnagoak
MF	Moss-side Farm (pollen site)
N	Northseat
SH	Skelmuir Hill site
TX	Tillybrex (former pit)
W	Westfield
WE	Warldsend pit
YB	Ythanbank

Glaciofluvial terraces and alluvium on Ugie Clay Formation

Flint-quartzite head

BG Buchan Ridge Gravel Member of Buchan Gravels Formation

Inland limit of tills of the Logie-Buchan Drift Group (LBDG)

Seaward limit of deposits of the East Grampian Drift Group (EGDG) at surface

Approximate inland limit of deposits of the Banffshire Coast Drift Group (BCDG) at depth

+ Locality mentioned in text

● Site mentioned in text (sections, pits etc)

◆ Important site, described in Appendix 1

▲ Summit of hill

Axis of large-scale glacial gouge (may occur thus —o—)

Axis of large-scale glacial flute (may occur thus —●—)

Glacial striae, arrow shows inferred direction of ice flow

Roche moutonnée with striae, arrow shows direction of ice flow

Glacial meltwater channel showing direction of flow

Subglacial meltwater channel with 'up-and-down' profile

Crestline of esker

Moraine, ticks on former ice side

Buried sediment-filled channel

Map 6   Glacial and glaciofluvial features and the distribution of glacigenic deposits on Sheet 87W Ellon.

Map 7
Glacial and glaciofluvial features and the distribution of glacigenic deposits on Sheet 87E Peterhead.

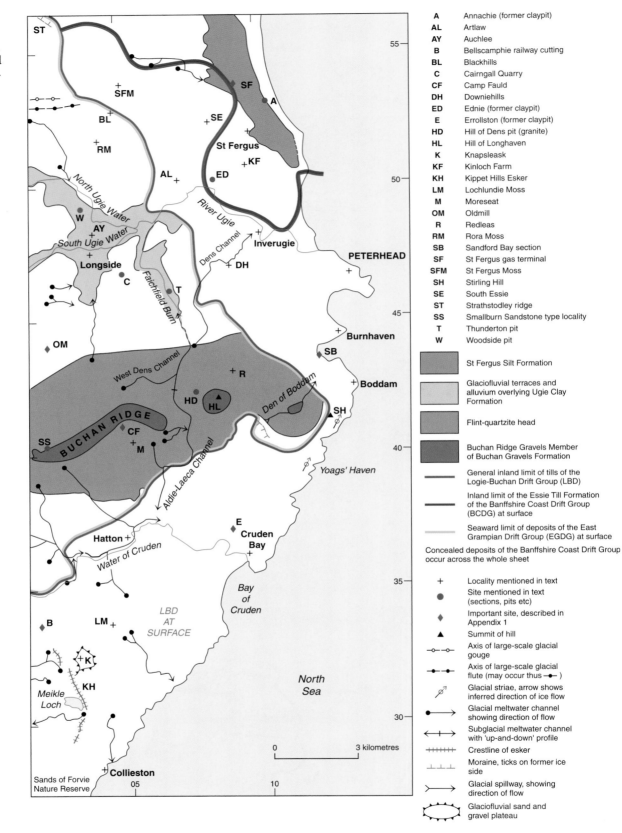

A	Annachie (former claypit)
AL	Artlaw
AY	Auchlee
B	Bellscamphie railway cutting
BL	Blackhills
C	Cairngall Quarry
CF	Camp Fauld
DH	Downiehills
ED	Ednie (former claypit)
E	Errollston (former claypit)
HD	Hill of Dens pit (granite)
HL	Hill of Longhaven
K	Knapsleask
KF	Kinloch Farm
KH	Kippet Hills Esker
LM	Lochlundie Moss
M	Moreseat
OM	Oldmill
R	Redleas
RM	Rora Moss
SB	Sandford Bay section
SF	St Fergus gas terminal
SFM	St Fergus Moss
SH	Stirling Hill
SE	South Essie
ST	Strathstodley ridge
SS	Smallburn Sandstone type locality
T	Thunderton pit
W	Woodside pit

St Fergus Silt Formation

Glaciofluvial terraces and alluvium overlying Ugie Clay Formation

Flint-quartzite head

Buchan Ridge Gravels Member of Buchan Gravels Formation

General inland limit of tills of the Logie-Buchan Drift Group (LBD)

Inland limit of the Essie Till Formation of the Banffshire Coast Drift Group (BCDG) at surface

Seaward limit of deposits of the East Grampian Drift Group (EGDG) at surface

Concealed deposits of the Banffshire Coast Drift Group occur across the whole sheet

+ Locality mentioned in text

● Site mentioned in text (sections, pits etc)

◆ Important site, described in Appendix 1

▲ Summit of hill

–○–○– Axis of large-scale glacial gouge

–●–●– Axis of large-scale glacial flute (may occur thus –●–)

↗ Glacial striae, arrow shows inferred direction of ice flow

●→ Glacial meltwater channel showing direction of flow

←→ Subglacial meltwater channel with 'up-and-down' profile

+++++++ Crestline of esker

⊥⊥⊥ Moraine, ticks on former ice side

⟩—→ Glacial spillway, showing direction of flow

Glaciofluvial sand and gravel plateau

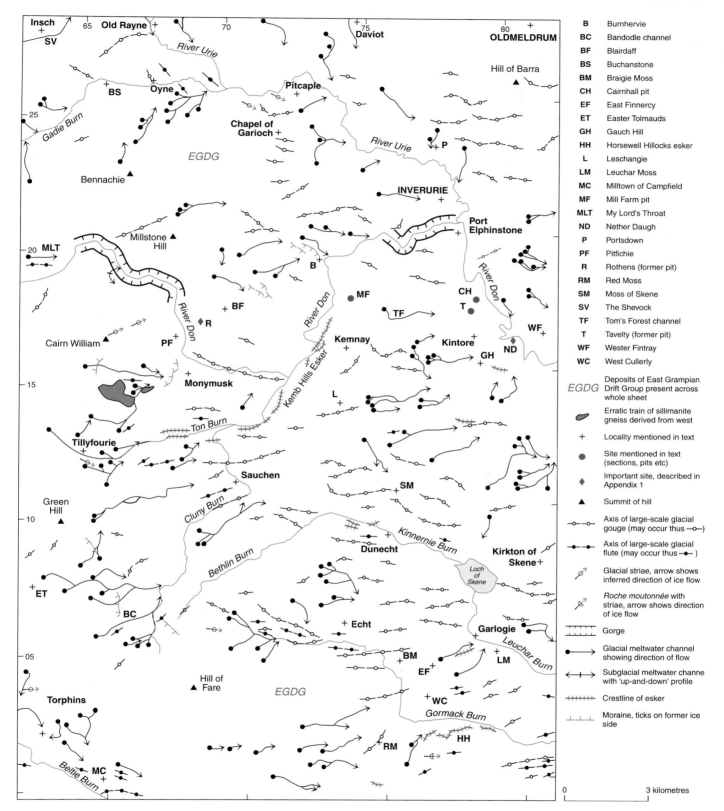

B	Burnhervie
BC	Bandodle channel
BF	Blairdaff
BS	Buchanstone
BM	Braigie Moss
CH	Cairnhall pit
EF	East Finnercy
ET	Easter Tolmauds
GH	Gauch Hill
HH	Horsewell Hillocks esker
L	Leschangie
LM	Leuchar Moss
MC	Milltown of Campfield
MF	Mill Farm pit
MLT	My Lord's Throat
ND	Nether Daugh
P	Portsdown
PF	Pitfichie
R	Rothens (former pit)
RM	Red Moss
SM	Moss of Skene
SV	The Shevock
TF	Tom's Forest channel
T	Tavelty (former pit)
WF	Wester Fintray
WC	West Cullerly

EGDG   Deposits of East Grampian Drift Group present across whole sheet

Erratic train of sillimanite gneiss derived from west

+   Locality mentioned in text

●   Site mentioned in text (sections, pits etc)

◆   Important site, described in Appendix 1

▲   Summit of hill

—o—o—   Axis of large-scale glacial gouge (may occur thus —o—)

—●—●—   Axis of large-scale glacial flute (may occur thus —●—)

Glacial striae, arrow shows inferred direction of ice flow

Roche moutonnée with striae, arrow shows direction of ice flow

Gorge

●——►   Glacial meltwater channel showing direction of flow

◄——►   Subglacial meltwater channel with 'up-and-down' profile

++++++   Crestline of esker

⊥⊥⊥   Moraine, ticks on former ice side

0   3 kilometres

Map 8   Glacial and glaciofluvial features and the distribution of glacigenic deposits on Sheet 76E Inverurie.

Legend

AF	Annfield pit
AH	Ardo House
BF	Brothersfield
BD	Blackdog (former day pit)
BM	Burreldale Moss
BL	Blairs pit
BS	Borrowstone (former pit)
C	Craibstone
CF	Cadgerford
CH	Cairdhillock
D	Drums
DM	Den of Murtle
F	Foggieton
GD	Greendams pit
HM	Harestone Moss
HL	Hillhead of Mundurno pit
LC	Little Clinterty
LH	Lochhills pit
K	Kingswells
M	Mill of Minnes
MD	Mill of Dyce pit
SA	South Anderson Drive site
SB	Strabathie pit
SL	Straloch
T	Tipperty (former claypit)
W	Waterside
WF	Westfield
WH	Westhills channel
YL	Ythan Lodge

+ Locality mentioned in text

● Site mentioned in text (sections, pits etc)

◆ Important site, described in Appendix 1

▲ Summit of hill

–○–○– Axis of large-scale glacial gouge (may occur thus –○–)

–●–●– Axis of large-scale glacial flute (may occur thus –●–)

Glacial striae, arrow shows inferred direction of ice flow

Roche moutonnée with striae, arrow shows direction of ice flow

Crag-and-tail

Glacial meltwater channel showing direction of flow

Glacial spillway, showing direction of flow

Subglacial meltwater channel with 'up-and-down' profile

+++++ Crestline of esker

⊥⊥⊥ Ice-contact slope, ticks on former ice side

Approximate margins of buried channel or valley

Inland limit of deposits of the Logie-Buchan Drift Group (LBDG)

Approximate Inland limit of the Banffshire Coast Drift Group (BCDG) at depth

Inland limit of deposits of the Mearns Drift Group (MDG)

Seaward limit of deposits of the East Grampian Drift Group (EGDG) at surface

Map 9 Glacial and glaciofluvial features and the distribution of glacigenic deposits on Sheet 77 Aberdeen.

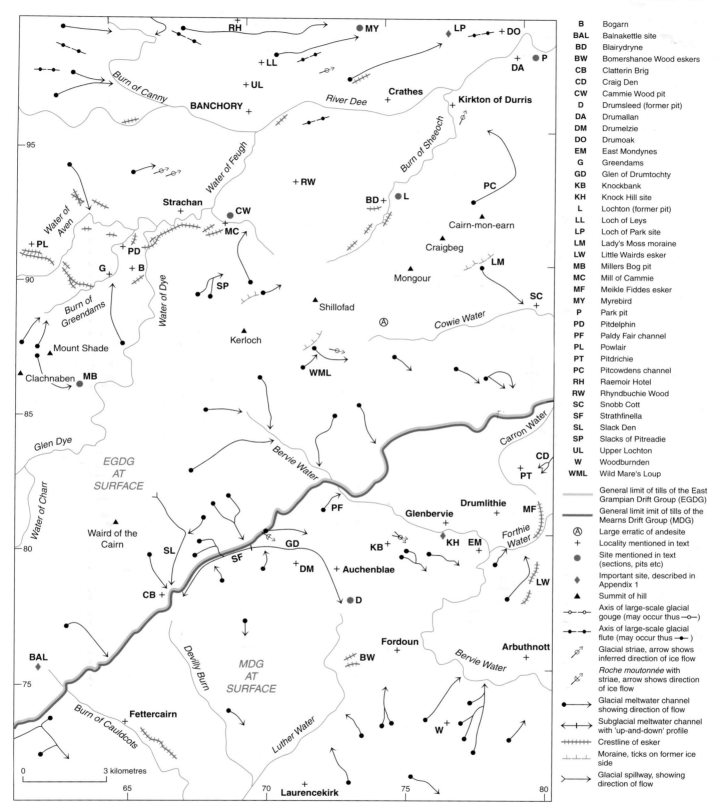

Map 10 Glacial and glaciofluvial features and the distribution of glacigenic deposits on Sheet 66E Banchory.

B	Bogarn
BAL	Balnakettle site
BD	Blairydryne
BW	Bomershanoe Wood eskers
CB	Clatterin Brig
CD	Craig Den
CW	Cammie Wood pit
D	Drumsleed (former pit)
DA	Drumallan
DM	Drumelzie
DO	Drumoak
EM	East Mondynes
G	Greendams
GD	Glen of Drumtochty
KB	Knockbank
KH	Knock Hill site
L	Lochton (former pit)
LL	Loch of Leys
LP	Loch of Park site
LM	Lady's Moss moraine
LW	Little Wairds esker
MB	Millers Bog pit
MC	Mill of Cammie
MF	Meikle Fiddes esker
MY	Myrebird
P	Park pit
PD	Pitdelphin
PF	Paldy Fair channel
PL	Powlair
PT	Pitdrichie
PC	Pitcowdens channel
RH	Raemoir Hotel
RW	Rhyndbuchie Wood
SC	Snobb Cott
SF	Strathfinella
SL	Slack Den
SP	Slacks of Pitreadie
UL	Upper Lochton
W	Woodburnden
WML	Wild Mare's Loup

Legend:

General limit of tills of the East Grampian Drift Group (EGDG)

General limit imit of tills of the Mearns Drift Group (MDG)

Ⓐ Large erratic of andesite

+ Locality mentioned in text

● Site mentioned in text (sections, pits etc)

◆ Important site, described in Appendix 1

▲ Summit of hill

—o—o— Axis of large-scale glacial gouge (may occur thus —o—)

—•—•— Axis of large-scale glacial flute (may occur thus —•—)

Glacial striae, arrow shows inferred direction of ice flow

Roche moutonnée with striae, arrow shows direction of ice flow

Glacial meltwater channel showing direction of flow

Subglacial meltwater channel with 'up-and-down' profile

Crestline of esker

Moraine, ticks on former ice side

Glacial spillway, showing direction of flow

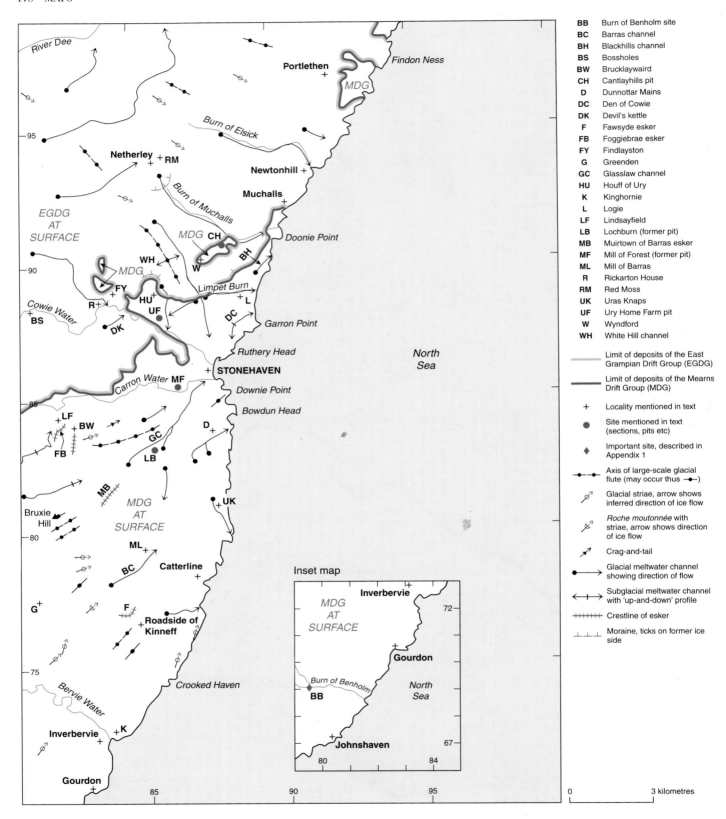

BB	Burn of Benholm site
BC	Barras channel
BH	Blackhills channel
BS	Bossholes
BW	Brucklaywaird
CH	Cantlayhills pit
D	Dunnottar Mains
DC	Den of Cowie
DK	Devil's kettle
F	Fawsyde esker
FB	Foggiebrae esker
FY	Findlayston
G	Greenden
GC	Glasslaw channel
HU	Houff of Ury
K	Kinghornie
L	Logie
LF	Lindsayfield
LB	Lochburn (former pit)
MB	Muirtown of Barras esker
MF	Mill of Forest (former pit)
ML	Mill of Barras
R	Rickarton House
RM	Red Moss
UK	Uras Knaps
UF	Ury Home Farm pit
W	Wyndford
WH	White Hill channel

Limit of deposits of the East Grampian Drift Group (EGDG)

Limit of deposits of the Mearns Drift Group (MDG)

+ Locality mentioned in text

● Site mentioned in text (sections, pits etc)

◆ Important site, described in Appendix 1

Axis of large-scale glacial flute (may occur thus —●—)

Glacial striae, arrow shows inferred direction of ice flow

Roche moutonnée with striae, arrow shows direction of ice flow

Crag-and-tail

Glacial meltwater channel showing direction of flow

Subglacial meltwater channel with 'up-and-down' profile

Crestline of esker

Moraine, ticks on former ice side

Map 11   Glacial and glaciofluvial features and the distribution of glacigenic deposits on Sheet 67 Stonehaven.